Spatial Hysteresis and Optical Patterns

## Springer

Berlin
Heidelberg
New York
Barcelona
Hong Kong
London
Milan
Paris
Tokyo

**Physics and Astronomy**

ONLINE LIBRARY

http://www.springer.de/phys/

# Springer Series in Synergetics

http://www.springer.de/phys/books/sssyn

Nikolay N. Rosanov

# Spatial Hysteresis
# and Optical Patterns

With 149 Figures

 Springer

Professor Nikolay N. Rosanov
Research Center "Vavilov State Optical Institute"
Research Institute for Laser Physics
12, Birzhevaya Liniya
199034 St. Petersburg, Russia

Library of Congress Cataloging-in-Publication Data

Rosanov, Nikolay N.
Spatial hysteresis and optical patterns/Nikolay N. Rosanov.
p.cm. – (Springer series in synergetics, ISSN 0712-7389)
Includes bibliographical references and index.

1. Optical bistability. 2. Hysteresis. I. Title. II. Series.
QC446.3.065 R67 2002  535'.2–dc21  2001055104

ISSN 0172-7389

ISBN 978-3-642-07672-5

Springer-Verlag Berlin Heidelberg New York
a member of BertelsmannSpringer Science+Business Media GmbH

http://www.springer.de

© Springer-Verlag Berlin Heidelberg 2010
Printed in Germany

Cover design: *design & production*, Heidelberg
Printed on acid-free paper

To dear Galya, Alexey, and Marianna

# Preface

This book addresses a subject at the crossing point of two lines of investigation – synergetics and modern nonlinear optics. First, the book is devoted to *optical patterns* that previously were mainly attributed to the display of instabilities of homogeneous field distributions. Second, we deal with *optical bistability* and *hysteresis*, which historically were studied predominantly for point (lumped, spatially nondistributed) systems.

In the book, which seems to be the first monograph devoted to optical patterns, I attempt to demonstrate that the variety of optical patterns and other synergetical phenomena is especially rich in spatially *distributed* bistable systems, and that instabilities of homogeneous states are not necessary for the formation of the patterns. Of special interest are patterns such as *dissipative optical solitons*, which individually have particle-like features and when combined resemble molecules, crystals, biological objects, and even social groups.

Another essential point is the key role of *inhomogeneities* of bistable systems. As for hysteresis, it has been studied mostly in point systems, in which, with fixed characteristics of the input signal, one of several sets of steady-state output signal characteristics can form, depending on the prehistory. In spatially distributed optical systems, the kinetics of hysteresis acquires the form of "*spatial hysteresis*", or hysteresis of spatial distributions of field characteristics. This problem of "stereoscopic vision" of such hysteresis was solved in optical bistability some 20 years ago, and the concepts proposed were later confirmed and expanded. Consistent description of spatial hysteresis is one of the basic subjects of this book. The concept of spatial hysteresis is fairly general, being applied to various systems with phase transition of the first kind. In optics, this and other types of inhomogeneities allow us to control pattern location and features, being thus of important applied significance.

Optics contributes also a new – *diffractive* – mechanism for coupling of spaced elements of a system, as compared with the *diffusion* coupling typical for chemical and biological systems. This mechanism, with field diffractive oscillations inherent in optics, is responsible for the presence of various spatial structures with unusual properties, which can be observed not only in optics but also in other "coherent" fields.

Of interest may also be the consideration of entirely confined, *three-dimensional optical solitons*, the discussion of the *quantum aspects* of optical bistability and the correspondence between the display of bistability in classical and quantum objects, as well as the possible applications of the phenomena of spatial distributivity for *optical information processing* given in the book.

The book material was used for teaching students specializing in quantum electronics and applied mathematics at the St. Petersburg State Institute of Fine Mechanics and Optics (Technical University). Knowledge of nonlinear optics is helpful, though not absolutely necessary, for the main notions, ideas and approaches are introduced in the book when required. For instance, in the first two appendices, more thorough and up-to-date as compared with standard textbooks, an analysis of certain fundamentals of nonlinear optics is given, including nonparaxial radiation propagation.

The book material is based on research carried out at the S.I. Vavilov State Optical Institute and at the Research Institute for Laser Physics (St. Petersburg). I am therefore grateful to E.B. Alexandrov and A.A. Mak for their support, and to V.E. Semenov, G.V. Khodova, A.V. Fedorov, S.V. Fedorov, A.G. Vladimirov, V.A. Smirnov and my other colleagues – coauthors of the articles – for joint work. During the work I had the pleasure of discussing various problems of bistability and optical patterns with S.A. Akhmanov, M.A. Vorontsov, F.V. Bunkin, F.V. Karpushko, G.V. Sinitsyn, A.V. Grigoryants, Yu.S. Balkarey, M.S. Soskin, H. Gibbs, G. Khitrova, W. Firth, L. Lugiato, M. Brambilla, F. Lederer, L. Torner, Yu. Kivshar, A. Akhmediev, A. Boardman, P. Mandel, C. Weiss, R. Kuszelewicz, and their colleagues.

I would particularly like to express my gratitude to Prof. H. Haken, who has done so much in the field of not only general synergetics, but optical synergetics as well, for his support for turning out an English translation of my book. The English version is a revised and expanded edition, as compared with my Russian book "Optical Bistability and Hysteresis in Distributed Nonlinear Systems" (Nauka, Moscow, 1997). Although many general aspects of the theory were clarified earlier, further experiments were still a desideratum. It would therefore seem fitting that "Spatial Hysteresis and Optical Patterns" appears with the onset of the new millennium, after the impressive experiments on dissipative optical solitons in semiconductor microcavities by C.O. Weiss, R. Kuszelevicz, and their colleagues in the year 2000.

Credit for the final English text version belongs to O.I. Brodovich and S.V. Voronin whose untimely death prevented him from seeing the book published. I am also very grateful to A.G. Samsonov and A.M. Kokushkin for their help in the typesetting of the manuscript.

St. Petersburg,                                                         *Nikolay N. Rosanov*
February 2002

# Contents

1. **Introduction to the Physics
   of Distributed Bistable Systems** ........................ 1
   1.1 Types of Optical Bistability, Nonlinearity,
       and Feedback ........................................ 1
       1.1.1 Bistability, Multistability, and Hysteresis ........... 1
       1.1.2 Absolute and Convective Bistability ................ 3
       1.1.3 Conditions of Bistability .......................... 4
       1.1.4 Types of Optical Bistability ....................... 5
       1.1.5 Increasing Absorption Optical Bistability ........... 6
       1.1.6 Electro-optical Schemes .......................... 7
       1.1.7 Nonlinear Cavities (Interferometers) ............... 7
       1.1.8 Nonlinear Reflection ............................. 7
       1.1.9 Bistable Lasers .................................. 8
   1.2 Transverse and Longitudinal Distributivity; Componency .... 8
   1.3 Why Study Distributed Bistable Systems? ................. 10
   1.4 First Investigations
       of Distributed Optical Bistable Systems .................. 13

2. **Increasing Absorption Bistability** ........................ 17
   2.1 Model and Initial Equations ............................ 18
   2.2 Bistability of Point Systems ............................ 21
       2.2.1 Stationary States ................................ 22
       2.2.2 Kinetics ........................................ 23
       2.2.3 Effect of Noise .................................. 27
   2.3 Steady-State Distributions
       in a One-Dimensional System ........................... 30
       2.3.1 Mechanical Analogy .............................. 30
       2.3.2 Stationary Inhomogeneous Distributions ............. 33
   2.4 Stability of Steady-State Distributions ................... 36
   2.5 Switching Waves ...................................... 38
       2.5.1 Switching Waves and the Mechanical Analogy ........ 38
       2.5.2 Stable and Unstable Switching Waves .............. 40
       2.5.3 Numerical Simulations ........................... 43
   2.6 Spatial Switching ..................................... 44

2.7   Wide Radiation Beam ................................. 48
      2.7.1  Asymptotic Analysis ........................... 48
      2.7.2  Spatial Bistability ........................... 52
2.8   Spatial Hysteresis ................................. 54
2.9   Systems with Sharp Transverse Inhomogeneity ........... 56
2.10  Longitudinal and Transverse Distributivity ............ 60
      2.10.1 Longitudinal Distributivity ..................... 61
      2.10.2 Joint Longitudinal and Transverse Distributivity ... 62
2.11  Other Factors; Bibliography ......................... 64
      2.11.1 Two-Dimensional Transverse Distributivity ........ 64
      2.11.2 Fluctuations ................................ 65
      2.11.3 Effect of Delay .............................. 66
      2.11.4 Isolated Loops .............................. 66
      2.11.5 Componency ................................ 66
      2.11.6 Experiments ................................ 67

3.  Hybrid Bistable Devices ................................. 69
    3.1   Stationary, Periodic, and Chaotic Regimes ............... 69
    3.2   Effect of Finite Feedback Bandwidth .................... 77

4.  Driven Nonlinear Interferometers ......................... 83
    4.1   Models of Nonlinear Interferometers ..................... 84
          4.1.1  Driven Ring Interferometers ..................... 84
          4.1.2  Model of Transverse Distributivity ................ 88
          4.1.3  Model of a Nonlinear Screen .................... 91
          4.1.4  Transversely Confined and Point Schemes .......... 92
          4.1.5  Model of Slow Nonlinearity ..................... 93
    4.2   Stability of Stationary Regimes ........................ 94
          4.2.1  Self-pulsations .............................. 95
          4.2.2  Modulation Instability ......................... 97
    4.3   Instabilities of Transverse Field Structure ............... 101
          4.3.1  Matrix Representation ......................... 101
          4.3.2  Cavities Without and with Magnification ........... 104
    4.4   Model of Threshold Nonlinearity ...................... 108
          4.4.1  General Solution ............................ 108
          4.4.2  Switching Waves ............................ 109
          4.4.3  Stability of Switching Waves .................... 111
          4.4.4  Switching Waves: Influence of Inhomogeneities ....... 113
          4.4.5  Single Localized Structures ..................... 114
          4.4.6  Stability of Dissipative Optical Solitons ........... 115
          4.4.7  Dissipative Solitons: Effect of Inhomogeneities ...... 118
          4.4.8  Cylindrically Symmetric Solitons ................ 121
          4.4.9  Combined Switching Waves ..................... 121
          4.4.10 Bound Dissipative Optical Solitons .............. 122
          4.4.11 The Case of Multistability ..................... 123

4.5 Structures for Other Nonlinearities ...................... 123
    4.5.1 Switching Waves .................................. 124
    4.5.2 Metastability and Kinetics of Local Perturbations .... 125
    4.5.3 Dissipative Optical Solitons......................... 127
4.6 Effect of Inhomogeneities ............................... 135
    4.6.1 Switching Waves .................................. 136
    4.6.2 Spatial Hysteresis ............................... 137
    4.6.3 Oblique Incidence ................................ 139
    4.6.4 "Mechanics" of Dissipative Solitons ................. 143
    4.6.5 Dissipative Solitons and Spatial Hysteresis........... 147
4.7 Switching Waves and Solitons
    in Conditions of Instability ............................ 148
    4.7.1 Plane-Wave Excitation ............................ 148
    4.7.2 Excitation with a Radiation Beam.................. 152
    4.7.3 Effect of Transverse Instability ..................... 155
4.8 Effect of Componency.................................... 156
    4.8.1 Model of a Magneto-optic Interferometer ............ 157
    4.8.2 Homogeneous Stationary and Nonstationary Regimes . 158
    4.8.3 Modulation Instability ............................ 161
    4.8.4 Polarized Dissipative Optical Solitons .............. 162
    4.8.5 Related Schemes ................................. 163
4.9 Bibliography; Experiments............................... 169
    4.9.1 Instabilities Related to Longitudinal Distributivity ... 169
    4.9.2 Wide-Aperture Semiconductor Microresonators....... 170
    4.9.3 Instabilities of the Transverse Field Structure ........ 170
    4.9.4 Switching Waves and Spatial Hysteresis ............. 172
    4.9.5 Dissipative Optical Solitons......................... 174

5. Nonlinear Radiation Reflection .......................... 177
5.1 Reflection of Plane Wave
    from Transparent Nonlinear Media ...................... 177
    5.1.1 Nonlinear Layer................................... 177
    5.1.2 Nonlinear Half-Space ............................. 181
5.2 A Weakly Absorbing Nonlinear Medium ................... 189
5.3 Nonlinear Reflection of Beams .......................... 196
    5.3.1 Boundary Conditions ............................. 196
    5.3.2 Paraxial Approach................................. 199
    5.3.3 Nonparaxial Approach: Bistability of Beam Reflection. 201
5.4 Discussion and References ............................... 203

6. Bistable Laser Schemes .................................. 207
6.1 Model and General Relations ............................ 207
    6.1.1 Laser Model ..................................... 207
    6.1.2 Maxwell–Bloch Equations ......................... 208
    6.1.3 Energy Balance .................................. 210

      6.1.4   Symmetry ....................................... 211
      6.1.5   Monochromatic Plane-Wave Solutions .............. 212
      6.1.6   Modulation Instability ........................... 213
  6.2  Transversely One-Dimensional Structures .................. 213
      6.2.1   Fast Nonlinearity ............................... 213
      6.2.2   Effect of Relaxation Rates ....................... 219
  6.3  Transversely Two-Dimensional Structures ................. 225
      6.3.1   Stripes ......................................... 225
      6.3.2   Cylindrically Symmetric Intensity Distribution ....... 226
      6.3.3   Asymmetric Rotating Laser Solitons ................ 227
      6.3.4   Nonparaxial Laser Solitons ....................... 230
  6.4  Interaction of Laser Solitons ........................... 231
      6.4.1   Conditions of "Galilean Symmetry" ................ 231
      6.4.2   Regimes of Weak Overlapping ..................... 232
      6.4.3   Multisoliton Structures ........................... 239
      6.4.4   Effect of Relaxation ............................. 240
  6.5  Effect of Smooth and Sharp Inhomogeneities .............. 242
      6.5.1   Smooth Inhomogeneities ......................... 242
      6.5.2   Sharp Inhomogeneities (Mirror Edges) .............. 243
  6.6  Laser Bullets – Three-Dimensional Laser Solitons .......... 246
      6.6.1   Bifurcation Analysis ............................. 247
      6.6.2   Formation of Laser Bullets ....................... 249
      6.6.3   Interaction of Laser Bullets ....................... 251
      6.6.4   Nonparaxial Effects .............................. 252

**7. Conclusion: Comparing Different Types
of Optical Patterns** ...................................... 255

**A. Paraxial and Nonparaxial
Radiation Propagation** .................................. 259

**B. Constitutive Equations
for Medium Nonlinear Polarization** ...................... 269

**C. Transverse Structures
and Digital Optical Computing** .......................... 273

**D. Bistability of the Quantum
Anharmonic Oscillator** .................................. 279

**E. Transverse Effects
for Squeezed States of Light** ............................ 287

**References** ................................................. 294

**Index** ..................................................... 307

# 1. Introduction to the Physics
# of Distributed Bistable Systems

The main goal of the book being the demonstration of the particular diversity of optical patterns under conditions of bistability, we start with the definitions and examples of bistable schemes, which will be studied in detail in Chaps. 2 to 6. We will distinguish the cases of "absolute" (in feedback systems) and "convective" (in systems without feedback) bistability. We consider the conditions of bistability and factors which influence the character and the kinetics of hysteresis. We then discuss the reasons for the interest in the effects of spatial distributivity in bistable systems and present a review of the first publications in this field.

## 1.1 Types of Optical Bistability, Nonlinearity, and Feedback

### 1.1.1 Bistability, Multistability, and Hysteresis

By "optical bistability" we usually term a property of a wide range of optical systems with fixed, time-independent characteristics (Fig. 1.1), which consists in their staying in one of two possible steady states with different characteristics of the output radiation. When the number of states is greater than two, the system is said to be multistable.[1] The selection of a particular variant of the state is determined by the system's prehistory. Steady states can be not only stationary, but also periodic in time, or even quasirandom (stochastic) under fixed characteristics of the system.

The systems we consider can be driven (illuminated with external radiation, as in Fig. 1.1) or active (laser schemes, external radiation not necessary). Characteristic examples of the transfer function (dependence of an output parameter on a control, or input parameter) for bistable and multistable driven systems are presented in Fig. 1.2. The intermediate branches depicted by the dashed lines correspond to unstable (and therefore unrealisable) states. The

---

[1] Initially bi- or multistable systems are able to change. Thus, a multistable system becomes bistable with a decrease in the control parameter range. In turn, systems bistable in the regular sense can have a great number of stable states when they are of a large spatial extent ("spatial multistability", see Chaps. 4 and 6).

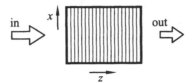

**Fig. 1.1.** An optical bistable system, input and output signals, longitudinal $(z)$ and transverse $(x)$ directions

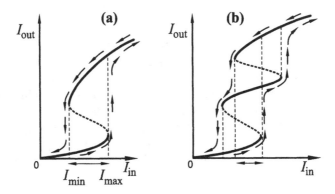

**Fig. 1.2.** Transfer function and hysteresis in driven bistable (**a**) and multistable (**b**) systems within the ranges indicated by *arrows* under axis $I_{in}$

arrows indicate the hysteresis character of temporal variation of output parameters with a slow (in comparison with transient times of regime settling) variation of control parameters. The most widely used control (input) parameter is the intensity (or power) of the incident radiation $I_{in}$. In the absence of external radiation ($I_{in} = 0$), the output radiation is also absent ($I_{out} = 0$). With the increase in $I_{in}$, the working point moves along the lower branch of the transfer curve up to its edge $I_{in} = I_{max}$, where $dI_{out}/dI_{in} = \infty$. With a further increase in $I_{in}$, the system switches "up" (from the lower to the higher branch). If $I_{in}$ decreases starting from high values, then the switching "down" (to the lower branch) takes place with a smaller value of intensity, $I_{in} = I_{min}$, which corresponds to the edge of the higher branch.

The described hysteresis is called "intensity hysteresis". If the optical system includes media with a nonlinear refractive index, intensity hysteresis is accompanied by hysteresis of the phase of the output signal (with respect to the phase of the input signal). Other parameters of incident radiation can be the control parameters in the same bistable system: frequency, polarization characteristics, angle of incidence, etc. Thus the frequency, polarization, angle, etc., hysteresis can, respectively, be observed.

The presence of bistability and hysteresis means that such an element can be used as a memory cell. With a certain variation of parameters of the

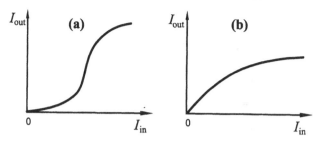

**Fig. 1.3.** Transfer function for regimes of differential gain (**a**) and power limiting (**b**)

same scheme, its transfer function can be radically changed into the single-valued one. The corresponding regimes of differential gain or of power limiting (Fig. 1.3) also present practical interest. As the modifications introduced into the scheme are only quantitative, the term "bistability" could be applied to any nonlinear optical system with feedback, in which, in accordance with the title of the book [117], light characteristics are controlled with light. We can also mention the possibility of bistability with modulated incident radiation [108]. Below we will predominantly use the initial, more narrow interpretation of optical bistability.

One more reservation is due to the presence of noise (fluctuations) in real systems and in the incident radiation. Their most important consequence is the possibility of spontaneous switchings between different steady (in the absence of the noise) states. Therefore the system has only limited memory time of its states, and it could be called "bimetastable". The finite memory time imposes restrictions on the rate of variation of the control parameter, as this rate should not be too small. (To obtain a regular hysteretic loop, the control parameters scanning velocity is bounded both above – by the reciprocal transient time of regime settling – and below – by the reciprocal memory time of the states). Therefore the hysteresis can be only dynamic. However, the noise intensity is very low in systems of practical interest. Fluctuation switchings occur with noticeable probability only in the near vicinity of the branch edges. The conditions when the dynamic character of hysteresis does not practically exhibit itself are quite attainable.

### 1.1.2 Absolute and Convective Bistability

Bistability means that there are a number of stable attractors with nonoverlapping domains of attraction in the system phase space, such that trajectories of system characteristics tend to these attractors during evolution (Fig. 1.4). The unstable state ($I = I_2$ in Fig. 1.4) separates the domains of attraction of stable states.

If we deal with temporal evolution (the evolution variable $\zeta$ is the time $t$), the bistability will be referred to as *absolute*. Then we have a number

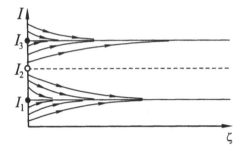

**Fig. 1.4.** The dynamics of a radiation characteristic $I$ for absolute ($\zeta$ is the time $t$) and convective ($\zeta$ is the longitudinal coordinate $z$) bistability; $I_{1,3}$ correspond to stable attractors, and $I_2$ represents an unstable state

of possible system states at the same point (e.g., intensity $I$ at the system output), settling with time in various ranges of initial parameters. This type of bistability is only possible in feedback systems (see below).

Now let us consider the following system without feedback: a continuous medium with saturating amplification and absorption decreasing with local intensity $I$. If the small-signal (unsaturated) absorption exceeds the unsaturated amplification, then the "nonlasing" regime with zero intensity will be formed with propagation distance $\zeta = z$ for sufficiently small intensities of input radiation ($I < I_2$, $I_1 = 0$ in Fig. 1.4). Under certain conditions (see below), for larger input intensities another regime with intensity $I = I_3 > 0$ will be formed for large $z$. This is the case of *convective bistability*, typical of laser schemes. Note also that these two types of bistability can be described by the same mathematical equations, distinctions being only in notation. This terminology is similar to that adopted for instabilities.

### 1.1.3 Conditions of Bistability

The necessary condition of optical bistability is the requirement of a nonlinear response of a system to the incident radiation. Radiation propagating in the system should change its optical characteristics. In the opposite case of a linear system, bistability is impossible, because then the parameters of the input and output radiation are connected unambiguously. In other words, the presence of phenomena of radiation self-action in the system is necessary for optical bistability. Note that the condition of the system's nonlinearity cannot be formulated in the general case as the requirement of nonlinearity of its governing equations. Indeed, it is easy to present examples of pure mathematical transforms which allow us to go from nonlinear differential equations to linear ones, and back. Therefore such a formal criterion would be controversial. It is also well known that quantum objects described by linear equations for the wave function or density matrix have a nonlinear response with respect to the electric field. Another necessary condition of *absolute* bistability is the requirement for the presence of "longitudinal" feedback. An important example of the feedback is given by systems in which a part of the transmitted radiation is applied again to the system entrance, e.g., by means of mirrors. In a number of cases the feedback is realized by means of spatial dispersion and different transport phenomena in a medium (e.g., in the bistability schemes with increasing absorption).[2]

To make this requirement clear, let us divide mentally the system into separate layers transverse to the direction of radiation propagation $z$, as shown in Fig. 1.1. The absence of "longitudinal" feedback means that characteristics of radiation in the medium at any section $z_0$ are determined by the same characteristics only in the previous layers $z < z_0$, but are not sensitive to variation of conditions at $z > z_0$. In particular, if we cut off (and remove) the part of the system which corresponds to $z > z_0$, in a scheme without longitudinal feedback this will not influence the values of the radiation field at $z < z_0$. Therefore, having fixed the parameters of input radiation and assuming that the state of the medium in an elementary layer is determined unambiguously by the local field of radiation, we can "pass" layer after layer (from the entrance to the exit) and unambiguously determine by that the field of radiation and medium characteristics in the entire scheme, including parameters of the output radiation. Hence, such a scheme without feedback can be only monostable (bistability is absent), though it is nonlinear. The presence of the feedback means that, at least in some pairs of the sections (layers) of the scheme $z_1$ and $z_2$ ($z_1 < z_2$), the state of the field or of the medium in the previous layer ($z_1$) depends also on the same characteristics in the subsequent layer ($z_2$). We are reminded that *convective* bistability does not necessitate longitudinal feedback.

In a macroscopic system whose elementary volumes are interconnected by radiation fluxes and by transport processes, bistability is not a local property of the medium – strictly speaking, it is a property of the system as a whole. When the feedback is weak (e.g., when the diffusion length is small as compared to the system sizes), one could speak, with approximation, of "local bistability".

Apart from these two bistability conditions, there are certain other important requirements, in which the factor of spatial distributivity of the system under consideration is manifested substantially. We will return to the discussion of these problems below.

### 1.1.4 Types of Optical Bistability

Bistable schemes can be classified in accordance with the character of nonlinearity, type of feedback, and optical materials. As has been noted, one can

---

[2] Extensive literature is also devoted to *mirrorless* schemes, where the feedback is caused by inter-atomic correlations and local field effects, see [117, 40, 210] and references therein.

differentiate between intensity, frequency, polarization, etc., by choosing the control parameter.

First of all, one can separate optical bistable systems into *all-optical* and *hybrid*. In the first ones optical nonlinearity is caused directly by radiation. If the radiation is transformed in the system to the fields of a different physical nature, which in their turn vary the optical characteristics of the medium with electro-optical, magneto-optical, acousto-optical, or other similar effects, such systems are called *hybrid* (*electro-optical, magneto-optical, acousto-optical*, etc.) systems. Depending on whether there is substantial nonlinear variation of only the absorption coefficient of the medium, or of its refractive index, or of both these values, *absorptive, dispersion*, or mixed (*absorptive–dispersion*) variants of bistability are distinguished.

Feedback can be external (a part of the transmitted radiation is applied to the feedback circuit by special elements) or intrinsic (special feedback elements are absent). External feedback is executed by means of mirrors in cavity bistable schemes – an example of concentrated feedback. Examples of distributed feedback are given below.

We restrict our choice in the book, except in Chap. 3, to wide-aperture systems with comparatively weak transverse coupling. In this section we briefly describe the schemes, which are considered in detail in the following chapters.

### 1.1.5 Increasing Absorption Optical Bistability

In this scheme, the medium has the absorption coefficient $\alpha$ increasing with the increase of radiation local intensity $I$. In the first theoretical works, the concentration $\alpha(n(I))$ [74, 81] and thermal $\alpha(T(I))$ [84] mechanisms of nonlinearity of the semiconductor absorption coefficient were proposed ($n$ is the carrier concentration, and $T$ is the temperature of the sample). When the sample is illuminated by radiation with intensity $I_{\text{in}}$, in a certain range of parameters, the sample transmission, depending on the initial state, is either high (and then only a small part of radiation power is absorbed in the sample, hence a high level of transmission is maintained) or low (in this case the part of radiation power absorbed in the sample is considerable, and the regime with low transmission is formed).

This scheme of bistability can be called all-optical, absorptive, with intrinsic distributed feedback realized by diffusion processes in the sample. Experimentally bistability was confirmed for the thermal mechanism of nonlinearity in [122], where the laser heating of a semiconductor (Ge) caused the increase of concentration of free carriers $n$ and thus of the absorption coefficient $\alpha$; so more precisely one should write $\alpha = \alpha(n(T(I)))$. Experimental studies of schemes with nonthermal (electronic) mechanisms of nonlinearity were made in a large number of the works, a review of which is presented in [117] (see also Sect. 2.11).

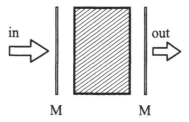

**Fig. 1.5.** A nonlinear Fabry–Perot interferometer driven by coherent radiation: M are the cavity mirrors, the nonlinear medium layer is hatched

### 1.1.6 Electro-optical Schemes

The simplest scheme of this type proposed and realized in [115] is the electro-optical modulator, in which a photodetector transforms part of the transmitted radiation into an electrical signal. This signal, after its amplification, is applied to the same electro-optical crystal, introducing corresponding variations in the polarization of radiation transmitted through it and in the transmission coefficient of the polarizing element in the scheme – the analyzer. This is a hybrid scheme of bistability (multistability) with external feedback. External radiation, as in the previous case, can be incoherent.

### 1.1.7 Nonlinear Cavities (Interferometers)

A nonlinear medium is placed inside a Fabry–Perot or ring cavity, and the cavity is driven externally by laser radiation (Fig. 1.5). The scheme can be bistable or multistable and both all-optical and hybrid. Weak nonlinearity presented by practically any known self-action phenomena is sufficient for realizing bistability, due to the resonance enhancement of the field intensity inside the cavity with a high $Q$-factor. The feedback is external, maintained by cavity mirrors. External radiation should be coherent. Experimentally this type of bistability was demonstrated in 1975 by McCall, Gibbs, Churchill, and Ventcatesan [225, 226].

### 1.1.8 Nonlinear Reflection

In this scheme radiation illuminates the interface of two media, one of which has a nonlinear refractive index. The feedback is performed first by the interface and second by the effective distributed nonlinear mirror induced by variation of field intensity in the nonlinear medium.

Bistability and hysteresis with reflection of electromagnetic radiation by a boundary of a nonlinear medium were predicted by Silin [360]. As applied to optics, the question of bistable reflection of a plane wave by a boundary of the medium with Kerr nonlinearity was formulated later in [52, 157]; the consistent solution to this problem was given in [178, 273].

### 1.1.9 Bistable Lasers

The laser scheme is close to the interferometer case (see Fig. 1.5) in that the feedback is caused by the cavity mirrors and there are a number of intracavity medium nonlinearities. Thus the necessary requirements of absolute bistability are satisfied. An important distinction is in the absence of external signal, not necessary here due to the medium gain and presence of self-oscillations (lasing).

A laser with an additional intracavity cell including a saturable absorber (absorption decreases with radiation intensity) served as one of the first examples of optical bistability realized experimentally [196]. The laser characteristics are chosen in such a way that, in a certain range of pumping, the small-signal gain is less than total losses. Then the nonlasing regime (radiation intensity $I_1 = 0$) is stable. The second requirement is that absorption decreases with intensity faster than gain. Therefore, for the same parameters of the laser and for a sufficiently large initiating signal (pulse of radiation), lasing with intensity $I_3 > 0$ will be formed (hard excitation of lasing). Thus there is bistability with two stable intensity levels $I_1$ and $I_3$; hysteresis variation of laser intensity occurs with slow variation of pumping. Very close to this scheme is a variant of the convective bistability discussed above.

## 1.2 Transverse and Longitudinal Distributivity; Componency

In the general case, both the nonlinear optical scheme and driving radiation (see Fig. 1.1) are spatially distributed. Therefore the transfer function (see Figs. 1.2 and 1.3) reflects the relation between certain integral characteristics of the input and output signals, this relation being also of integral type over spatial variables. It can be said that the transfer functions describe a point scheme, meaning a system whose dimensions can be neglected within the scope of problems we are interested in; compare the notion of "mass point" in mechanics [185]. The notion of a point system is asymptotic, reached at the limit of the decrease in the dimensions of a spatially distributed system. The cost of such simplification in bistability analysis is, however, the difficulty in interpretation of the feedback mechanism: the entrance and exit of a "point" object are spatially indistinguishable. This indeterminateness is resolved by taking into account the non-zero extent – however small – of the system.

Radiation propagation provides the preferential direction in the bistable scheme; thus, it is useful to separate phenomena of longitudinal and transverse (with respect to the beam axis or to the system axis, which is close to it) distributivity. In certain cases one of these types of distributivity is insubstantial, and this allows us to simplify the description of hysteresis phenomena. Let us write the conditions for neglecting the longitudinal and transverse distributivity of the system respectively in the form

$$r_\| \ll r_{\|,\mathrm{cr}} , \quad r_\perp \ll r_{\perp,\mathrm{cr}} . \tag{1.2.1}$$

Here $r_\|$ and $r_\perp$ are the system's dimensions (or the scales of variation of radiation characteristics[3] and medium state) in the longitudinal and transverse directions, and $r_{\|,\mathrm{cr}}$ and $r_{\perp,\mathrm{cr}}$ are the corresponding critical dimensions. The system can be considered as a point system when both inequalities (1.2.1) are satisfied simultaneously. Note also that regimes with different scales of spatial variations can be formed in the system depending on the initial conditions. Therefore for some of these regimes the description in terms of a point system is justified, whereas for others the use of more complicated distributed models is needed. Besides, the transverse distributivity can be one-dimensional (distributivity along the second transverse dimension is insubstantial) or two-dimensional.

In the electro-optical scheme described above the transverse distributivity is not essential, as the photodetector integrates radiation intensity over the beam cross-section. Thereby the scheme presents an example of the system distributed only longitudinally. Further simplification with transition to a point system is connected to neglect by the delay time corresponding to the system's longitudinal extent $\tau_{\mathrm{del}} = r_\|/c$ ($c$ is the light speed). It is allowable if the delay time is small in comparison with the transient time of regime settling $\tau_{\mathrm{st}}$ in the point system. Therefore the critical longitudinal dimension is $l_{\mathrm{del}} = c\tau_{\mathrm{st}}$.

In order to neglect the longitudinal distributivity of a bistable optical device with increasing absorption, the variations of field characteristics and of medium parameters have to be small over its length. The structure of a collimated radiation beam can vary because of the effect of the linear (diffraction or scattering of radiation) and nonlinear factors considered in Appendix A. In particular, for the beam with width $r_\perp = w_b$ the diffraction length $l_{\mathrm{dfr}} = w_b^2/\lambda$. There can be diffusion processes in the medium which are characterized by the diffusion coefficient $D$ and the lifetime (relaxation time) of diffusing particles or quasi-particles $\tau_{\mathrm{rel}}$; the corresponding free path is $l_{\mathrm{dfs}} = (D\tau_{\mathrm{rel}})^{1/2}$. The minimum of these values ($l_{\mathrm{dfr}}, l_{\mathrm{dfs}}, l_{\mathrm{del}}$, etc.) will be the critical longitudinal dimension entered in (1.2.1).

The critical transverse dimensions for the schemes of optical bistability with increasing absorption are determined by the same factors. Thus, for diffraction and diffusion

$$r_{\mathrm{dfr}} = \sqrt{\lambda r_\|} , \quad r_{\mathrm{dfs}} = l_{\mathrm{dfs}} = \sqrt{D\tau_{\mathrm{rel}}} . \tag{1.2.2}$$

A number of additional factors act in nonlinear cavities (interferometers and lasers) besides those indicated above. A serious problem is the development of instabilities of different kinds, which change the scale of spatial

---

[3] The critical dimensions usually exceed the light wavelength $\lambda$, so here we speak about the characteristic scales of variation of the field envelope (slowly varying amplitude, see Appendix A).

field structures. This can also limit the possibilities for neglect by one or another type of spatial distributivity. At the same time, partial simplifications are also possible here (see Chap. 4). When the transverse field structure is constant, which is true for single-mode (with respect to transverse indexes) regimes or at strong discrimination of the transverse modes because of difference in their losses, transition to pure longitudinal distributivity (or even to the point model) is possible. In wide-aperture systems, when the number of the modes is great, a description in mode terms becomes ineffective.

One more important factor strongly influencing the character of the effects of spatial distributivity is the number of degrees of freedom of the bistable system. The widely accepted term "the number of degrees of freedom" is, however, used in quite a number of different senses. We have thus proposed [332] the term "componency", designating the number of independent values that fully describe the state of the corresponding point model. With the growth of the componency, the diversity of spatio-temporal patterns in the distributed bistable systems widens.

## 1.3 Why Study Distributed Bistable Systems?

From the practical point of view, the optical bistability is interesting because of its applications for controlling the characteristics of laser radiation (spatio-temporal light modulators), and, especially, because of opportunities for its use in optical information processing [117]. Differential gain, intensity limiting, multilevel memory (multistability), and transformation of incident radiation into repetitively pulsed radiation have already been demonstrated in optical bistable elements. Therefore bistable elements can act as optical analogues of electronic triggers, transistors, and other basic computer elements.

Even in one-channel (close to point) schemes, analysis of their distributivity not only allows us to reveal the nature of the feedback, but also determines, to a considerable extent, the bistability conditions themselves, the characteristics of switching and the limits for miniaturization of the schemes. The spatial distributivity of the system also provides opportunity to realize multichannel schemes (if the input signal substantially varies over the aperture). For these schemes, the phenomena of spatial distributivity are responsible for the interaction of the channels and for the ultimate density of their package. Some of these questions are considered in Appendix C. Here we note only the important role of the effects of spatial (and especially transverse) distributivity in the prospects of parallel information processing inherent in optics.

The other reason for interest in spatially distributed bistable systems is connected to the novelty of the problems arising here and with their broad implications for various fields. We mean here the problem of the character and the kinetics of hysteresis in a nonlinear system with substantial spatial

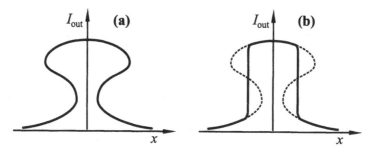

**Fig. 1.6a,b.** Determination of the transverse profile of the output signal

distributivity, i.e., the "stereoscopic vision" of hysteresis. Under conditions (1.2.1), when the system is close to the point one, spatial variation of the system parameters is small. In this case "spatial bistability" and "spatial hysteresis", meant as the bistability and hysteresis of the spatial distributions of the system parameters, do not fundamentally differ from notions corresponding to point systems. However, these notions become deficient when conditions (1.2.1) are violated. Thus, for transversely distributed systems the presence of even weak interaction of scheme elements spaced in the transverse direction is fundamental. Indeed, if it is absent, a wide collimated radiation beam can be divided into separate noninteracting ray tubes, and the transfer function for a point scheme (see Fig. 1.2) can be applied to each separate tube. With such a procedure we would obtain the shape of the intensity profile of the output radiation $I_{out}(x)$ depicted in Fig. 1.6a by a dashed line which has no physical meaning (in the case when the maximum intensity of the input radiation exceeds the intensity of switching up for the point scheme $I_{max}$). Its cause is in the fact that for the distributed system the condition of spatial switching between the branches of the transfer function is not known beforehand, and this condition cannot coincide with the criterion of temporal switching of the point scheme (the branch edges). If in a distributed system we take into account even weak transverse interaction of ray tubes, we obtain a profile that does have a correct physical meaning (solid line in Fig. 1.6b); this profile includes a sharp spatial transition between the states, which correspond to different branches of the transfer function (see Chap. 2). In this case both bistability conditions and the kinetics of spatial hysteresis differ fundamentally from those known for point schemes.

Systems of this kind are also present in other fields of physics, chemistry, and biology, but this problem was raised [179, 274] and first solved [323, 276] precisely in the field of optical bistability. The reason could be the following: Spatial hysteresis is an example of the effect of inhomogeneities; in the case of spatial hysteresis we have inhomogeneity in the characteristics of external signal (intensity transverse variation in the case of radiation beams). Such inhomogeneities are inherent in optics, and they open up new possibilities for pattern control (see Appendix C). Although the kinetics of the spatial hysteresis in the volume is substantially richer than in point schemes, its basic features can be revealed in sufficiently general form.

Further presentation is devoted predominantly to a consistent analysis of the display of spatial distributivity in optical bistable systems. To describe them, I do not propose a unified model, but a hierarchy of models. In the first place, we analyze the simplest models, in which particular phenomena of spatial distributivity can be shown more clearly. Thus, in Chap. 2 increasing absorption bistability is the model of a one-component system with only transverse distributivity; the refinement in description of this type of bistability, taking into account the longitudinal distributivity and a number of other additional factors, is discussed in Sects. 2.10 and 2.11. Similarly, the electro-optical scheme of bistability with considerable delay in the feedback contour (Chap. 3) is important for us as a model of the system with pure longitudinal distributivity. The effects described by the simplest models are reproduced under certain conditions also in more complex schemes – in nonlinear cavities. At the same time, the higher componency of the cavity schemes causes new, additional features, including such striking, to our mind, phenomena, as dissipative optical solitons (Chaps. 4 and 6). This relates also to the schemes with nonlinear reflection of radiation (see Chap. 5), the consideration of which gave rise to lively discussions in the literature. I also note here that a new (diffractive) mechanism of coupling of spaced system elements is present in optics, as compared with the diffusion coupling considered earlier. The "diffractive field oscillations" yield an exceptional variety of spatial structures (see Chaps. 4 and 6). Such "particle-like" field structures can be formed not only in optical, but also in acoustical, hydrodynamical, and other wave systems. Different types of structures are compared in Chap. 7.

Fundamental is also the problem of the relation between the classical and quantum descriptions of microobjects and the possibility of bistability for elementary quantum objects. This subject deserves separate discussion, which is beyond the framework of this book. Therefore in Appendix D I present a brief, but, as I believe, didactic consideration of resonance excitation of a quantum anharmonic oscillator, whose classical analogue is the well-known sample of a bistable object [185]. Intensively developing investigations of transverse phenomena for the *quantum field* are illustrated by a simple example in Appendix E.

I strove to accompany the theoretical constructions with a review of experimental results, as far as this was possible. Today the basic concepts of the spatial hysteresis have been confirmed experimentally. This justifies the discussion in Appendix C of the opportunities for the application of wide-aperture bistable schemes in the design of optical computer systems with nonstandard architecture applicable for parallel operations and allowing the advantages of discrete and analogous methods of computations to be wed.

More specific questions, predominantly those of method, are considered in Appendices A and B.


## 1.4 First Investigations of Distributed Optical Bistable Systems

The first investigations of optical bistability in passive nonlinear optical systems and the comprehension of the importance of this trend date back to 1969. The history of the question is presented in [117] and in the first review on this theme [205]. Here we will briefly discuss the earliest works which analyze passive bistable systems with substantial spatial distributivity.

Note that hereinafter we consider predominantly wide-aperture bistable systems, which have the property of bistability for very wide incident beams (in theory, in the limit of excitation by a plane wave). The schemes of so-called transverse optical bistability [117], which began to be studied somewhat later [44, 45, 159, 160], strongly differ from them. It is sufficient to place only one feedback mirror in those schemes, besides the layer of a nonlinear medium. The presence of a gradient of incident radiation intensity and a variation of the transverse profile of the radiation intensity of counterpropagating beams under their propagation in the medium with self-focusing or self-defocusing nonlinearity is fundamental for this type of bistability. Accordingly, bistability in such schemes is absent under excitation by a plane wave. Different one-channel schemes of transverse optical bistability have been proposed and studied in literature. In contrast to wide-aperture systems, the organization of the transverse transport of information (see Appendix C) is hindered in these schemes. Apparently, the first indication as to the possibility of bistability with self-focusing of counterpropagating beams can be found in [197].

Generally speaking, using the expansion of the field into a series of modes with fixed spatial structure, one could reduce the problem to determination of time-dependent amplitudes of these modes, i.e., to pass from the distributed model to the point one [205]. However, the number of modes required for adequate description of wide-aperture bistable systems is so great that it makes such an approach unrealistic and not appropriate for the study of optical patterns.

A consistent analysis of stationary modes of a longitudinally distributed bistable scheme corresponding to the nonlinear reflection of a plane wave was, as it has been noted, first performed by Silin [360]; the role of transverse phenomena was not considered in this work. Discussion of the possibility of bistability and hysteresis in a system with substantial transverse distributivity began in the literature, as far as I know, with discussion [179, 274]. Both these papers were devoted to the analysis of beam reflection by the interface of media with a nonlinear refractive index (see Chap. 5). The use of the paraxial approximation (slowly varying envelope), justified for smooth

field variation in the transverse direction $x$ (see Fig. 1.1), led the authors of
[179] to the conclusion stating the fundamental impossibility of bistability
for transversely distributed radiation beams. This conclusion was not made
in our article [274], submitted for publication before [179] was published; it
was shown in [274] that this opinion is unjustified due to the inapplicabil-
ity of the paraxial approximation under conditions of sharp hysteresis jumps
(see Fig. 1.6b). Considerations of the possibility and the general character of
bistability for the distributed system considered were also stated in [274].

These conclusions were confirmed in the works published in 1980 and devoted to excitation of a nonlinear interferometer by a wide beam of external
radiation [323] and to heating (also by a wide beam) of the semiconductor sample with the absorption coefficient increasing with temperature [276].
The similarity of the kinetics of spatial hysteresis[4] in these greatly differing
schemes of bistability (with a mechanism of transverse coupling that was
diffractive in [323] and diffusive in [276]) allowed us to extend these conclusions to the general case of wide-aperture bistable systems [277].

Instability of the transverse field structure, which is characteristic for a
medium with a nonlinear refractive index, was found in [323]. The spatial
hysteresis has been demonstrated with instability suppression by spatial filtration. The specific features of the oblique (with respect to the interferometer axis) incidence of the radiation beam were revealed. Geometrical drift of
radiation arising in this case during the light round-trip through the interferometer causes the disappearance of bistability with the angles of incidence
exceeding the critical value,[5] and also, in terms of [7], the two-dimensional
feedback and large-scale nonlocality.

The analytical theory of spatial hysteresis was proposed in [276] for the
scheme of bistability with increasing absorption (thermal mechanism of nonlinearity). We used a graphical "mechanical analogy" to determine the localization of the fronts of sharp spatial switching between the branches of
the transfer function (see Fig. 1.6b). It was indicated that the switching
"up" takes place in the distributed system not simultaneously over the entire
beam cross-section, but only in its narrow zone – the propagating boundary
layer. Computer analysis of characteristics of similar single switching waves
and of their interaction was performed in [278].

In [277] attention was paid to convenience of separation of the spatial
distributivity phenomena into longitudinal and transverse ones. The scheme
considered in [276] was interpreted in [277] as a model of a transversely distributed system. The relation of "propagating boundary layers" (switching
waves) of [323, 276] to "trigger waves" known earlier for problems of another
nature [395] has been revealed. The model of purely transverse distributiv-

---

[4] In the first works [323, 276] we spoke about the hysteresis of the intensity profile.
The more general term "spatial hysteresis" was proposed by Firth and Wright
[102].
[5] Accordingly, the conclusion of [179] about the absence of bistability for limited
beams is justified at sufficiently large angles of radiation incidence.

ity was proposed in [277] also for the case of the nonlinear interferometer, in which the transverse coupling was carried out by diffusion in a nonlinear medium (but not by diffraction of radiation, as in the preceding work [323]). The analysis of spatial hysteresis in ring interferometers with diffractive transverse coupling and different types of nonlinear medium was carried out in [324]; examples of simulations for the kinetics of spatial hysteresis for pulses of incident radiation consistent with conceptions developed in [323, 276, 277] were also presented. Experimental investigations of spatial hysteresis in the wide-aperture interferometer with thermooptical nonlinearity [23] confirmed the basic conclusions of [323, 276, 277].

Delay is one of the consequences of the system's longitudinal distributivity. As Ikeda has shown [146], the delay in a nonlinear interferometer (cavity) driven by monochromatic external radiation can cause the formation of periodic regimes or even regimes stochastic (quasirandom) in time. We noted in [277] that the interesting results of [146] need additional justifications when the transverse field structure and its instabilities are taken into account; the possibility of periodic or stochastic time variation of the beam profile was also pointed out. As proposed in [275, 147], pure delay phenomena (in the absence of transverse phenomena) can be revealed by using hybrid schemes with delay in the feedback contour; this variant of "optical turbulence" was first experimentally realized in [119].

A review of these and other initial investigations of hysteresis and stochastic phenomena in spatially distributed nonlinear optical systems was given in [279]. Further important progress in the field will be discussed in the subsequent chapters.

# 2. Increasing Absorption Bistability

In schemes of optical bistability with increasing absorption, radiation propagates through a sample whose absorption coefficient increases with intensity, for instance, due to the sample heating. With a sufficiently small sample, a point model is justified, describing bistability with an S-like transfer function. For larger samples, the simplest model, which takes into account one-dimensional transverse distributivity, corresponds to thin rod samples, whose characteristics can be averaged in the directions normal to the rod axis.

The basic problem here is to determine the character and kinetics of hysteresis under the effect of a radiation beam. Analyzed locally, a wide beam is similar to a plane wave; therefore we begin by analyzing the effect of a radiation plane wave on a wide sample (unbounded in the transverse directions). In this case the transversely homogeneous temperature distributions correspond to the states of a point scheme. In the case of bistability, along with these distributions, the *mechanical analogy* reveals stable and unstable transversely inhomogeneous distributions. The most important ones are stable *switching waves*, which are the spatial transition between two stable states of the point scheme (different branches of the transfer function). The velocity of the switching wave front is determined by the intensity of incident radiation and changes its sign in the bistability range. At a certain value of intensity the switching wave is motionless. This case corresponds to spatial coexistence of two different transversely homogeneous states. We call this intensity value *Maxwell's intensity value*, referring to the analogy with Maxwell's condition of phase equilibrium in systems with a phase transition of the first kind. Maxwell's intensity value separates the branches of the transfer function, which describe transversely homogeneous states, into intervals corresponding to globally stable and metastable states. Local temperature perturbations (overshoots) arising on the background of metastable state are separated into those that are undercritical (disappearing with time) and those that are supercritical (switching the scheme to the stable state). The overshoot parameters (their width and maximum temperature value) can change nonmonotonically with time.

The presence of switching waves determines effects of smooth in space and time *inhomogeneities*, including *spatial bistability* (the existence and stability of two different transverse distributions of the sample characteristics and of the intensity profiles of the transmitted radiation) and the kinetics of a *spatial hysteresis* with slow temporal variation of the incident radiation power. If these conditions are met, one of the stationary profiles is smooth and consists locally of stable states corresponding to the lower branch of the transfer function; the second profile is combined, with spatial switching between the branches in the region where the local value of incident radiation intensity is equal to Maxwell's value. Switching up (with the radiation power increasing) starts in the beam center. Then, even with constant incident power, the boundaries of the switched region move apart in the form of switching waves with gradually decreasing velocity; finally, the wave front stops in the neighbourhood of Maxwell's value of the local intensity. Switching down (with incident radiation power decreasing) takes place in the form of the switched region's gradual narrowing.

The mechanical analogy allows us also to analyze stationary states of piecewisely homogeneous rod bistable schemes. Inhomogeneity of the medium or incident radiation intensity can, depending on the inhomogeneity characteristics, either stop the front of the incoming switching wave or initiate the formation of the switching waves that successively switch on the entire scheme.

In a purely longitudinally distributed scheme (in which transverse variations are inessential) bistability is possible if the medium layer thickness is less than a certain critical value. If diffusion is weak, sharp switching fronts and high-temperature domains can be formed in the medium. Manifestation of joint longitudinal and transverse distributivity with sample heating by a focused beam is the periodic birth and travel of high-temperature domains, which results in temporal pulsations of the output radiation power. Some other factors, including fluctuations and the system's componency, may also be of importance.

## 2.1 Model and Initial Equations

In this chapter we will describe the apparently simplest transversely distributed scheme of optical bistability, whose point variant was discussed above, in Sect. 1.1. Let us consider the incidence of a radiation beam on the surface of a plate (Fig. 2.1) with intensity distribution over the surface $I_{\text{in}}(x, y, t)$. We assume the plate medium to be isotropic and denote its volume absorption coefficient as $\alpha$.

Let the plate thickness $d$ be sufficiently small compared to the characteristic diffraction length of radiation beam (see Appendix A), which allows us to describe radiation in terms of geometrical optics.[1] The beam is assumed to be collimated, so the light rays are parallel to the $z$-axis, which coincides

---

[1] Otherwise it is necessary to take into account the effects of diffraction [390].

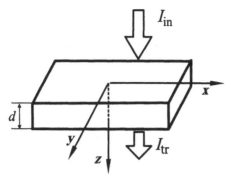

**Fig. 2.1.** Increasing absorption bistable scheme

with the normal to the entrance $(z = 0)$ and exit $(z = d)$ plate surfaces. The transfer equation for radiation intensity $I$ has the form:

$$\frac{\partial I}{\partial z} = -\alpha I \quad (0 < z < d) . \tag{2.1.1}$$

The entrance facet of the plate is characterized by the intensity reflection coefficient $R_0$, and the exit facet is assumed to be nonreflecting (e.g., with antireflective coating). Then it follows from (2.1.1) that

$$I(x, y, z, t) = (1 - R_0)I_{\text{in}} \exp\left(-\int_0^z \alpha \, dz\right) .$$

In particular, the intensity of the radiation transmitted through the plate is

$$I_{\text{tr}} = I(x, y, d, t) = (1 - R_0)I_{\text{in}} \exp\left(-\int_0^d \alpha \, dz\right) . \tag{2.1.2}$$

The absorption coefficient $\alpha$ is assumed to increase with the intensity rise. The nonlinearity can conveniently be considered thermal, corresponding to the temperature dependence of the absorption coefficient (see Sect. 1.1):

$$\alpha = \alpha(T) . \tag{2.1.3}$$

The temperature distribution inside the plate may be determined by the heat conductivity equation

$$\varrho_0 c_{\text{v}} \frac{\partial T}{\partial t} = \nabla(\Lambda \nabla T) + w_{\text{v}} . \tag{2.1.4}$$

Here $\varrho_0$ is the medium specific density, $c_0$ is the specific heat (at constant volume), $\Lambda$ is the thermal conductivity and $w_{\text{v}}$ is the heat release density due to radiation absorption:

$$w_{\mathrm{v}} = -\partial I/\partial z = \alpha I .$$

The boundary conditions have the standard form:

$$\Lambda \left.\frac{\partial T}{\partial z}\right|_{z=0} = h(T_{z=0} - T_{\mathrm{amb}}) , \quad \Lambda \left.\frac{\partial T}{\partial z}\right|_{z=d} = -h(T_{z=d} - T_{\mathrm{amb}}) , \qquad (2.1.5)$$

where $h$ is the heat exchange coefficient and $T_{\mathrm{amb}}$ is the ambient temperature.

To gain bistability, it is sufficient to have only the absorption coefficient $\alpha$ [see (2.1.3)] temperature-dependent; therefore the heat conductivity coefficient can be assumed to be constant. Then instead of (2.1.4) we have

$$\varrho_0 c_{\mathrm{v}} \frac{\partial T}{\partial t} = \Lambda \left(\frac{\partial^2 T}{\partial x^2} + \frac{\partial^2 T}{\partial y^2} + \frac{\partial^2 T}{\partial z^2}\right) - \frac{\partial I}{\partial z} . \qquad (2.1.6)$$

Let us average this equation over the plate thickness integrating both sides of (2.1.6) with respect to $z$ from 0 to $d$ and dividing by $d$. Introducing the average temperature

$$\langle T \rangle = d^{-1} \int_0^d T(x, y, z, t) \mathrm{d}z$$

and taking into account the boundary conditions (2.1.5), we obtain

$$\varrho_0 c_{\mathrm{v}} \frac{\partial \langle T \rangle}{\partial t} = \Lambda \left(\frac{\partial^2 \langle T \rangle}{\partial x^2} + \frac{\partial^2 \langle T \rangle}{\partial y^2}\right)$$
$$- \frac{2h}{d}\left(\frac{T_{z=0} + T_{z=d}}{2} - T_{\mathrm{amb}}\right) + \frac{1}{d}[(1 - R_0)I_{\mathrm{in}} - I_{\mathrm{tr}}] . \qquad (2.1.7)$$

Using (2.1.2), the last terms on the right-hand side of (2.1.7) can be written as follows:

$$(1/d)[(1 - R_0)I_{\mathrm{in}} - I_{\mathrm{tr}}] = (1/d)(1 - R_0)\,I_{\mathrm{in}}\left[1 - \exp\left(-\int_0^d \alpha\,\mathrm{d}z\right)\right]$$
$$\approx (1/d)(1 - R_0)\,I_{\mathrm{in}}\left\{1 - \exp[-\alpha(\langle T \rangle)d]\right\} . \qquad (2.1.8)$$

The final version of the equation has been derived at the cost of the approximate substitution $\alpha \to \alpha(\langle T \rangle)$, which is justified for small temperature differences over the plate thickness:

$$d \ll \Lambda/h ; \qquad (2.1.9)$$

in this case relative variation of radiation intensity over the plate thickness may be not small. Similarly, one can approximately substitute the terms describing heat exchange in the right-hand side of (2.1.7):

$$(T_{z=0} + T_{z=d})/2 - T_{\mathrm{amb}} \approx \langle T \rangle - T_{\mathrm{amb}}.$$

The above approximations allow us to obtain the closed equation for temperature $\langle T \rangle$ averaged over the plate thickness. Omitting below the averaging sign, $\langle T \rangle \to T$, we write this equation in the form:

$$\varrho_0 c_v \frac{\partial T}{\partial t} = \Lambda \triangle_\perp T - F(T) \,. \qquad (2.1.10)$$

Here

$$\triangle_\perp T = \frac{\partial^2 T}{\partial x^2} + \frac{\partial^2 T}{\partial y^2} \,, \quad F(T) = -P(T)I_{in} + H(T - T_{amb}) \,,$$

$$P(T) = [(1 - R_0)/d]\{1 - \exp[-\alpha(T)d]\} \,, \quad H = 2h/d \,. \qquad (2.1.11)$$

The $P(T)$ value characterizes the fraction of radiation power absorbed in the plate. At sufficiently small thicknesses ($\alpha d \ll 1$)

$$P(T) \approx (1 - R_0)\alpha(T) \,. \qquad (2.1.12)$$

Changing the scales of time and coordinates,

$$t' = t/\varrho_0 c_v \,, \quad x' = x/\Lambda^{1/2} \,, \quad y' = y/\Lambda^{1/2} \,, \qquad (2.1.13)$$

we arrive at a somewhat simplified form of (2.1.10):

$$\frac{\partial T}{\partial t'} = \triangle'_\perp T - F(T) \,. \qquad (2.1.14)$$

The *reduced heat conductivity equation* (2.1.10) allows us to determine temperature distributions $T(x, y, t)$. Using (2.1.2), the latter determine the distributions of transmitted radiation $I_{tr}(x, y, t)$ in a single-valued form. A more complete analysis without using averaging over the plate thickness and that of condition (2.1.9) is given in Sect. 2.10.

## 2.2 Bistability of Point Systems

Let incident radiation be a plane wave, so the intensity $I_{in}$ does not depend on the coordinates $x$ and $y$. In this case the solutions of (2.1.10) include transversely homogeneous ones:

$$\partial T/\partial x = \partial T/\partial y = 0 \,.$$

For them, temperature evolution is described by the ordinary differential equation

$$\varrho_0 c_v \frac{dT}{dt} = -F(T) \,. \qquad (2.2.1)$$

Equation (2.2.1) can also be derived from the initial heat conduction equation (2.1.6), averaging it not only over the longitudinal direction ($z$), but also over the transverse ($x, y$) directions, which is justified for samples with small dimensions as compared with the characteristic length of heat conduction. Thus the system described by (2.2.1) belongs to the class of point (spatially nondistributed) systems, which have been well investigated in the theory of nonlinear oscillations.

## 2.2.1 Stationary States

Let the intensity of the incident radiation be constant $(dI_{in}/dt = 0)$. Then temperature temporal dependence $T(t)$ is implicitly determined by the relation

$$\int_{T_0}^{T} F^{-1}(T)\, dT = -(\varrho_0 c_v)^{-1}(t - t_0) , \qquad (2.2.2)$$

where $T_0 = T(t_0)$ is the initial temperature value. The stationary temperature values $\Theta = T(\infty)$ correspond to equilibrium balance of heat (2.1.11):

$$F(\Theta) = 0 . \qquad (2.2.3)$$

For these stationary states, the heat release $(PI_{in})$ is compensated by the heat removal $H(\Theta - T_{amb})$. This condition can be written in the form

$$I_{in} = H(\Theta - T_{amb})/P(\Theta) \qquad (2.2.4)$$

or

$$P(\Theta) = (H/I_{in})(\Theta - T_{amb}) . \qquad (2.2.5)$$

In practice, it is convenient to vary $\Theta$ in (2.2.4) in order to find the dependence $I_{in}(\Theta)$, which, although single-valued, can be nonmonotonic (N-like). The inverse dependence $\Theta(I_{in})$ will then be three-valued (S-like).

The stationary temperature values $\Theta$ are graphically found from the intersection points of the curve $P(T)$ [left-hand side of (2.2.4)] with the straight line [right-hand side of (2.2.4)] coming out the point $\Theta = T_{amb}$ with the slope $H/I_{in}$ (see Fig. 2.2). It is necessary for bistability that the dependence $P(\Theta)$ has the point of inflection $\Theta = \Theta_{in}$, in which $d^2P/dT^2 = 0$. Denote as $T^*$ the point of intersection of the $T$-axis with the tangent to the curve $P(\Theta)$ in the inflection point $T_{infl}$ (see Fig. 2.2). With high ambient temperatures, there is only one point of intersection of the curve $P(\Theta)$ with the indicated curve, i.e., there is only one stationary value of temperature $\Theta$ (monostability). Critical conditions $T_{amb} = T^*$ and $I_{in} = H/(dP/dT)_{T=T^*}$ correspond to a triple root of (2.2.5), $\Theta_1 = \Theta_2 = \Theta_3$. If $T_{amb} < T^*$, then in the intensity range

$$I_{min} < I_{in} < I_{max} \qquad (2.2.6)$$

(boundary values correspond to tangency, see Fig. 2.2a) there are three such points. Accordingly, at the same value of incident radiation intensity $I_{in}$, there are three stationary solutions: $\Theta_1$ (the "low-temperature", or lower branch), $\Theta_2$ (the intermediate branch), and $\Theta_3$ (the "high-temperature", or upper branch). Here the lower, or "thermodynamic" branch refers to regimes that transfer to the trivial state with $T = T_{amb}$ with intensity decrease ($I_{in} \rightarrow 0$). The dependence of the stationary temperature and transmitted intensity values on holding intensity at $T < T_{amb}$ is shown in Fig. 2.3. At more complex (with several points of inflection) temperature dependence $P(\Theta)$ realized in experiments [371], multistability is also reached [a greater number of stable stationary solutions (2.2.3), see Fig. 2.2b].

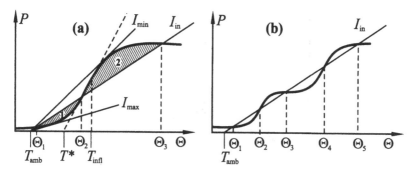

**Fig. 2.2.** Graphical determination of stationary regimes: (a) bistability, (b) multistability

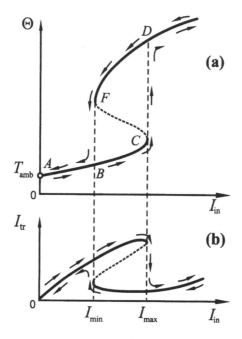

**Fig. 2.3.** Temperature (a) and intensity (b) bistability and hysteresis in a point scheme; the intermediate branches (*dashed lines*) correspond to unstable states

## 2.2.2 Kinetics

The typical temperature dependence of the resulting heat balance $F(T)$ (with necessary condition of bistability $T^* > T_{amb}$) is presented in Fig. 2.4 for different values of intensity. When only one stationary solution is present, the temperature approaches $T = \Theta$ monotonously with time (Fig. 2.5a). Hence a (single) stationary state is stable with respect to arbitrary (including large) deviations of temperature. The characteristic transient time to the stationary

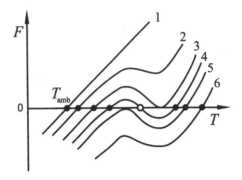

**Fig. 2.4.** Temperature dependence of the thermal balance function $F(T)$. 1: $I_{in} = 0$; 2: $I_{in} < I_{min}$; 3: $I_{in} = I_{min}$; 4: $I_{min} < I_{in} < I_{max}$; 5: $I_{in} = I_{max}$; 6: $I_{in} > I_{max}$

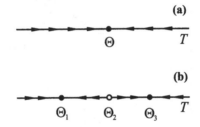

**Fig. 2.5.** Temperature kinetics in the case of monostability (**a**) and bistability (**b**) for the point scheme; *arrows* show changes of temperature with an increase in time

state $\tau_s$ is obtained by linearization of (2.1.1) in the vicinity of $T = \Theta$ with respect to small deviations of temperature $\delta T = T - \Theta$:

$$\varrho_0 c_v \frac{d\delta T}{dt} = -V_\Theta \delta T , \qquad (2.2.7)$$

where

$$V_\Theta = \left(\frac{dF}{dT}\right)_{T=\Theta} . \qquad (2.2.8)$$

Hence

$$\delta T = \delta T_0 \exp(-t/\tau_s) \qquad (2.2.9)$$

and

$$\tau_s = \varrho_0 c_v / V_\Theta . \qquad (2.2.10)$$

More interesting is the case when three stationary values of temperature (2.2.6) are present. As is seen from Fig. 2.4, the sign of $F(T)$ is such that at initial temperature $T < \Theta_2$ the "low-temperature" stationary state $T = \Theta_1$ forms, while at $T > \Theta_2$ the final state is "high-temperature" with $T = \Theta_3$, see Fig. 2.5b. Correspondingly, if the system has been driven to a state close to the unstable state ($T_0 \approx \Theta_2$), then the time of settling of a stable state ($T = \Theta_1$ or $\Theta_3$) anomalously increases (logarithmically with respect to deviation $|T_0 - \Theta_2|$). Such system behaviour is sometimes called uncritical slowing down [43, 358, 232]. The stationary solution $T = \Theta_2$ turns out to be unstable (and therefore the intermediate branch in Fig. 2.3 is shown by a point curve). This means that with the smallest deviations of temperature from the value $\Theta_2$ ($\delta T_0 = T - \Theta_2$) we come with time to the state $\Theta_1$ (if $\delta T_0 < 0$) or $\Theta_3$ (if $\delta T_0 > 0$). The physical meaning of the unstable solution $T = \Theta_2$ is, thus, in fact, that the value of $\Theta_2$ serves as the "watershed", i.e., the boundary of the domains of attraction for two stable states $\Theta_1$ and $\Theta_3$. The time of their settling can be again evaluated with (2.2.10), substituting into it $\Theta_1$ or $\Theta_3$ instead of $\Theta$. At $T = \Theta_2$, (2.2.10) gives a negative value for $\tau_s$, which is a reminder of the instability of such solutions. Note also that the value $V_\Theta$ for the stationary state, corresponding to the edge of the relevant branch of the dependence $\Theta(I_{in})$ (Fig. 2.3), approaches zero when intensity approaches $I_{min}$ or $I_{max}$. Therefore the time of settling of such a regime increases unrestrictedly (the effect of *critical slowing down*, see, e.g., [117]).

This consideration shows the presence of "temperature bistability", i.e., the existence of two stable states of a system with different temperature values under the fixed intensity of incident radiation. Hence, according to (2.1.2), "optical bistability" also follows, i.e., the presence of two different values of transmitted radiation intensity under the same conditions (Fig. 2.3b).

How does the hysteresis transition take place between different stable states? Let the intensity of incident radiation increase slowly from low values. Then, while $I_{in}(t) < I_{max}$, the state is described by the lower (low-temperature) branch $\Theta_1$ (see Fig. 2.3, branch $ABC$). When $I_{in}$ exceeds $I_{max}$, the state changes over to the only stationary regime remaining under such intensities, that which corresponds to the upper (high-temperature) branch ($CDE$). When the intensity of incident radiation decreases, temporal variations follow another path ($EDFBA$). Thus we obtain the hysteresis loop $BCDF$.

Switching between the branches of the hysteresis loop also takes place with the slow variation of other parameters, for instance, the temperature of the external medium $T_{amb}$. "Slowness" is understood here in the scale of settling time for the stationary states $\tau_s$ (2.2.10). However, in the vicinity of the branch edges variation of parameters with any fixed rate ceases to be slow because of the effect of critical slowing down ($\tau_s \to \infty$). This matter deserves somewhat more attention.

For definiteness, consider temperature evolution along the lower branch caused by a slow increase in $T_{amb}$. At fixed intensity $I_{in}$ the temperature values $\Theta_1 = \Theta_2 = \Theta_b$, $T_{amb} = T_b$ correspond to the edge of the lower branch, whereas the temperature for the high-temperature state $\Theta_3$ is far from $\Theta_b$. Thus we exclude the domain of the critical conditions, in which $T_{amb} \approx T^*$, and all three branches merge ($\Theta_1 = \Theta_2 = \Theta_3$).

We assume the deviations of temperatures from the values corresponding to the branch edge to be small:

$$T = \Theta_{\mathrm{b}}[1 + \delta T(t')] , \quad T_{\mathrm{amb}} = T_{\mathrm{b}}[1 + \delta T_{\mathrm{amb}}(t')] , \quad \delta T^2 \ll 1 . \quad (2.2.11)$$

The time $t'$ was introduced in (2.1.13). Retaining the terms up to the quadratic ones in an $F(T)$ expansion in $\delta T$, we transform (2.2.1) to the form

$$\frac{\mathrm{d}\delta T}{\mathrm{d}t'} = m\delta T^2 + H\delta T_{\mathrm{amb}}(t') , \quad m = \frac{1}{2}\Theta_{\mathrm{b}}I_{\mathrm{in}}\left(\frac{\mathrm{d}^2 P}{\mathrm{d}T^2}\right)_{T=\Theta_{\mathrm{b}}} > 0 . \quad (2.2.12)$$

In the case $\delta T_{\mathrm{amb}} = \mathrm{const.} < 0$, (2.2.12) has two stationary solutions:

$$\delta T^{(\pm)} = \pm(-H\delta T_{\mathrm{amb}}/m)^{1/2} . \quad (2.2.13)$$

The solution $\delta T^{(+)}$ is unstable (the intermediate branch) and the solution $\delta T^{(-)}$ is stable (the lower branch) with respect to small perturbations. The nonlinear equation of the first order (2.2.12) is the Riccati equation and can be reduced to a linear equation of the second order [155]:

$$\frac{\mathrm{d}^2 u}{\mathrm{d}t'^2} + mH\delta T_{\mathrm{amb}}(t')u = 0 , \quad (2.2.14)$$

where

$$u(t') = \exp\left(-mH\int^{t'}\delta T\,\mathrm{d}t\right) , \quad \delta T(t') = -\frac{1}{mH}\frac{\mathrm{d}\ln u}{\mathrm{d}t'} . \quad (2.2.15)$$

We will consider the rate of $T_{\mathrm{amb}}$ variation to be a constant,

$$H\delta T_{\mathrm{amb}}(t') = \gamma t' . \quad (2.2.16)$$

Then the solutions of (2.2.14) are the cylinder functions with the index $1/3$. By requiring that the asymptotic at $t' \to -\infty$ is the stable solution (2.2.13) $\delta T^{(-)} = -(-\gamma t'/m)^{1/2}$, we find

$$u(t') = (-t')^{1/2}K_{1/3}[(2/3)(-m\gamma t'^3)^{1/2}] , \quad (2.2.17)$$

where $K_{1/3}$ is the cylinder function of imaginary argument, $t' < 0$. The deviation of temperature $\delta T(t')$ is expressed through $u(t')$ by means of (2.2.15). At $\gamma \to 0$ the value, which corresponds the branch edge, is $\delta T^{(-)}(0) = 0$. At the finite rate $\gamma$ of the $T_{\mathrm{amb}}$ variation, the measure of the critical slowing down is the value $\delta T(0)$. According to (2.2.15) and (2.2.17),

$$\delta T(0) = -q\gamma^{1/6}m^{1/3}/H , \quad q = 12^{1/3}\pi^{1/2}/\Gamma(1/6) \approx 1.05 \quad (2.2.18)$$

(here $\Gamma$ is the gamma function).

Although $\delta T(0)$ decreases with a decrease in the rate $\gamma$, this decrease is extremely slow. Thus, to decrease the value $\delta T(0)$ by a factor of 10, the rate $\gamma$ should be decreased by a factor of $10^6$. Much more slowing down should be expected when the critical conditions are approached.

The initial stage of switching from the lower branch to the upper one at $t > 0$ is also described by means of (2.2.14) and (2.2.16). The solution is expressed through Bessel's functions. At the small scanning rate $\gamma$ the system is "stuck" in the vicinity of the lower branch edge $T \approx \Theta_b$ during the characteristic time $t'_{st} \propto (\gamma m)^{-1/3}$. The corresponding value of deviation of the ambient temperature needed for transition to the upper branch, is $\delta T_{amb} \propto \gamma t'_{st}/H \propto \gamma^{2/3}/m^{1/3}H \to 0$ at $\gamma \to 0$. Similar conclusions also are true for switching down.

### 2.2.3 Effect of Noise

The presence of noise caused by thermodynamic fluctuations in the sample and external medium and by fluctuations of intensity of incident radiation is inevitable for any scheme. Large – and therefore rare – fluctuations will induce switching between the stationary (in the absence of fluctuations) states with temperatures $\Theta_1$ and $\Theta_3$. The probability of the switching is small in the case of weak noise because of stability of the stationary states with respect to weak perturbations. Therefore the system's residence time in a state with temperature $\Theta_1$ or $\Theta_3$ is, though finite, very great [outside a small vicinity of the bistability range edges (2.2.6)]. Correspondingly, the system under consideration should be called not bistable, but rather "bimetastable".

The description of a system subjected to noise needs stochastic language. It is necessary to determine the density of probability $w(T, t)$ of the temperature value $T$ at the moment $t$. Among the initial conditions, an exact value of the initial temperature $T_0$ can also be initialized:

$$w_{init}(T) = w(T, 0) = \delta(T - T_0), \tag{2.2.19}$$

where $\delta$ denotes the delta function. Let the basic noise source be intensity fluctuations:

$$I_{in}(t) = I_{st} + \xi(t). \tag{2.2.20}$$

Here $I_{st}$ is the stationary (average) intensity value and $\xi(t)$ is the random process, which, for the sake of simplicity, we consider to be white noise:

$$\langle \xi \rangle = 0, \quad \langle \xi(t)\xi(t + \tau) \rangle = 2D_n\delta(\tau) \tag{2.2.21}$$

(angular brackets denote statistical averaging). Such an approximation is justified if the width of the noise spectrum exceeds considerably the inverse stabilization time for the stationary state in the absence of noise $1/\tau_s$ (2.2.10). Taking into account (2.2.21) and changing the time scale (2.1.13), we will represent (2.2.1) in the form

$$\frac{dT}{dt'} + F(T) - P(T)\xi(t') = 0 , \qquad (2.2.22)$$

where the stationary value of intensity $I_{\mathrm{st}}$ is substituted into the expression for $F$ (2.1.11). Stochastic equation (2.2.22) is reduced in a standard way to the Fokker–Planck equation for the probability distribution function [350, 114, 156]:

$$\frac{\partial w}{\partial t'} = -\frac{\partial (Aw)}{\partial T} + D_n \frac{\partial^2 (P^2 w)}{\partial T^2} , \qquad (2.2.23)$$

where

$$A(T) = -F(T) + D_n P(T) \frac{dP(T)}{dT} . \qquad (2.2.24)$$

The stationary solution of (2.2.23) has the form

$$w_{\mathrm{st}}(T) = \mathrm{const.} \, [P(T)]^{-1} \exp\left(-D_n{}^{-1} \int^T F(T) P^{-2}(T) \, dT\right) . \qquad (2.2.25)$$

The constant factor is determined from the normalizing condition:

$$\int w \, dT = 1 . \qquad (2.2.26)$$

For the case we are interested in, that of weak noise ($D_n \to 0$), the stationary distribution function differs from zero only in the near vicinity of the maxima of the exponent in (2.2.25). In the conditions of monostability there is only one such maximum. It is reached at $T = \Theta$, where $\Theta$ is the stationary temperature value in the absence of noise, defined by (2.2.3), and $V(\Theta) = (dF/dT)_{T=\Theta} > 0$. Replacing the integrand by the first terms of its expansion in $T - \Theta$, we obtain

$$w_{\mathrm{st}}(T) = (\pi^{1/2} a)^{-1} \exp[-(T - \Theta)^2 / a^2] , \quad a^2 = 2D_n[P(\Theta)]^2 / V(\Theta) . \qquad (2.2.27)$$

The probability distribution function is represented by a peak of a Gaussian shape with the maximum at $T = \Theta$ and the width $a \propto D_n^{1/2}$ (Fig. 2.6a). In the case of bistability the exponent has maxima at $T = \Theta_1$ and $\Theta_3$ and a minimum at $T = \Theta_2$. The stationary distribution function is calculated similarly. It consists of two narrow (with widths of order of $D_n^{1/2}$) peaks located in the vicinities of $T = \Theta_1$ and $\Theta_3$ (Fig. 2.6b). The ratio of the areas of the left and the right peaks gives the relative time of the system's residence in the low-temperature and high-temperature states. The system spends the main part of the time in the low- or high-temperature state near the lower or upper edge, respectively, of the bistability interval (2.2.6) ($I_{\mathrm{in}} \approx I_{\mathrm{min}}$ or $I_{\mathrm{in}} \approx I_{\mathrm{max}}$).

The stationary probability distribution function (2.2.25) does not depend on the initial state of the system, and its settling requires a time that exceeds the mean time of fluctuational transition between the low- and high-temperature states. As we have seen, the temperature value $T = \Theta_2$ of the

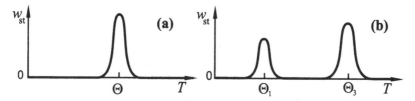

**Fig. 2.6.** Stationary distribution functions for monostable (**a**) and bistable (**b**) point schemes

unstable state is the boundary of the attraction domains for the stable stationary regimes ($T = \Theta_1$ and $\Theta_3$) in the absence of noise. Therefore if the sample temperature, the initial value of which is $T(0) = T_0$, approaches the level $\Theta_2$ under the effect of fluctuations, then spontaneous state transition will take place (with finite probability, for a return to the initial state is also possible). Thereby the mean time of fluctuational transition is expressed through the mean time $\bar{t}'$ the boundary $T = \Theta_2$ is reached for the first time [350]. To determine this, it is convenient to replace $T$ in (2.2.22) by a new variable

$$S = \int^T \frac{dT}{P(T)} = 1 , \qquad (2.2.28)$$

which obeys a simpler equation

$$\frac{dS}{dt'} = -G(S) + \xi(t'), \quad G = F/P . \qquad (2.2.29)$$

Stationary, in the absence of noise, states correspond to the condition $G(S) = 0$.

Stochastic equation (2.2.29) has been studied in detail, including studies in the field of optical bistability [77, 3, 55]. The results allow us a pictorial presentation with the introduction of the potential

$$U_S = \int^S G \, dS , \qquad (2.2.30)$$

which has a two-well character under the condition of bistability (2.2.6) (two minima at $S = S_1$ and $S_2$, that is $T = \Theta_1$ and $\Theta_3$, and a maximum at $S = S_2$, $T = \Theta_2$). The system spends the principal part of the time near the bottom of one or the other well (mainly near the deepest one). Fluctuational transition can be regarded as overcoming the potential barrier during noise-induced transition from one well to the other. Time of transition from state 1 to 3 is

$$t_{1\to3} \approx \frac{\pi}{\sqrt{|U_2''||U_1''|}} \exp\left(\frac{U_2 - U_1}{D_n}\right) , \quad U_n = U_S\big(S(\Theta_n)\big) . \qquad (2.2.31)$$

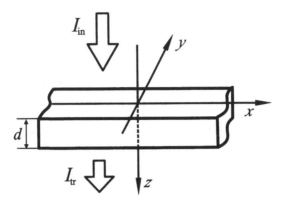

**Fig. 2.7.** Rod increasing absorption bistable scheme

It follows from (2.2.31) that the probability of spontaneous transitions between the states is exponentially small within the whole range of bistability (2.2.6) except at its edges. We will not discuss here these concepts thoroughly analyzed in [114, 156]. Additional peculiarities of bistable system response arise if incident radiation is subject not only to noise, but simultaneously to a regular (periodical) modulation [80]. Then "stochastic resonance" takes place: a sharp increase and subsequent decrease in the signal-to-noise ratio with an increase in noise intensity (the signal-to-noise parameter is determined as the ratio of intensity of the fluctuation spectral density at regular modulation frequency at the system output to this intensity value in the absence of modulation).

The scheme considered is characteristic for well-studied point nonlinear systems. In the following sections we will see that spatial distributivity of a nonlinear system introduces new nontrivial features into hysteresis phenomena.

## 2.3 Steady-State Distributions in a One-Dimensional System

The simplest distributed bistable system is a one-dimensional (rod) system (Fig. 2.7). We will start our acquaintance with phenomena of spatial (more precisely, transverse) distributivity from the analysis of its stationary regimes, following [277].

### 2.3.1 Mechanical Analogy

Equation (2.1.10) for this problem is reduced to

$$\Lambda \frac{\mathrm{d}^2 T}{\mathrm{d}x^2} = F .$$ (2.3.1)

We still consider incident radiation as a plane wave. Then $F$ does not depend explicitly on $x$: $F = F(T)$, and (2.3.1) is integrated by quadratures for the arbitrary type of nonlinearity [see (2.3.7) below]. The mechanical analogy is convenient for graphic presentation of solutions. Within its framework we interpret (2.3.1) as Newton's equation for mechanical motion of a material point with mass $m$ under the effect of the force $F$:

$$m \frac{\mathrm{d}^2 X}{\mathrm{d}t^2} = F(X) .$$ (2.3.2)

The correspondence of the variables in (2.3.1) and (2.3.2) is shown in Table 2.1.

**Table 2.1.** Variables in the mechanical analogy

| Heat equation | | Mechanical equation | |
|---|---|---|---|
| Thermal conductivity | $\Lambda$ | Mass | $m$ |
| Temperature | $T$ | Coordinate | $X$ |
| Coordinate | $x$ | Time | $t$ |
| Heat balance | $F(T)$ | Force | $F(T)$ |

"Potential energy" $U$ is constructed with the "force" $F$,

$$U = - \int^T F(T) \, \mathrm{d}T .$$ (2.3.3)

Taking into account the form of $F(T)$ [see (2.1.11)], we obtain

$$U = A(T) I_{\mathrm{in}} - B(T) ,$$ (2.3.4)

where

$$A(T) = \int^T P(T) \, \mathrm{d}T , \quad B(T) = (H/2)(T - T_{\mathrm{amb}})^2 + \mathrm{const.}$$ (2.3.5)

The determination of the stationary temperature distribution $T(x)$ is thus reduced to the solution of the classical problem of one-dimensional motion of a material point in a potential field [185]. It is convenient to use the integral of motion, which corresponds to the mechanical energy conservation law

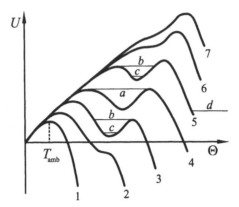

**Fig. 2.8.** "Potential curves" $U(\Theta)$ for different values of incident radiation intensity. 1: $I_{in} = 0$; 2: $I_{in} = I_{min}$; 3: $I_{min} < I_{in} < I_0$; 4: $I_{in} = I_0$; 5: $I_0 < I_{in} < I_{max}$; 6: $I_{in} = I_{max}$; 7: $I_{in} > I_{max}$

$$\frac{\Lambda}{2}\left(\frac{dT}{dx}\right)^2 + U(T) = W = \text{const.} \tag{2.3.6}$$

The first summand on the left-hand side corresponds to the kinetic energy. We find the general solution to (2.3.1) from (2.3.6):

$$x = \pm(\Lambda/2)^{1/2} \int^T [W - U(T)]^{-1/2}\, dT . \tag{2.3.7}$$

The temperature distribution $T(x)$ is obtained by inversion of dependence (2.3.7).

According to (2.1.11) and (2.3.4), the "force" $F$ and the "potential" $U$ are linear functions of the incident radiation intensity. The series of dependencies of the "force" $F(T)$, characteristic for bistability conditions (Sect. 2.2), is depicted in Fig. 2.4 for different values of intensity $I_{in}$, and the corresponding series of potential curves $U(T)$ is presented in Fig. 2.8.

The extremes of the potential curves correspond to the transversely homogeneous states of the previous section $T = \Theta$, defined by (2.2.3) (temperature does not depend on coordinate $x$). Outside the bistability range, at $I_{in} < I_{min}$ or $I_{in} > I_{max}$, the potential curves are "single-humped" (one maximum), and inside this interval [condition (2.2.6)] they are "two-humped" (two maxima). As we have seen in Sect. 2.2, the solution $T = \Theta_2$ corresponding to the minimum of the potential curve [intermediate branch of dependency $\Theta(I_{in})$, Fig. 2.3], is unstable. Stability of the regimes $\Theta_1$ and $\Theta_3$ (the maxima of the potential curve) with respect to small perturbations with arbitrary spatial structure will be shown in Sect. 2.4.

The left hump of the potential curve is higher than the right one, if the incident radiation intensity is in the interval

$$I_{\min} < I_{\text{in}} < I_0 \tag{2.3.8}$$

and is lower if

$$I_0 < I_{\text{in}} < I_{\max} . \tag{2.3.9}$$

The intensity $I_0$ corresponding to equality of heights of the two humps $[U(\Theta_1) = U(\Theta_3)]$ is determined by the relation

$$\int_{\Theta_1}^{\Theta_3} F(T)\,dT = 0 . \tag{2.3.10}$$

This rule can also be interpreted geometrically as the condition of equality of areas of the two hatched domains in Fig. 2.2 (1 and 2). Another interpretation of this important relation will be given in Sect. 2.5.

### 2.3.2 Stationary Inhomogeneous Distributions

The mechanical analogy and Fig. 2.8 show the presence of inhomogeneous stationary temperature regimes $T(x)$ as well as the transversely homogeneous states $T = \Theta_1$, $\Theta_2$, and $\Theta_3$. According to (2.3.6), they correspond to the intervals of the horizontal straight lines $W = \text{const.}$ in classically allowed domains $W > U(T)$ in Fig. 2.8. In this section, we will limit ourselves by analysis of the asymptotically settling regimes (in the transverse direction), which satisfy the natural boundary conditions

$$\left(\frac{dT}{dx}\right)_{x\to\pm\infty} = 0 . \tag{2.3.11}$$

In the language of the mechanical analogy, condition (2.3.11) means that the trajectories – the horizontal intervals in Fig. 2.8 depicting temperature distributions $T(x)$ – should originate ($x = -\infty$) and terminate ($x = +\infty$) at the maxima of the potential curves. Such trajectories can be of two types:

(i) Trajectory $a$ in Fig. 2.8 exists only under condition (2.3.10), i.e., for $I_{\text{in}} = I_0$. As is shown in Fig. 2.9a, it corresponds to a smooth transition with variation of the coordinate between two stable stationary states $\Theta_1$ and $\Theta_3$ – the lower and higher branches of hysteresis dependence $\Theta(I)$ for the transversely homogeneous problem (Fig. 2.3). Such regimes have the following asymptotic behaviour:

$$T_{x\to-\infty} = \Theta_1 , \quad T_{x\to+\infty} = \Theta_3 , \tag{2.3.12}$$

or

$$T_{x\to-\infty} = \Theta_3 , \quad T_{x\to+\infty} = \Theta_1 . \tag{2.3.13}$$

These regimes are named motionless switching waves, according to the reasons stated in Sect. 2.5.

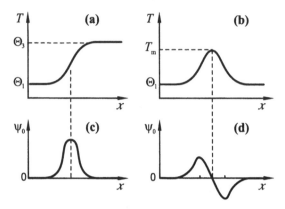

**Fig. 2.9.** Stationary temperature distributions for trajectories $a$ and $b$ in Fig. 2.8 (**a**, **b**) and the corresponding eigenfunctions ("neutral modes") for the perturbations (**c**, **d**)

The presence of the indicated stationary temperature distribution of the transient type means spontaneous violation of spatial symmetry, allowable because of the character of the temperature dependence of the absorption coefficient $\alpha(T)$. The colder part of the rod ($T \approx \Theta_1$) is not heated because of the small heat release within it, while higher temperature is maintained in the hotter part of the rod ($T \approx \Theta_3$) because of an increase in the absorption coefficient and, accordingly, of heat release. Spatial coexistence of two asymptotically homogeneous temperature regimes ($T = \Theta_1$ and $\Theta_3$) resembles the coexistence of two phases (for instance, vapour and liquid), and the condition of such existence (2.3.10) is "Maxwell's rule" known in the theory of phase transitions of the first kind [194]. According to this reason, we will denote the intensity value $I_0$ defined by (2.3.10) as Maxwell's value. Analogy with phase transitions is traced in more detail in [283, 284].

(ii) Trajectory $b$ in Fig. 2.8 exists at any value of intensity inside the bistability range (2.2.6), except Maxwell's value, $I_{in} = I_0$. It corresponds to a symmetric (with respect to the coordinate of extreme value of temperature $T_m$) distribution $T(x)$ (Fig. 2.9**b**). Accordingly, the asymptotic values of temperature under $x \to +\infty$ and $x \to -\infty$ coincide for them:

$$T_{x \to \pm\infty} = \begin{cases} \Theta_3, & I_{in} < I_0 , \\ \Theta_1, & I_{in} > I_0 . \end{cases} \qquad (2.3.14)$$

When the intensity approaches Maxwell's value $I_0$, the temperature distribution resembles the domain of the stable state against a background of the metastable state; in this case the width of such a stationary domain increases unrestrictedly. The dependence of the extreme [minimum for (2.3.8) and maximum for (2.3.9)] value of $T_m$ on the radiation intensity $I_{in}$ is shown in Fig. 2.10. Because at $I_{in} \neq I_0$ the value $T_m$ does

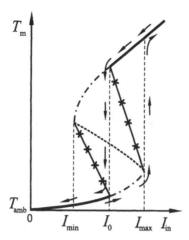

**Fig. 2.10.** Characteristics of stationary states for a distributed system. Metastable states are indicated with *dash-dotted lines* in the lower and upper branches of the transfer function. The *lines with crosses* are extreme values for critical nuclei

not reach the stationary temperature value $\Theta$ for another branch, one can call such distributions "regimes of incomplete switching" or, according to considerations presented in Sect. 2.6, "*critical nuclei*". Note that on the phase plane of (2.3.1) the stationary transversely homogeneous regimes $dT/dx = 0$, $T = \Theta_1$ and $\Theta_3$ correspond to saddles, switching waves represent *heteroclinic trajectories* exiting one saddle and entering another one, and critical nuclei correspond to *homoclinic trajectories* exiting and entering the same saddle. However, it is more convenient for us here to use not the phase plane, but "potential portraits" shown in Fig. 2.8.

As the problem considered here is invariant with respect to a shift along the $x$-axis, both types of solutions are also invariant with respect to this operation. This is exhibited also in the form of (2.3.7), according to which $x$ is determined with an accuracy up to an arbitrary constant. In other words, if some solution of (2.3.1) is known,

$$T = T(x) \,, \tag{2.3.15}$$

then there is also solution of (2.3.1) of the form

$$T = T(x + \delta x) \,, \tag{2.3.16}$$

where $\delta x = $ const. Note that for a motionless switching wave there is the distribution $T(-x)$ with boundary condition (2.3.13) along with the distribution $T(x)$ with boundary condition (2.3.12). This circumstance corresponds to the equivalence of directions $x$ and $-x$.

To find the explicit form of the stationary distributions $T(x)$, one has to specify dependence $F(T)$. In a number of cases the integral in (2.3.7) is expressed through elementary functions [395], e.g., for the case of the piecewise-linear approximation of $F(T)$.

## 2.4 Stability of Steady-State Distributions

The stability analysis is based on the following nonstationary heat conduction equation

$$\varrho_0 c_v \frac{\partial T}{\partial t} = \Lambda \frac{\partial^2 T}{\partial x^2} - F(T) \, . \tag{2.4.1}$$

The stability of the stationary solution $T_{st} = T_{st}(x)$ will take place if the temperature distribution close to it at initial time moment $t = 0$,

$$T(x, 0) = T_{st}(x) + \delta T(x, 0) \, , \tag{2.4.2}$$

is represented for large times in the form

$$T(x, t) = T_{st}(x + \delta x) + \delta T(x, t) \, , \tag{2.4.3}$$

where $\delta x = $ const. and $\delta T(x, t) \to 0$ for $t \to +\infty$. The initial perturbation is considered to decrease sufficiently quickly at $x \to \pm\infty$ $[\delta T(\pm\infty, 0) = 0]$. Note that the constant shift $\delta x$ of stationary solution along the coordinate induced by initial perturbation (see (2.4.3)), does not mean instability of the initial solution [420].

Because perturbations are small, one can linearize (2.4.1) and obtain the following equation for deviation $\delta T(x, t)$:

$$\varrho_0 c_v \frac{\partial \delta T}{\partial t} = \Lambda \frac{\partial^2 \delta T}{\partial x^2} - V(x) \delta T \, , \tag{2.4.4}$$

where

$$V(x) = \left( \frac{\mathrm{d}F}{\mathrm{d}T} \right)_{T=T_{st}(x)} = -\left( \frac{\mathrm{d}^2 U}{\mathrm{d}T^2} \right)_{T=T_{st}(x)} \tag{2.4.5}$$

[after calculation of the derivatives in (2.4.5), value $T$ has to be replaced by the stationary distribution $T_{st}(x)$ investigated for its stability]. Separating the variables in (2.4.4), we come to exponential temporal variation of the perturbation

$$\delta T(x, t) = \psi(x) \exp(-\gamma t) \tag{2.4.6}$$

and to the equation of the Schrödinger type

$$\Lambda \frac{\mathrm{d}^2 \psi}{\mathrm{d}x^2} + [\varrho_0 c_v \gamma - V(x)] \psi = 0 \tag{2.4.7}$$

with the boundary condition

$$\psi_{x \to \pm \infty} = 0 \ . \tag{2.4.8}$$

This is the thoroughly investigated equation for eigenvalues $\gamma$ and eigenfunctions $\psi$, i.e., the Storm–Liouville boundary problem [382]. Let us mark separate solutions of this problem by index $n$. Then the general solution for (2.4.7) is the superposition

$$\delta T(x, t) = \sum_n C_n \psi_n(x) \exp(-\gamma_n t) \ . \tag{2.4.9}$$

The stationary solution $T_{st}(x)$ will be unstable if even one negative eigenvalue is found among the eigenvalues $\gamma_n$, and it will be stable if there are no negative eigenvalues. The constants $C_n$ are determined by the form of the initial perturbation $\delta T(x, 0)$ and do not influence the character of asymptotic stability of the stationary solution.

It follows from (2.4.5) that $V = V_{1(3)} > 0$ for the transversely homogeneous stationary regimes $T_{st} = \Theta_1$ or $\Theta_3$ (the humps of the potential curve). Therefore the condition $\gamma_n > V_{1(3)}/\varrho_0 c_v > 0$ holds for all $n$ (in the opposite case the eigenfunction would increase unrestrictedly with an increase in $x$). Hence the states $T = \Theta_1$ and $\Theta_3$ are stable with respect to small perturbations with an arbitrary transverse profile. Naturally, instability of the states $T = \Theta_2$ (intermediate branch) remains.

The analysis for stability of the transversely inhomogeneous distributions found in Sect. 2.3 is more complex; it is based on the invariance to the shift of coordinate $x$ both for the stationary (2.3.1) and nonstationary (2.4.1) heat conduction equations [420]. Indeed, because both $T_{st}(x)$ and $T_{st}(x + \delta x)$ are solutions of (2.4.1), then the "neutral mode", their linear combination,

$$\psi_0(x) = \lim_{\delta x \to 0} \{ [T_{st}(x + \delta x) - T_{st}(x)]/\delta x \} = \frac{dT_{st}(x)}{dx} \ , \tag{2.4.10}$$

will be a solution for (2.4.7) with the eigenvalue $\gamma = 0$. This conclusion is applicable to the stationary distributions both of the transient type (motionless switching wave) and of the type of critical nucleus. The difference between these two solutions is in the character of coordinate dependency $\psi_0(x)$. For a motionless switching wave, $T_{st}(x)$ is a monotone function; therefore $\psi_0 = dT_{st}(x)/dx$ is a bell-like function, which has no nodes (zeros), besides $x = \pm\infty$ (Fig. 2.9c). In this case $\gamma = 0$ is the lower eigenvalue for (2.4.7). The rest of the eigenvalues in expansion (2.4.9) can only be positive, and this implies the stability of the stationary distribution of the transient type. For the critical nucleus the dependency $T_{st}(x)$ is bell-like; therefore $\psi_0(x)$ has a node (zero) at the coordinate $x = x_m$, corresponding to extreme temperature $T(x_m) = T_m$ (Fig. 2.9d). Therefore a lower negative eigenvalue should exist which corresponds to the eigenfunction without nodes. Thus the critical nucleus is unstable. Note that in this case there is only one negative eigenvalue $\gamma < 0$ in expansion (2.4.9), whereas the number of positive eigenvalues of the discrete spectrum depends on the specific form of potential $U(T)$ [i.e., on $F(T)$].

Similar considerations show the instability of stationary distributions with periodic transverse variation of temperature $T_{st}(x)$ in an unlimited rod (trajectories $c$ in Fig. 2.8). Indeed, in the Storm–Liouville problem with the periodic boundary conditions, the solution $\psi_0 = dT_{st}/dt$ corresponding to $\gamma = 0$ has a node inside each period of $T_{st}(x)$. Therefore, as well as in the case of a critical nucleus, there should also be an eigenfunction $\psi(x)$ without nodes with a negative eigenvalue ($\gamma < \gamma_0 = 0$), which means instability of the distribution considered. The trajectories of the type $d$ in Fig. 2.8 have no physical meaning, as temperature increases unrestrictedly with an increase in the coordinate for them. Therefore the solutions obtained in Sect. 2.3, which satisfy the boundary conditions (2.3.9), actually exhaust all possible stationary regimes of a homogeneously heated unlimited rod.

## 2.5 Switching Waves

### 2.5.1 Switching Waves and the Mechanical Analogy

As we determined in Sect. 2.3, even at a constant intensity of incident radiation (a plane wave, no dependence on the transverse coordinates), a stable steady-state temperature distribution can be transversely inhomogeneous, if the condition $I_{in} = I_0$ is satisfied ["Maxwell's rule", see (2.3.10)]. This naturally brings up the following question: What will happen with such a temperature profile (of the transient kind) if the intensity of incident radiation differs from $I_0$, i.e., $I_{in} \neq I_0$? The answer to this problem leads us to an important class of regimes, namely, *switching waves* [277].

As the distributions we are interested in cannot be stationary [all stationary distributions with natural boundary conditions (2.3.9) have already been found in Sect. 2.3], to analyze them we have to refer to the nonstationary heat conduction equation (2.4.1):

$$\varrho_0 c_v \frac{\partial T}{\partial t} = \Lambda \frac{\partial^2 T}{\partial x^2} - F(T) , \qquad (2.5.1)$$

with the boundary conditions

$$T_{x \to -\infty} = \Theta_n , \quad T_{x \to +\infty} = \Theta_m , \quad n \neq m . \qquad (2.5.2)$$

The function of heat balance $F(T)$ in conditions of bistability (2.2.6) has three zeros (see Fig. 2.4). As noted in Sect. 2.2, they correspond to two stable ($T = \Theta_1$ and $\Theta_3$) and one unstable ($T = \Theta_2$) transversely homogeneous states.

Equations of the form (2.5.1), with similar behaviour of the function $F(T)$, have been investigated in detail in problems of the diffusive propagation of threshold chemical reactions and of flames, and also in a number of adjacent problems [395]. The theory of the stationary and nonstationary regimes described by (2.5.1) is expounded in [420], where extensive literature is also presented. Below we give a summary of results on switching waves, employing the graphical mechanical analogy.

The presence of switching waves is attributed to invariance of (2.5.1) with respect to a shift of time $t$ and coordinate $x$ and to the function $F(T)$ features. They are characterized by the stationary temperature profile propagating along the $x$-axis with a constant velocity $v$, i.e., depending on the only coordinate $\xi$:

$$T(x,t) = T(\xi), \quad \xi = x - vt, \tag{2.5.3}$$

with (2.5.2) asymptotic. What follows for such profiles from (2.5.1) is

$$\Lambda \frac{d^2 T}{dx^2} + \varrho_0 c_v v \frac{dT}{dx} - F(T) = 0. \tag{2.5.4}$$

At $v = 0$ the ordinary differential equation (2.5.4) coincides with (2.5.1). Therefore the stationary solution of transient type found in Sect. 2.3 is a switching wave with zero velocity, which is possible only for $I_{in} = I_0$ (*Maxwell's intensity value*).

The term in (2.5.4) proportional to velocity $v$ (and corresponding to "friction" in the context of the mechanical analogy) results in a violation of conservation of "energy" $W$ introduced by (2.3.6):

$$\frac{dW}{d\xi} = -\varrho_0 c_v \, v \left( \frac{dT}{d\xi} \right)^2. \tag{2.5.5}$$

In parallel with the switching wave with velocity $v$ and asymptotic behaviour (2.5.2), there is also the wave with velocity $-v$ and inverted boundary conditions (2.5.2) with $n \leftrightarrow m$ at the same value of intensity $I_{in}$, because of invariance of (2.5.1) with respect to substitution $x \to -x$. For definiteness we will consider $v > 0$. Then it follows from (2.5.5) that the value $W$ should decrease with an increase in $\xi$; its decrease ends at values $W = U(\Theta_n)$, corresponding to one of the transversely homogeneous states $T = \Theta_n = $ const. On the plane $W, T$ trajectories $T(\xi)$ are represented by curves with decreasing (more precisely, not increasing with an increase in $\xi$) "energy" $W$, located inside classically allowed regions $W > U(T)$ (Fig. 2.11). Between turning points where $U(T) = W$, the trajectory is determined by

$$\frac{dW}{dT} = \mp \varrho_0 c_v v \left\{ (2/\Lambda)[W - U(T)] \right\}^{1/2}. \tag{2.5.6}$$

The upper sign is taken for the trajectory sections where temperature increases with coordinate $\xi$; the lower sign is taken when temperature decreases.

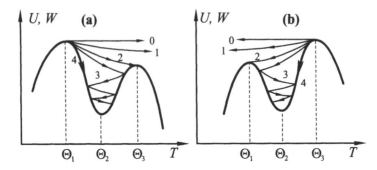

**Fig. 2.11.** "Potential curves" and trajectories as solutions of (2.5.4) with boundary conditions $T_{\xi \to -\infty} = \Theta_1$ (a) and $\Theta_3$ (b). Value $v$ increases from trajectory 0 ($v = 0$) to 4 ($v = \infty$); the stable switching wave corresponds to trajectory 2

### 2.5.2 Stable and Unstable Switching Waves

**Stable Switching Waves.** First we consider waves of switching between stable transversely homogeneous states $\Theta_1$ and $\Theta_3$ with the condition (2.3.8), when the left hump of the potential curve is higher than the right one. A family of trajectories starting from the top of a potential curve at $T = \Theta_1$ is shown in Fig. 2.11a. Near the top $U(T) = U(\Theta_1) - |U''|(T - \Theta_1)^2/2$, where $U'' = \mathrm{d}^2 U/\mathrm{d}T^2 < 0$ at $T = \Theta_1$. In this case the trajectory has the parabolic form

$$W(T) = U(\Theta_1) - \alpha(T - \Theta_1)^2 \, , \quad \alpha = (\beta^2 + \beta|U''|)^{1/2} - \beta \, , \quad \beta = \varrho_0 c_v v/(4\Lambda) \, . \tag{2.5.7}$$

It follows from (2.5.6), with the upper sign in the right-hand side, that the trajectory is a horizontal straight line at $v = 0$; for small $v$

$$W(T) = U(\Theta_1) - \varrho_0 c_v v (2/\Lambda)^{1/2} \int_{\Theta_1}^{T} [U(\Theta_1) - U(T)]^{1/2} \, \mathrm{d}T \, , \tag{2.5.8}$$

and for large velocities ($v \to \infty$) the trajectory approaches a slope of potential shape $W(T) = U(T)$, $\Theta_1 \le T \le \Theta_2$. Because "energy" $W$ decreases monotonically with $v$ at fixed temperature, only one trajectory hits the top $T = \Theta_3$ at $\xi \to \infty$. It means that waves of switching between states with $T = \Theta_1$ and $T = \Theta_3$ with positive velocity $v$ exist within intensity range (2.3.8), and there is only one value of the velocity at a fixed intensity.

Similarly, within intensity range (2.3.9) when the right hump of the potential curve is higher than the left one (Fig. 2.11b), switching waves with positive velocity exist only at the inverted boundary conditions $T(\xi \to -\infty) = \Theta_3, T(\xi \to +\infty) = \Theta_1$, because "energy" $W$ decreases with $\xi$. Now if the form of the boundary conditions remain, the wave velocity turns negative. Therefore switching waves of the type considered exist within the whole bistability range (2.2.6), and their velocity depends uniquely on the incident radiation

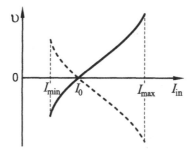

**Fig. 2.12.** Dependence of the switching wave velocity $v$ on the incident radiation intensity $I_{in}$; the *dashed line* corresponds to the inverted boundary conditions

intensity. At the switching wave front, the temperature varies monotonically from one asymptotic value $(T = \Theta_1)$ to another $(T = \Theta_3)$ when the coordinate $\xi$ varies from $-\infty$ to $+\infty$. A typical dependence of velocity $v$ on intensity $I_{in}$ is shown in Fig. 2.12.

Demonstration of stability of the switching waves with the formed temperature profile $T_{sw}(\xi)$ comes down to that given above in Sect. 2.4. Indeed, let us assume

$$T(x,t) = T_{sw}(\xi) + \sum_n C_n \psi_n(\xi) \exp(-\gamma_n t) . \tag{2.5.9}$$

Substitution of (2.5.9) into (2.5.1), linearized with respect to small perturbations, gives an equation for functions $\psi_n$ in the form

$$\Lambda \frac{d^2 \psi_n}{d\xi^2} + [\varrho_0 c_v \gamma_n - V(\xi)]\psi_n + \varrho_0 c_v v \frac{d\psi_n}{d\xi} = 0 , \tag{2.5.10}$$

where, in accord with (2.4.5),

$$V(\xi) = \left(\frac{dF}{dT}\right)_{T=T_{sw}(\xi)} = -\left(\frac{d^2 U}{dT^2}\right)_{T=T_{sw}(\xi)} . \tag{2.5.11}$$

The additional, as compared with (2.4.7), term on the left-hand side of (2.5.10), is eliminated by the substitution $\psi_n(\xi) = \chi(\xi) \exp(-\varrho_0 c_v v \xi / 2\Lambda)$:

$$\Lambda \frac{d^2 \chi}{dx^2} + [\varrho_0 c_v \gamma_n - V(\xi) - (\varrho_0 c_v v)^2 / (4\Lambda)]\chi = 0 . \tag{2.5.12}$$

Now the nodeless eigenfunction with zero eigenvalue $\gamma_n = 0$ is $\chi = (dT_{sw}/d\xi)$ $\exp(\varrho_0 c_v v\xi/2\Lambda)$. Therefore the other perturbation eigenfunctions have positive eigenvalues, so that the switching wave is stable.

**Unstable Switching Waves.** More precisely, the structures considered should be called *stable switching waves*. The mechanical analogy shows the existence, parallel with them, of unstable waves of switching between stable and unstable transversely homogeneous states. These structures include a semi-infinite interval of unstable state with $T = \Theta_2$ (asymptotically, at $\xi \to -\infty$ or $\xi \to +\infty$, $T \to \Theta_2$) and will be called *unstable switching waves.*[2] Despite their instability, these waves are needed for the description of local perturbation kinetics (Sect. 2.6).

Let us return to the case shown in Fig. 2.11a [incident radiation intensity within the range of (2.3.8)]. As we have seen, the trajectory exiting the top of the left hump of the potential curve hits the top of the right hump at only one value of stable switching wave velocity, $v = v_{st}$. At greater velocities the trajectory goes lower. It reaches the slope of the potential curve at $T < \Theta_3$, and after a number of reflections from the slopes, it approaches the potential minimum ($T \to \Theta_2$ at $\xi \to \infty$). Therefore, there are unstable waves of switching between the states with $T = \Theta_1$ and $T = \Theta_2$, characterized by a continuous velocity spectrum $v_{st} < v < \infty$. For every value of velocity, there is a definite temperature profile that can include oscillations. If, however, the trajectory exits the top of the lower hump with $T = \Theta_3$, it hits, as is evident from Fig. 2.11a, the potential minimum at $T = \Theta_2$ for any (positive) velocity. Therefore the velocity range for unstable waves of switching between states with $T = \Theta_3$ and $T = \Theta_2$ is different: $0 < v < \infty$.

Similarly, it is not difficult to see from Fig. 2.11b for range (2.3.9) of incident radiation intensity that unstable waves of switching between states with $T = \Theta_3$ (at $\xi \to -\infty$) and $T = \Theta_2$ (at $\xi \to +\infty$), have the continuous velocity spectrum $v_{st} < v < +\infty$, while for switching between states with $T = \Theta_1$ (at $\xi \to -\infty$) and $T = \Theta_2$ (at $\xi \to +\infty$), the spectrum is $0 < v < +\infty$. Note that there are no motionless unstable switching waves for any incident radiation intensity. The velocity sign is such that the boundary of the domain occupied by stable states always moves toward the domain of unstable states (extent of system domain corresponding to the state with $T = \Theta_2$ reduces with time). The temperature profile of an unstable switching wave can include oscillations.

The consideration presented is also valid in the case of a greater number of function $F(T)$ zeros, when the potential has a number of maxima and minima (multistability). Then mechanical analogy shows the existence of stable and unstable waves of switching between an increased number of transversely homogeneous states. An example is given in Fig. 2.13, where three types of stable switching waves are presented (between pairs of states with $T = \Theta_1$ and $\Theta_3$, $\Theta_1$ and $\Theta_5$, $\Theta_3$ and $\Theta_5$). Again, it would not be difficult to determine ranges of variation in unstable switching wave velocity.

---

[2] Such waves were first discovered by Kolmogorov, Petrovskii, and Piskunov [174] in the context of a biological problem.

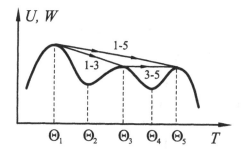

**Fig. 2.13.** "Potential curves" in the case of tristability and stable waves of switching between different transversely homogeneous states with $T = \Theta_n$

### 2.5.3 Numerical Simulations

For a number of model functions $F(T)$, the velocity of a stable switching wave can be found analytically (see, e.g., [395]). For an arbitrary "force" $F(T)$ it can be determined by perturbation theory in the vicinity of "Maxwell's intensity value" $I_0$. However, the most general way, valid for practically any case, is the numerical solution of nonstationary heat conduction equation (2.5.1). Because for stable switching waves their characteristics do not depend on smaller variations of initial conditions, the latter can be taken in a step-like form:

$$T_{x,0} = \begin{cases} \Theta_1, & x < 0, \\ \Theta_3, & x > 0. \end{cases} \qquad (2.5.13)$$

Boundary conditions (2.3.11) are formulated at finite distance:

$$\left(\frac{\partial T}{\partial x}\right)_{x=\pm A} = 0 . \qquad (2.5.14)$$

Numerical simulation of switching waves upon laser heating of a Ge sample [278] confirms the alternation of switching wave velocity $v$ depending on incident radiation intensity $I_{\text{in}}$.

Let us now return to the main case of bistability and stable switching waves. The sign of switching wave velocity is such that within intensity range (2.3.9) the rod "hot" section widens, i.e., what occurs is successive switching of the rod zones from the low-temperature state ($\Theta_1$) to the high-temperature state ($\Theta_3$). In contrast, a cooling wave propagates in the system under condition (2.3.8).

If the temperature profile includes two steps at the initial moment, e.g.,

$$T_{x,0} = \begin{cases} \Theta_1, & |x| < a, \\ \Theta_3, & |x| > a, \end{cases} \qquad (2.5.15)$$

then two switching waves will be formed and propagate for a sufficiently large width of the switched section $2a$. The waves are independent, and their interaction is negligibly weak if the distance between their fronts exceeds the

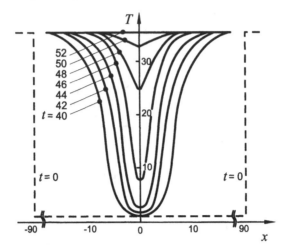

**Fig. 2.14.** Collision of two switching waves: temperature profiles at different time moments $t$ [278]

front width. At $I_{in} < I_0$ the switching wave velocity is such that the initial overshoot of type (2.5.15) widens with time. Therefore the entire rod will be switched gradually to the low-temperature state $(T = \Theta_1)$. If $I_{in} > I_0$, the central overshoot will shrink and disappear after the collision of counterpropagating switching waves (Fig. 2.14); in this case the entire rod will occur in the high-temperature state $(T = \Theta_3)$. The kinetics of initial overshoots of the type (2.5.15) with the exchange $\Theta_1 \leftrightarrow \Theta_3$ is similar.

The temperature gradients at the fronts of switching waves cause noticeable gradients of pressure and mechanical stress waves. Special increase in mechanical stress is to be anticipated under conditions when the switching wave velocity approaches the sound velocity in the medium.

## 2.6 Spatial Switching

Features of the stable switching waves discussed above allow conclusions as to the stability or metastability of stationary transversely homogeneous states. Due to fluctuations, there is a finite probability of large local temperature perturbations (i.e., overshoots), which determine the subsequent kinetics of the system. Because the probability is small, one can consider the large overshoots to be separated in time. Then for the states shown in Fig. 2.10 by dash-dotted lines [section of lower branch at (2.3.9) and of upper branch at (2.3.8)], overshoots of the form of (2.5.15) with inversion of $\Theta_1$ and $\Theta_3$ will widen. The system will be switched with time to the state corresponding to the other stable branch. Therefore the initial transversely homogeneous states (dash-dotted lines in Fig. 2.10) should be attributed to metastable ones, unlike the stable states, indicated in Fig. 2.10 by solid lines [281].

Separation of states into those that are stable and those that are metastable is intrinsic to spatially distributed (wide-aperture) bistable systems (in the strict sense, the latter should be called not bistable, but bimetastable due to the metastability of one of the two states). Generally speaking, fluctuational switching from the stable state to the metastable state is also possible. However, it requires a fairly large initial temperature overshoot over almost entire system aperture, which should be treated as an improbable event.

It would be natural to denote overshoots that switch the system from a metastable state to a stable state as overcritical, the term taken from the theory of phase transitions of the first kind [194]. Because metastable states are stable with respect to weak perturbations, the latter will dissolve (disappear) with time; this means that they are undercritical. Note that initial overshoots can be initiated not only by fluctuations, but artificially too. To do this, one can overheat some section of the rod by an additional source and then switch off the source sharply. This raises the following question: What is the kinetics of different initial overshoots (local perturbations), and how are we to separate them into overcritical and undercritical? For large overshoots, it is realistic to obtain the full answer for point systems only (see Sect. 2.2 and Fig. 2.5). For spatially distributed systems, the nonstationary heat conduction equation (in its dimensionless form)

$$\frac{\partial T}{\partial t} = \frac{\partial^2 T}{\partial x^2} - F(T) \qquad (2.6.1)$$

describes much more diverse overshoot kinetics, which we consider below, following [316].

We assume for definiteness the case (2.3.9), then states with temperature $\Theta_1$ (lower branch) are metastable. If the maximum temperature in an initial overshoot does not surpass $\Theta_2$ (intermediate unstable branch)

$$T^{(m)}(0) = \max T(x,0) < \Theta_2 , \qquad (2.6.2)$$

then the overshoot of any form and width will be undercritical (dissolving). When (2.6.2) is violated in a sufficiently wide domain, a transition to the stable regime, which corresponds to the upper branch, takes place inside the domain, and switching waves are formed at the domain boundaries. The width of the domain of "stable phase" $R$ grows with time linearly:

$$T^{(m)} \to \Theta_3 , \quad \dot{R} \to v(I_{in}) , \quad t \to \infty , \qquad (2.6.3)$$

where $v$ is the velocity of the switching wave (see Sect. 2.5). Evidently, such overshoots are overcritical.

A narrow initial local perturbation (overshoot) can also be overcritical, if its maximum temperature value $T^{(m)}(0)$ is sufficiently large. At the first stage of development of narrow overshoots, the first term on the right-hand side of (2.6.1) prevails. Then, neglecting the contribution of $F(T)$, we come to the linear heat conduction equation with self-similar solution:

$$T(x,t) - T_\infty \propto t^{-1/2} \exp(-x^2/4t) , \quad T_\infty = \text{const.} , \quad t > 0 . \tag{2.6.4}$$

Relation (2.6.4) describes the overshoot spread and decrease in the maximum temperature. According to (2.6.4), the overshoot "height" $T^{(m)}$ and width $R$ are connected by the following relation:

$$(T^{(m)} - T_\infty)R = \text{const.} \tag{2.6.5}$$

When the perturbation width is considerable, both terms on the right-hand side of (2.6.1) are comparable, and initially narrow overshoots can be both undercritical and overcritical.

To pass on to the general case, it is necessary to simplify the problem. Abstracting from the description of the detailed spatial form of the overshoot, let us characterize it by two values only – the extreme temperature $T^{(m)}(t)$ and the overshoot width at some level $R(t)$. If we restrict ourselves to overshoots with the bell-like (unimodal) spatial temperature profile, then the specific form of the profile does not influence the qualitative character of the switching kinetics. The latter is determined by the presence of the transversely inhomogeneous distributions described above – the regime of critical nucleus and switching waves.

The scheme integrating the indicated elements is presented in Fig. 2.15. The arrows' directions correspond to an increase in time. The stationary point with the parameters $(T_m, R_{cr})$ represents the critical nucleus, a localized stationary regime with saddle-type instability. Trajectory 1 corresponding to self-similar solution (2.6.4) and trajectory 2 changing over into an unstable switching wave at $t \to -\infty$ serve as separatrices entering the saddle. These two trajectories separate the domains of the overshoot parameters into the overcritical (hatched) and undercritical. Two trajectories also exit from the saddle – separatrices 3 and 4. For a large amount of time, the regimes presented by separatrix 3, as well as by all other trajectories for overcritical perturbations, transform to a stable switching wave. Separatrix 4 "ends" (at $t \to +\infty$) in the domain corresponding to the metastable state ($T = \Theta_1$).

An interesting feature of the kinetics of overshoots with initial parameters close to the separatrices entering the saddle (curves 1 and 2) is the nonmonotonic temporal variation of their parameters. It is illustrated in Fig. 2.15 by trajectories 5–8. For instance, for trajectory 7 the narrowing of the local perturbation will be observed first (an unstable switching wave); then, in the vicinity of the parameters of the critical nucleus, the maximum temperature will abruptly increase up to a level close to $\Theta_3$, and after that widening in the form of a stable switching wave will take place. Nonmonotonic variation takes place for overshoot width (along trajectories 5 and 8) and for the maximum temperature (along curves 6 and 8).

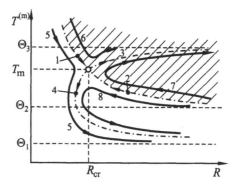

**Fig. 2.15.** Evolution of large local perturbations over the background of the metastable state (the domain of overcritical parameters is *hatched*; $I_0 < I_{in} < I_{max}$)

Note that we could obtain a scheme close to Fig. 2.15 with another approach, if we would assume $T^{(m)}$ and $R$ as the parameters of the initial temperature overshoot with fixed (e.g., Gaussian) shape:

$$T(x, 0) = \Theta_1 + (T^{(m)} - \Theta_1) \exp(-x^2/R^2) . \qquad (2.6.6)$$

The curve close to separatrices *1* and *2* would arise as the line which divides the domain of parameters $(T^{(m)}, R)$ of overshoots, which, with time, switch on the system to stable state $\Theta_3$ or else disappear.

One more aspect of Fig. 2.15 is connected with the conditions for formation of stable switching waves. Strictly speaking, a switching wave occupies the entire rod length ($-\infty < x < +\infty$); therefore the description of temperature distributions in terms of two switching waves has an approximate (asymptotic) character. As seen from Fig. 2.15, one can use the conceptions of switching waves only in the case when the width of the corresponding overshoot exceeds the width of the critical nucleus $R_{cr}$.

Now consider the formation of two switching waves from initial wide symmetric temperature overshoot in more detail. Assume $I_{in} > I_0$, so two switching waves, propagating apart, must be formed, their interaction being weak because of the spacing of fronts of the waves. Let us denote the coordinates of the fronts $R(t)$ and $-R(t)$ by relation $T\left(\pm R(t), t\right) = \Theta_2$. Multiplying (2.6.1) by $\partial T/\partial x$ and integrating over $x$ from 0 to $\infty$, we obtain

$$\dot{R} = v - F'_{T=\Theta_3} (\Theta_3 - T_{x=0})^2 \left[ 2 \int_0^\infty \left( \frac{\partial T}{\partial x} \right)^2 dx \right]^{-1} . \qquad (2.6.7)$$

The value

$$v = - \int_{\Theta_1}^{\Theta_3} F(T) \, dT \left[ \int_0^\infty \left( \frac{\partial T}{\partial x} \right)^2 dx \right]^{-1} \qquad (2.6.8)$$

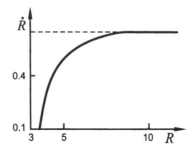

**Fig. 2.16.** Dependence of the velocity of switching front propagation on local perturbation width

is the velocity of a single switching wave. The second term on the right-hand side of (2.6.7) is caused by the finite width of the central domain of the high-temperature phase in the overshoot, where $T \approx \Theta_3$. It leads to a dependence of the velocity of front propagation $\dot{R}$ on the width $R$, which indicates that $\dot{R}$ increases together with $R$, asymptotically approaching $v$ at $R \to \infty$. For large $R$, the correction term in (2.6.7) is small, and one can obtain the explicit form of $\dot{R}$ dependence on time or on the overshoot width. However, it is easier and more reliable to find this dependence numerically. The result of such a calculation is presented in Fig. 2.16.

A similar picture for the kinetics of local perturbations is obtained also for (2.3.8), when the high-temperature states $(T = \Theta_3)$ are metastable. The corresponding scheme is presented in Fig. 2.17, and it does not seem to need any comment. Note that the nonmonotonic character of temperature overshoot parameters has been confirmed experimentally in optical bistable schemes [133]. Similar results for kinetics of artificial overshoots in thermodynamic systems with a phase transition of the first kind have as yet been found only in theory [283, 284].

## 2.7 Wide Radiation Beam

### 2.7.1 Asymptotic Analysis

The previous sections treated the "ideal" system obeying translation symmetry. In the rest of this chapter we will study important effects of the system inhomogeneities, starting with the case of transverse inhomogeneity of incident radiation intensity.

Consider an incident radiation beam with characteristic scale $w_b$ of transverse variation of intensity, for instance, a Gaussian beam $I_{in}(x) = I_m \exp(-x^2/w_b^2)$. The stationary profiles of rod temperature $T(x)$ are still determined by (2.3.1). However, now the "force" $F$ depends on the coordinate explicitly:

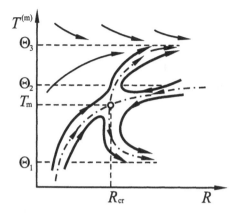

**Fig. 2.17.** Kinetics of large perturbations ($I_{\min} < I_{\text{in}} < I_{\max}$)

$$F = F(T, x) = -P(T)I_{\text{in}}(x) + H(T - T_{\text{amb}}) , \qquad (2.7.1)$$

where the form of $P(T)$ is given by (2.1.11). Therefore the "energy" $W$ [see (2.3.6)] is not conserved. As one can see,

$$\frac{dW}{dx} = P(T)\frac{dI_{\text{in}}}{dx} . \qquad (2.7.2)$$

The results of the previous sections are valid for $dI_{\text{in}}/dx = 0$, when $W = $ const. follows from (2.7.2). The "energy" $W$ increases (decreases) for the beam in the intervals of increase (decrease) in intensity $I_{\text{in}}(x)$ with a rate proportional to the gradient of incident radiation intensity.

Now we make an important assumption. We will assume that the radiation beam is wide, i.e., the width of the bell-like intensity profile $w_{\text{b}}$ substantially exceeds the width of the switching wave front $w_{\text{fr}}$:

$$w_{\text{b}} \gg w_{\text{fr}} . \qquad (2.7.3)$$

Under (2.7.3), one can obtain the solution for the stationary heat conduction equation (2.3.1) with an asymptotic (singular perturbations) technique [65]. To do that, we will use the dimensionless coordinate $x' = x/w_{\text{fr}}$. Then (2.3.1) takes the form:

$$\mu\frac{d^2T}{dx'^2} = F(T, x') , \qquad (2.7.4)$$

where a small parameter, $\mu = \Lambda/w_{\text{fr}}^2$, appears corresponding to a small "particle mass" – in the spirit of the mechanical analogy.

The asymptotic ($w_{\text{b}} \to \infty, \mu \to 0$) solution to (2.7.1) is constructed from the external expansions joined by an internal expansion, i.e., the boundary layer. The external expansion corresponds to the solution of (2.7.4) in the form of a series in terms of $\mu$ integer powers:

$$T^{\text{ext}}(x') = \sum_{s=0}^{\infty} \mu^s T_s^{\text{ext}}(x') \, . \tag{2.7.5}$$

Substituting (2.7.5) into (2.7.4) and equating the terms with equal powers of $\mu$, we obtain algebraic (not differential) relations for the determination of $T_s^{\text{ext}}$. In the lower approximation ($\mu = 0$) we thus find

$$F(T_0^{\text{ext}}, x') = 0 \, , \tag{2.7.6}$$

which coincides with (2.2.3) for transversely homogeneous states. The corresponding profile $T_0^{\text{ext}}(x')$ is determined by inversion of dependence equivalent to (2.2.4):

$$I_{\text{in}}(x') = H \left[ T_0^{\text{ext}}(x') - T_{\text{amb}} \right] / P \left( T_0^{\text{ext}}(x') \right) \, , \tag{2.7.7}$$

where the $x'$ coordinate plays the role of a parameter. To find the profile, it is actually sufficient to use the dependence $T(I_{\text{in}})$ constructed for transversely homogeneous regimes (see Fig. 2.3). This corresponds to the scheme with separation of the radiation beam into noninteracting ray tubes (Chap. 1).

In the case of monostability, the external expansion is sufficient for the construction of the temperature profile $T(x')$ for any smooth profile of the incident radiation intensity $I_{\text{in}}(x')$. The external expansion is sufficient also in the case of bistability, if maximum intensity $I_{\text{m}} = \max_{x'} I(x')$ is smaller than the threshold one for the lower branch (although we will see below that the solutions with spatial switching between branches, which cannot be described by the external expansion, are also possible under this condition). At $I_{\text{m}} > I_{\text{max}}$ the external expansion is insufficient, as sharp transverse temperature jumps of the type of boundary layer are then unavoidable. But the indicated external expansion (approximation of noninteracting ray tubes) is valid "almost everywhere" for such temperature profiles too. In this case the external expansion is constructed in the periphery of the beam with the lower branch of the hysteresis curve of Fig. 2.3, while in the beam's central part it is necessary to use the upper one. Note that within the framework of the mechanical analogy the humps of the potential curve (Fig. 2.8) correspond to the external expansion. Here we have to find the solution with the spatial transition between the two humps.

For the consequent construction of the temperature distributions we are interested in, we have to use the concept of the internal (boundary) layer situated in the vicinity of coordinate $x_0'$ not known beforehand. To describe the sharp variation of temperature in this domain, we introduce a new ("fast") coordinate $\tilde{x} = (x' - x_0')/\mu^{1/2}$. Let us expand the intensity profile $I_{\text{in}}(\tilde{x})$ into Taylor's series in the vicinity of $x_0'$:

$$I_{\text{in}}(\tilde{x}) = I_0 + I_1 \mu^{1/2} \tilde{x} + I_2 \mu \tilde{x}^2 / 2! + \dots \, , \tag{2.7.8}$$

where $I_0 = I(x_0')$ and $I_n = (\text{d}^n I / \text{d}x'^n)_{x'=x_0'}$. Taking into account (2.1.11), we present (2.7.4) in the following form:

$$\frac{d^2T}{d\tilde{x}^2} = -\left(I_0 + I_1\mu^{1/2}\tilde{x} + \frac{1}{2!}I_2\mu\tilde{x}^2 + \dots\right)P(T) + H(T - T_{\text{amb}}) . \quad (2.7.9)$$

The solution of (2.7.9) is constructed in the form of series in powers of $\mu^{1/2}$ [whereas the external expansion (2.7.5) includes only integer powers of $\mu$]:

$$T^{\text{int}}(\tilde{x}) = \sum_{s=0}^{\infty} \mu^{s/2} T_s^{\text{int}}(\tilde{x}) . \quad (2.7.10)$$

Substituting (2.7.10) into (2.7.9) and equating once more the terms with equal powers of the small parameter $\mu$, we obtain the equation for the determination of $T_s^{\text{int}}(\tilde{x})$. Joining of the external and internal expansions plays the role of boundary conditions. In the lower approximation in $\mu^{1/2}$ the leading term $T_0^{\text{int}}(x)$ is found from

$$\frac{d^2T_0^{\text{int}}}{d\tilde{x}^2} = -I_0 P(T_0^{\text{int}}) + H(T_0^{\text{int}} - T_{\text{amb}}) = F(T_0^{\text{int}}) . \quad (2.7.11)$$

Taking into account (2.7.1), we come to the initial equation (2.7.4) with substitution $\mu \to 1$. However, now $I_{\text{in}} = I_0 = \text{const.}$, and the "force" $F$ does not depend explicitly on the coordinate $x$. Thus we reduced the problem to the one considered in Sect. 2.3, although the intensity value $I_0$ at the front of the boundary layer is not yet known. Therefore, the boundary layer in the language of the mechanical analogy (see Fig. 2.8) is presented by the interval of the horizontal straight line $W = \text{const.}$ Our task is reduced, in view of (2.7.2), to the construction of a solution that is represented by the motion over the left humps [$I_{\text{in}}(x) < I_0$, the beam peripheral part], "horizontal" transition from the left hump to the right one at $I_{\text{in}} \approx I_0$, and further motion over the right humps [$I_{\text{in}}(x) > I_0$, the beam's central part]. It is clear from that stated above that the transition between the humps of the potential curves is possible only with their heights equal. This allows us to identify the value $I_0$ of the present section with "Maxwell's intensity value" $I_0$ (2.3.10). Indeed, precisely at this condition there are solutions corresponding to the transition between the branches under variation of the $x$ coordinate.

As a result, the position of the boundary layer (coordinate $x_0$) is determined from

$$I(x_0) = I_0 , \quad (2.7.12)$$

where the value $I_0$ is found from (2.3.10). Moving along the boundary layer (motionless switching wave) away from the front to $\tilde{x} \to \pm\infty$, we come to the top of the left ($T = \Theta_1$) or the right ($T = \Theta_3$) humps. With an increase in $|\tilde{x}|$, the deviation of $T_0^{\text{int}}(\tilde{x})$ from these asymptotic values ($\Theta_1$ or $\Theta_3$) decays exponentially, which allows us to join the internal expansion with the external one.

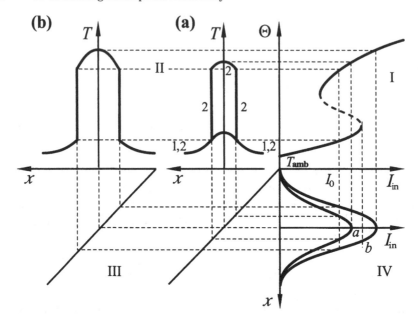

**Fig. 2.18.** The scheme for the construction of temperature profiles: the transfer function (quadrant I); the incident radiation intensity profile (quadrant IV); and sample temperature distributions (quadrant II): (a) spatial bistability, (b) spatial monostability

### 2.7.2 Spatial Bistability

When the width of the boundary layer $w_{\mathrm{fr}} \sim (\Lambda/H)^{1/2}$ is negligible as compared with the radiation beam width $w_{\mathrm{b}}$, we come to the scheme for the determination of the stationary temperature profile given in Fig. 2.18. Quadrant I presents the S-like dependence of the temperature $T = \Theta$ on the radiation intensity $I_{\mathrm{in}}$ for the transversely homogeneous distributions, which coincides with Fig. 2.3. The bell-like intensity profile $I_{\mathrm{in}}(x)$ is shown in quadrant IV. With these two dependencies and the known value of $I_0$, we find the profiles for temperature $T(x)$ by the construction depicted in the figure (quadrant II).

If the intensity maximum over the beam cross-section $I_{\mathrm{m}} = I_{\mathrm{in}}(0)$ is lower than $I_0$, i.e., $I_{\mathrm{m}} < I_0$, then there is the only temperature profile $T(x)$ completely determined by the lower branch of the hysteresis curve. Note that in this case the left part of the upper branch of the hysteresis curve is not realized [condition (2.3.8), the dash-dotted line in Fig. 2.10]. This part is realized for radiation beams with an intensity dip up to a value exceeding $I_{\mathrm{min}}$ (Fig. 2.19). The solution at $I_{\mathrm{m}} > I_{\mathrm{max}}$ is also unique. However, in this case the temperature profile is not smooth, but combined. It consists of the peripheral part determined by the lower branch [$I_{\mathrm{in}}(x) < I_0$] and the central part corresponding to the upper branch [$I_{\mathrm{in}}(x) > I_0$], with sharp switching between

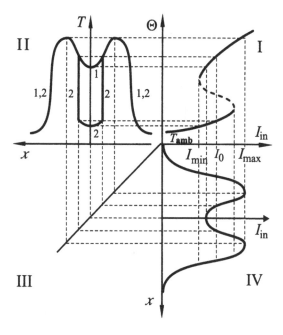

**Fig. 2.19.** Spatial bistability in the case of an incident beam with an intensity dip

the branches in vicinity of Maxwell's intensity value $I_{in} \approx I_0$ (Fig. 2.18b). Bistability of the temperature profile exists only in the following interval of intensities:

$$I_0 < I_m < I_{max} . \tag{2.7.13}$$

Under this condition, for fixed parameters of the beam with the bell-like profile, there are two possible temperature profiles: smooth profile 1 and combined profile 2 with switching between the branches (quadrant II, Fig. 2.18a). It is evident that bistability of the temperature profile is accompanied by bistability of the profile of intensity of transmitted radiation. The corresponding scheme is presented in Fig. 2.20.

The presented theory is asymptotic; it is valid for sufficiently wide radiation beams (2.7.3). As its lower approximation has already allowed us to find the solution for the problem of spatial bistability, it is not necessary to determine subsequent approximations. For comparatively narrow beams, when the condition (2.7.3) is violated, it is easier and more reliable to solve the problem numerically.

Note that the indicated temperature profiles are composed of elements of the transversely homogeneous problem, stable with respect to small perturbations. This serves as an argument for the stability of the profiles of sufficiently wide radiation beams. The analysis of kinetics of spatial hysteresis confirms this statement (next section).

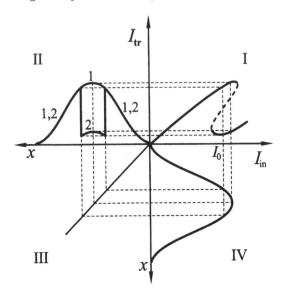

**Fig. 2.20.** Construction of intensity profiles for the transmitted radiation (quadrant II)

## 2.8 Spatial Hysteresis

Now we consider how the temperature profile varies in a rod with a slow temporal variation of the maximum intensity of incident radiation: $I_m = I_m(t)$. For definiteness, we will assume the coordinate dependence of intensity to be fixed (bell-like), $I_{in}(x, t) = I_m(t)f(x)$, $\max f(x) = 1$, and the beam width to be sufficiently large (2.7.3). Then we can speak about switching waves with a slowly varying front propagation velocity, determined by local radiation intensity in the beam at the front location $I_{fr}$: $v = v(I_{fr})$.

With a slow temporal increase in intensity $I_m$ from small values up to the intensity of breakdown of the lower branch $I_{max}$, the temperature profile $T(x, t)$ will be smooth at each time moment (of type 1 in Fig. 2.21a), corresponding to the lower branch of the hysteresis curve. At the moment when $I_m$ exceeds $I_{max}$, a narrow and sharp temperature overshoot – local perturbation – appears in the centre of the beam (profile 2 in Fig. 2.21a). Then it will gradually widen in the form of two stable switching waves propagating in opposite directions even at the stabilization of $I_m$. The velocity of propagation of these waves is close for wide beams to the velocity $v = v(I_{fr})$, determined above for external radiation in the form of a plane wave (see Fig. 2.12). Therefore for the considered bell-like beams the propagation of the front will decelerate (because of the decrease in the local radiation intensity value $I_{fr}$). Finally, the front will stop at a position where the local intensity is $I_{fr} = I_0$ (note that $I_0$ is determined by Maxwell's rule (2.3.10), i.e., $v(I_0) = 0$). Thus the hysteresis transition takes place not simultaneously over

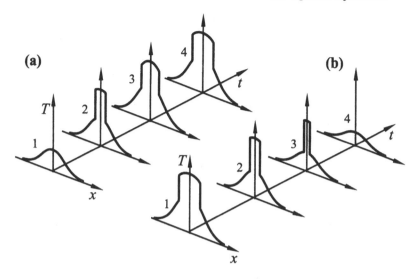

**Fig. 2.21.** Kinetics of hysteretic variations of the transverse profile of the output signal at the incident beam power increase (**a**) and decrease (**b**)

the entire beam cross-section, but only in its narrow zone – the propagating front of the switching wave (Fig. 2.21**a**). Correspondingly, the duration of the hysteresis transition is determined by the time of the transverse propagation of the switching waves.

If the maximum value of intensity starts to decrease further, the kinetics of variation of the temperature profile will be the following: In the first stage, the temperature profile will still be of the combined type, with sharp spatial switching in the domain $I_{fr} \approx I_0$ (profile 1 in Fig. 2.21**b**). But with a decrease in $I_m$ the central domain of the beam (switched to the upper state) will gradually narrow, and at $I_m \approx I_0$ the central local perturbation will entirely disappear (profile 4 in Fig. 2.21**b**). Therefore in the intensity range (2.7.13), in agreement with the results of Sect. 2.7, bistability and hysteresis of the temperature and transmitted radiation intensity profiles will take place (as well as of profiles of pressure and mechanical stress). Although one of the two profiles (smooth at $I_m > I_0$) is metastable, the probability of its fluctuational switching into the stable state (combined profile with the central part switched into the upper state) is negligible for wide radiation beams outside a small vicinity of the branch edge ($I_m \approx I_{max}$). Here, therefore, metastability is practically indistinguishable from stability.

The kinetics of the hysteresis described here manifests itself in the form of waves of switching from the lower branch to the upper one ($I_{in} > I_0$), but not vice versa. This is connected with the assumed bell-like profile of incident radiation intensity. The waves of switching from the upper branch to the lower one ($I_{in} < I_0$) can be observed for beams with a sufficiently flat peak in the intensity profile or for beams with an intensity dip in their centre. Also note that, with a smooth variation of beam radiation intensity, the position of the motionless switching wave front in the vicinity of Maxwell's value $I_0$ can be either stable (if signs of gradients of intensity at the front of the switching wave and in the considered beam domain coincide) or unstable (in the opposite case).

If the incident radiation intensity profile includes several spaced spatial oscillations around $I_0$, at each such oscillation (irrespective of the others) individual spatial hysteresis can be realized. We obtain here multichannel memory on the basis of a single (but wide-aperture) bistable element [344], which may be of practical interest (see Appendix C).

## 2.9 Systems with Sharp Transverse Inhomogeneity

The mechanical analogy (see Sect. 2.3) is also instructive in the analysis of piecewise-homogeneous systems, for which sharp boundaries or inhomogeneities of incident radiation beam and/or system physical characteristics are important [309, 285]. The stationary heat conduction equation for temperature $T$ averaged over the rod cross-section has the form of (2.3.1), where $F$ is given in (2.1.11), and the heat conduction coefficient can be taken as $\Lambda = 1$.

The simplest – and most revealing – problem is stationary heating of a semi-bounded rod ($x > 0$), a fixed temperature value being maintained at its edge:

$$T_{x=0} = T^* . \tag{2.9.1}$$

The presence of inhomogeneities (and boundaries) of the system is not so important if they are situated in the peripheral part of the beam, where the intensity of incident radiation is small ($I_{\text{in}} < I_{\text{min}}$). We are interested here in the opposite case, when $I_{\text{in}}$ does not vary within the entire system. At $x \to +\infty$ boundary condition (2.3.11) remains,

$$\left(\frac{\mathrm{d}T}{\mathrm{d}x}\right)_{x \to +\infty} = 0 . \tag{2.9.2}$$

Therefore the asymptotic value of temperature $T_{x \to +\infty} = T_\infty$ must coincide with the temperature $\Theta$ of the transversely homogeneous problem (see Sect. 2.2),[3] i.e.,

$$F(T_\infty) = F(\Theta) = 0 . \tag{2.9.3}$$

In the language of the mechanical analogy, conditions (2.9.2) and (2.9.3) mean that the "trajectories" $T(x)$ (intervals of the horizontal straight lines 1–2 and 3–4 in Fig. 2.22) must arrive at the top of the hump of the potential

---

[3] In what follows we do not consider solutions with a periodic temperature oscillation at $x \to +\infty$ because of their instability, as pointed out in Sect. 2.4.

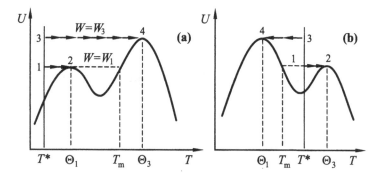

**Fig. 2.22.** "Potential curves": (a) $I_{\min} < I_{\text{in}} < I_0$, (b) $I_0 < I_{\text{in}} < I_{\max}$

curve $U(T)$ corresponding to a chosen value of intensity $I_{\text{in}}$ at $x \to \infty$. This determines the "energy" $W = U(T_\infty)$. Because of (2.9.2) the trajectory must originate from the point of the indicated straight line with abscissa $T^*$ (Fig. 2.22). To determine the temperature profile $T(x)$ we have

$$x = \pm \int_{T^*}^{T} \{2[W - U(T)]\}^{-1/2} \, dT \ . \tag{2.9.4}$$

The upper sign in (2.9.4) is taken if $T^* < T_\infty$, while for $T^* > T_\infty$ the lower sign is to be used. For a semi-bounded rod (2.9.1), there is no translation invariance and no transversely homogeneous states except the trivial one with $T = T^*$ (at $T^* = T_\infty = \Theta$). Therefore, switching waves are here only an approximate solution and an asymptotic notion.

One could anticipate that boundary condition (2.9.1) influences the temperature distribution only near the rod edge (the boundary layer at $x = 0$). Then $T_\infty = \Theta_1$ or $T_\infty = \Theta_3$ (the low- and high-temperature branches, respectively) would be the temperature asymptotic values far from the edge $(x \to \infty)$, independent of the boundary temperature $T^*$, in the bistability range for a point system (2.2.6). However, it is not so – because of the problem's nonlinearity. Let us consider the case when the influence of the boundary condition spreads over the entire rod.

Let, for instance, the intensity of radiation incident on the rod be in the interval (2.3.9). Then the potential curve $U(T)$ is two-humped (bimodal) with the higher right maximum $[U(\Theta_3) > U(\Theta_1)$, see Sect. 2.5]. As seen from Fig. 2.22a, two different temperature distributions with asymptotic $T_\infty = \Theta_1$ and $T_\infty = \Theta_3$ exist only under condition $T^* < T_m$, where $T_m$ is the extreme temperature for the unstable state of incomplete switching ("critical nucleus", see Sect. 2.3). These two distributions correspond to the segments of the horizontal straight lines 1–2 and 3–4 with "energies" $W = W_1$ and $W = W_3$ indicated in Fig. 2.22a. The dependence $T(x)$ is depicted in Fig. 2.23 for different boundary temperatures $T^*$. Similarly, for (2.3.8) two different stationary distributions exist only at $T^* > T_m$ (see Fig. 2.22b). The general

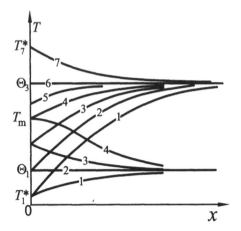

**Fig. 2.23.** Stationary temperature distributions with different temperature values at the boundary, $T_1^* < T_7^*$

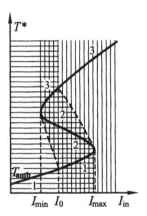

**Fig. 2.24.** Domains of existence of the low-temperature (*horizontal hatching*) and high-temperature (*vertical hatching*) states; curves 1, 2, and 3 are the branches of the transfer function for a point system

result of separation of the plane of parameters $I_{in}, T^*$ into domains of existence of two indicated regimes is presented in Fig. 2.24.

The stability analysis for the stationary distributions found needs some revisions as compared with that made for an unbounded rod (see Sect. 2.4), for which the translation invariance was essential. For a semi-bounded rod, it follows from (2.9.1) for perturbation $\psi$ that $\psi_{x=0} = 0$. The latter condition excludes the even eigenfunction with negative eigenvalue $\gamma$, and eigenvalue $\gamma = 0$ is the lowest one. Therefore the stationary temperature distributions for a semi-bounded rod with (2.9.1) constructed above are stable, and the scheme is really bistable.

Now let us elaborate on the character of corresponding hysteresis variations. Assume that the temperature of the rod edge $T^*$ increases at a fixed intensity of incident radiation $I_{in}$ (2.3.9). Let us trace the variation of the initial "low-temperature" $(T_\infty = \Theta_1)$ distribution. According to Fig. 2.23, such a distribution remains as long as $T^* < T_m$. When the value $T_m$ is exceeded, switching to the only stationary ("high-temperature", $T_\infty = \Theta_3$) distribution is inevitable. As in the case of heating of an unlimited rod by a wide radiation beam (see Sect. 2.7), this switching takes place not simultaneously over the entire rod length, but within a limited region only. The coordinate of the region increases with time. For a great distance from the rod edge, a stationary switching wave will be formed (see Sect. 2.5). Thus, the conditions at the boundary initiate the switching wave, which is not localized near the rod edge, but propagates over the entire rod. This is close to the kinetics of ignition of a combustible gas mixture by a hot surface [420].

With a further decrease in the edge temperature $T^*$, the high-temperature profile will also remain in the range $T^* < T_m$; so hysteresis variation of temperature distribution takes place. However, reverse switching to the low-temperature profile does not happen. The complete hysteresis loop is present with other kinds of parameter variation, e.g., of radiation intensity $I_{in}$ at fixed temperature $T^*$.

The use of the mechanical analogy also allows us to consider other types of piecewisely homogeneous schemes [309, 285]. An interesting aspect of the problem of inhomogeneous rod switching is connected to the initiation of switching by the system inhomogeneities. Even one "dangerous" local inhomogeneity is able to switch with time the entire wide-aperture scheme from a metastable state to a stable one.

Assume that the inhomogeneity is located in the central part of the rod $|x| < l$, and has absorptance increased by a value $|\Delta_a|$, the radiation intensity being in the range of (2.3.9) (then the state of a homogeneous rod with $T = \Theta_1$ is metastable). For fixed intensity $I_{in}$ fine inhomogeneities only disturb the temperature distribution in the central part of the rod, while overheating of sufficiently large inhomogeneities ($l > l_{max}$) induces the breakdown of the stationary distribution. When intensity increases up to the critical value $I_{cr}$ ($I_{cr} < I_{max}$) in a bistable scheme with a large inhomogeneity, waves of switching from the metastable state to the stable state will arise and run away from the inhomogeneity. The qualitative dependence of the breakdown critical intensity on characteristics of inhomogeneity is presented in Fig. 2.25, and a quantitative analysis is given in [87]. Estimates show that for a Ge rod with thickness $d = 2$ mm and an absorbing inhomogeneity with parameters $|\Delta_a| = 15$ cm$^{-1}$ and $l = 2$ mm, one has $I_{cr} = 20$ W/cm$^2$ (at $I_0 = 7.3$ W/cm$^2$ and $I_{max} = 28.2$ W/cm$^2$).

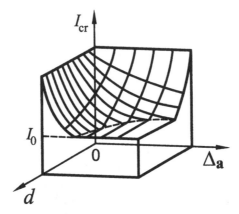

**Fig. 2.25.** Dependence of critical intensity of switching on inhomogeneity characteristics [309]

## 2.10 Longitudinal and Transverse Distributivity

As mentioned in Chap. 1, it is convenient to differentiate transverse and longitudinal (with respect to the direction of radiation propagation) distributivity of bistable systems. In systems distributed in transverse directions only (when their longitudinal distributivity is not essential) excited by transversely homogeneous radiation, the fundamental spatio-temporal structures are switching waves (see Sect. 2.5). Knowledge of them allows us to describe spatial hysteresis and switching at slow variation of power of the external wide radiation beam.

Longitudinal distributivity of bistable systems can lead to similar structures. Thus an intensity decrease in the longitudinal direction due to radiation absorption is, in a sense, analogous to transverse intensity variation in the radiation beam.[4] Therefore under conditions typical of an electron mechanism of nonlinearity, the formation of kinks – longitudinal analogs of a motionless switching wave – is possible in sufficiently expanded samples. In this case the location of the switching front is also determined by "Maxwell's rule" (2.3.10) [172, 142]. The presence of sample facets is of fundamental importance. A kink is stable only if it is localized inside the sample at sufficient distance from the facets [184]. The existence of stable kinks involves the appearance of a new branch of transfer function ("tristability") [142]. If input intensity increases linearly with time (and not too slowly as compared with the time of kink formation [184]), then the kinks move periodically in the direction of radiation propagation, and saw-tooth temporal dependence of the

---

[4] Practically a full equivalence with the transverse variant of spatial hysteresis (Sect. 2.7) can be reached with radiation focused inside the sample, when local intensity is maximum near the focus and decreases with distance towards both the front and back facets.

output radiation power is observed [118, 195]. There are also similar fronts of temperature or carrier concentration that move, at fixed radiation power, towards the radiation (to the sample front facet) [332].

Joint transverse and longitudinal distributivity of a bistable scheme leads to new additional phenomena, including formation and complex dynamics of absorption domains [370]. The physics is close to optical discharge in gases [264]. In particular, under certain conditions and constant incident radiation intensity, periodic motion of absorption domains is possible, which results in the pulsation of output radiation power. Note that self-oscillations of output radiation intensity can also be caused by other mechanisms, e.g., by competition of different mechanisms of nonlinearity [223, 137].

For the mirrorless schemes of increasing absorption bistability [84, 122, 69, 276], joint longitudinal and transverse distributivity is essential if their sizes exceed the typical length of heat conduction. The effects of longitudinal distributivity were studied in [141, 57, 332]. A further detailed analysis of purely longitudinal effects and an approximate account for transverse phenomena are given in [335]. In this section we will briefly discuss the effects of longitudinal and joint longitudinal and transverse distributivity [332, 335].

### 2.10.1 Longitudinal Distributivity

At large plate thickness $d$, averaging over the sample thickness (see Sect. 2.1) becomes incorrect. An account of temperature longitudinal variation is, in this case, essential.

For incident radiation in the form of a plane wave, stationary regimes are described by

$$\Lambda \frac{\mathrm{d}^2 T}{\mathrm{d}z^2} = \frac{\mathrm{d}I}{\mathrm{d}z} = -\alpha(T)I \ , \tag{2.10.1}$$

with boundary conditions [omitting inessential factor $(1 - R_0)$]

$$\Lambda \left( \frac{\mathrm{d}T}{\mathrm{d}z} \right)_{z=0} = h(T_{z=0} - T_{\mathrm{amb}}) \ , \quad I_{z=0} = I_{\mathrm{in}} \ , \tag{2.10.2}$$

$$\Lambda \left( \frac{\mathrm{d}T}{\mathrm{d}z} \right)_{z=d} = -h(T_{z=d} - T_{\mathrm{amb}}) \ .$$

First let us demonstrate that bistability is absent for sufficiently thick plates. It follows from (2.10.1):

$$I - \Lambda \frac{\mathrm{d}T}{\mathrm{d}z} = C = \mathrm{const.} \tag{2.10.3}$$

At $d = \infty$ (half-space), $C = 0$. Invoking boundary conditions (2.10.2), we get a single-valued – and, moreover, linear – dependence of the boundary temperature $T_{\mathrm{b}} = T_{z=0}$ on the incident radiation intensity $I_{\mathrm{in}}$:

$$T_{\mathrm{b}} = T_{\mathrm{amb}} + I_{\mathrm{in}}/h \ . \tag{2.10.4}$$

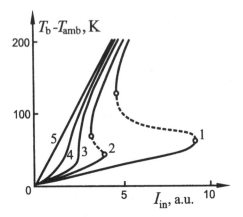

**Fig. 2.26.** Dependence of temperature at the sample front facet on intensity of incident radiation for stationary regimes: $d = 0.3$ (1) and 0.5 (2), $d = d_{cr} = 1.4$ (3), 3.5 (4) and $\infty$ (5) mm [332]

Knowing the value $T_b$, one can find the temperature profile $T(x)$ over the whole plate thickness in a single-valued way. Therefore bistability is absent in the heating of a half-infinite medium with an arbitrary temperature dependence of absorption coefficient $\alpha(T)$. It arises if the plate thickness becomes smaller than a certain critical value, $d_{cr}$. With a further decrease in $d$, the hysteresis loop width and the critical intensity of switching increase, as shown in Fig. 2.26.

Bistability of the temperature longitudinal distribution is illustrated in Fig. 2.27. With a slow increase in incident radiation intensity $I_{in}$ near its value $I_{in} = I_{max}$, a fairly fast switching of the temperature profile occurs [332, 167] that can be related to the longitudinal motion of the large absorption domain found in experiments [370].

### 2.10.2 Joint Longitudinal and Transverse Distributivity

In the experiment in [370] a semiconductor CdS sample was a parallel-sided plate and the incident cylindrically symmetric radiation was a Gaussian beam focused behind the plate. Near the back facet, an additional absorption could be introduced by ion implantation.

Numerical simulation of this scheme reveals the various regimes illustrated in Fig. 2.28. After radiation is switched on, the sample is heated mainly near its back facet. If the relation between power absorption and heat removal is such that the maximum temperature does not exceed 130°C, sample transmission remains high. Otherwise the sample portion near its back facet transits to the state of high absorption (curve **a** in Fig. 2.28, $t = 40$ μs). Then for about 75 μs the sample is overheated further, and a high-absorbing domain

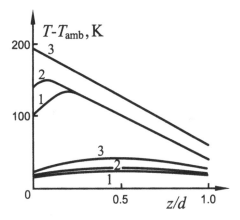

**Fig. 2.27.** Stationary longitudinal distributions of temperature for incident radiation intensities $I_{in1} < I_{in2} < I_{in3}$ [332]

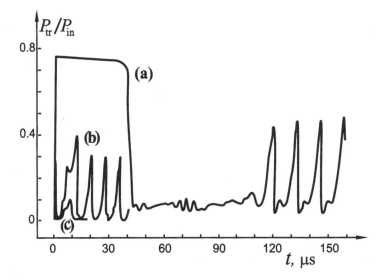

**Fig. 2.28.** Temporal dependence of transmission of a bistable device at different powers of incident radiation: $P_{in} = 0.16$ (*curve* **a**), 0.49 (*curve* **b**) and 1.97 W (*curve* **c**) [335]

forms. The domain is nonstationary; its position oscillates in the longitudinal direction.

With an increase in incident radiation intensity (Fig. 2.28, curve **b**), the sample is overheated faster, and the amplitude of the domain oscillations increases. If the intensity increases more (Fig. 2.28, curve **c**), the domain reaches the front facet, and the process is finished by heating of the entire sample up to the stationary state corresponding to sample low transmission.

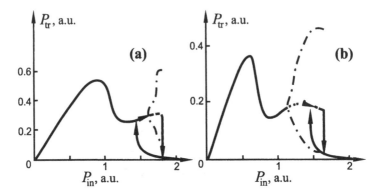

**Fig. 2.29.** Hysteresis of transmitted radiation power $P_{tr}$ with the variation of the power of incident radiation $P_{in}$. *Dashed lines* show power values averaged in time, and *dash-dotted lines* show power maximum and minimum values. $F =$ (**a**) 600 and (**b**) 400 µm

These results allow us to explain self-pulsation of output radiation as follows: When the absorbing domain moves towards focused radiation, it enters an area with lower and lower radiation intensity. If the domain reaches the area where radiation absorbed by it does not compensate heat removal, it stops and cools. Then shading of the sample area near its back facet by the domain terminates, giving way to its heating, and the process repeats.

The presence of pulsations changes the hysteresis character, because now co-existence of stationary and periodic regimes is possible. Examples are given in Fig. 2.29. With a slow increase in radiation power, pulsation regimes break up in conditions when the domain originated near the back facet reaches the front facet and stops there. With a further decrease in radiation power, pulsations can be absent entirely (weak focusing, Fig. 2.29b) or take place in a narrower range of incident radiation intensity (Fig. 2.29a).

Of even greater diversity is the scheme dynamics under conditions of temporal modulation of incident radiation intensity [335]. Note also that similar phenomena with an essentially reduced time scale are anticipated in media with a sharp dependence of absorption on the free carrier concentration [165].

## 2.11 Other Factors; Bibliography

### 2.11.1 Two-Dimensional Transverse Distributivity

With the transition from a rod sample (Fig. 2.7) to a plate (Fig. 2.1), an additional term with a radial derivative arises in the reduced heat conductivity equation for cylindrically symmetric distributions of intensity and temperature:

$$\frac{\partial T}{\partial t'} = \frac{\partial^2 T}{\partial r'^2} + \frac{1}{r'}\frac{\partial T}{\partial r'} - F(T) \, . \qquad (2.11.1)$$

This term does not influence the transversely homogeneous states considered in Sect. 2.2. In the context of the mechanical analogy for stationary regimes, this term plays the role of friction with an $r$-dependent friction coefficient. Regimes of "critical nuclei" exist as before [316]. As for the switching waves, it is essential that the radial coordinate $r$ changes in the semi-bounded range $0 < r < \infty$. It is similar to the case of the semi-bounded rod (Sect. 2.9); even more complete is an analogy with the kinetics of overshoots with finite width (see Sect. 2.6). As in the latter case, it is possible to use this notion only approximately, if the switching front radial coordinate $r_{fr}$ exceeds the width of switching wave front $w_{fr}$ [316, 346]:

$$r_{fr} \gg w_{fr} \, . \qquad (2.11.2)$$

If this condition is fulfilled, the additional term of (2.11.1) is only a small correction, and the notion of cylindrical switching waves being locally close to plane waves is valid. Note, however, that the velocity of front propagation will depend on the front coordinate $r_{fr}$ and will approach the constant velocity of the plane switching wave $v$ at $r_{fr} \to \infty$, as for the case shown in Fig. 2.16.

For the condition (2.11.2) the "quasi-one-dimensional" description of spatial hysteresis is also valid. Moreover, if both (2.11.2) and (2.7.3) are valid, the radial correction term does not participate in the lowest-order approximation of the singular perturbation approach used in Sect. 2.7. Therefore the scheme of spatial hysteresis (Fig. 2.18) also holds for wide cylindrical beams. Note that for the "one-component" bistable schemes considered here, there are no azimuthal instabilities of cylindrical switching waves.[5]

Three-dimensional spatial hysteresis with the same kinetics can be realized in weakly absorbing media, e.g., in high-frequency plasma heating [282].

### 2.11.2 Fluctuations

The effect of noise was considered for point schemes in Sect. 2.2. A new factor arising for wide-aperture (transversely distributed) systems is the metastability of one of two states stable without noise; corresponding spatial switching from a metastable state to a stable one is analogous to those known in the theory of phase transitions of the first kind (see Sect. 2.6 and [403]).

The effect of weak fluctuation is most pronounced near the critical conditions, where the differential gain of the point scheme is extremely large (see Sect. 2.2). Then, similar to the effect of critical opalescence [86, 187], it is natural to anticipate a sharp increase in the small-angle scattering of light on the temperature fluctuations [281].

In a phenomenological description of the fluctuational switching of a distributed bistable system, it is possible to introduce the probability density

---

[5] They can arise in two- or more component systems [420].

of birth of "dangerous" (overcritical) nuclei for unit time and space. Then, depending on the relations between this density, the switching wave velocity, and the system size, its switching will be caused by the growth of a single nucleus ("monocrystal") or numerous nuclei ("policrystal") [281, 346].

### 2.11.3 Effect of Delay

These effects were absent in our previous discussion. The reason is the use of the standard heat conductivity equation (2.1.4) corresponding to an infinitely large velocity of perturbation propagation. Of course, this velocity must be finite and depends on the medium microstructure. It can be introduced phenomenogically by means of the temperature relaxation time $\tau_{rel}$. Then the modified heat conductivity equation takes the form [208]

$$\varrho_0 c_v \left( \frac{\partial T}{\partial t} + \tau_{rel} \frac{\partial^2 T}{\partial t^2} \right) = \nabla(\Lambda \nabla T) + w_v . \qquad (2.11.3)$$

In the framework of (2.11.3), there is feedback delay proportional to the system's size. In this case instabilities of a stationary regime can arise, known in optical bistability as Ikeda instabilities [146]. This, of course, is not the only reason for instabilities in these schemes (see Sect. 2.10 and the effects of componency below in this section).

### 2.11.4 Isolated Loops

Absorption of some materials depends not only on temperature but also on radiation intensity: $\alpha = \alpha(T, I_{in})$. If absorption is saturated with an increase in intensity, then the transfer function is not only S-shaped, but can include isolated loops (Fig. 2.30). Conclusions arrived at in Sect. 2.5 on waves of switching between the states with temperatures $\Theta_1$ and $\Theta_3$ hold good here; however, the following distinction takes place: The velocity of switching waves does not change its sign if areas $S_{1,2}$ of the two domains shaded in Fig. 2.30a (1 and 2) differ at any intensity $I_{in}$ (e.g., $S_1 > S_2$). Then all the states corresponding to the loop are metastable, and spatial hysteresis is not realized, though the point scheme is bistable.

### 2.11.5 Componency

The medium state can be characterized not only by the temperature, but also by the concentration of different carriers. Then the kinetics is described by a system of coupled equations of diffusion type [395]. An important example is the system of two equations for temperature $T$ and carrier concentration $n$:

$$\tau_T \frac{\partial T}{\partial t} = L_T^2 \triangle T - F(T, n) , \quad \tau_n \frac{\partial n}{\partial t} = L_n^2 \triangle n - G(T, n) . \qquad (2.11.4)$$

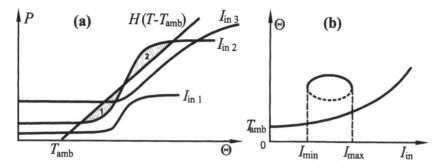

**Fig. 2.30.** Determination of stationary regimes (a) and an isolated loop of the transfer function (b)

Here $\tau_T$ and $\tau_n$ are typical relaxation times for temperature and concentration, $L_T$ and $L_n$ are the corresponding diffusion lengths, function $F$ describes the thermal flux, and $G$ represents carrier generation and recombination.

Now even in a point scheme [averaging of (2.11.4) over coordinates],

$$\tau_T \frac{dT}{dt} + F(T,n) = 0 \ , \quad \tau_n \frac{dn}{dt} + G(T,n) = 0 \ , \qquad (2.11.5)$$

not only stationary, but also periodic regimes exist for stationary incident radiation; these relaxation oscillations were demonstrated in optical bistability in [370]. A description with only the heat conductivity equation (2.1.4) is valid in the absence of these instabilities for regimes with sufficiently smooth temporal variation (for duration of fronts $\tau_{fr} \gg \tau_n$).

A vast body of literature is devoted to the analysis of the reaction–diffusion equations (2.11.4), see, e.g., [395]. In particular, they describe, in the case of transversely homogeneous excitation, not only switching waves, but also motionless and moving stable localized structures (see [166] and Chap. 4), nonstationary pulsing distributions, spiral waves, and many other spatio-temporal structures. Note that similar structures arise with laser initiation of different chemical reactions [61, 162].

### 2.11.6 Experiments

The first experiments in increasing absorption bistability were performed in [122]. The sample was the semiconductor Ge, the nonlinearity being of thermal nature, and the increase in absorption with increase in intensity of the $CO_2$-laser connected with the increase in free carrier concentration. Switching intensities are about tens $W/cm^2$. Although the scheme kinetics is very slow, it is convenient for model experiments.

A review of theoretical and experimental studies of mechanisms in CdS electronic nonlinearity is given in [171]. In [246, 245] temperature transverse profiles were studied for CdS crystals; some numerical simulations were presented there also. The conclusions are in qualitative agreement with the theory given above. Numerical simulations and experiments were realized for the same scheme in [170, 227]. Spatial hysteresis was not found in these papers, presumably because of insufficient beam width as compared with the switching wave front width. However, even under the conditions considered in [170, 227], there were sharp radial structures in the transmitted radiation intensity that could be associated with the switching waves.

Nevertheless, such effects of transverse distributivity as switching waves and spatial hysteresis were not investigated experimentally in detail. The reason may be that cavity bistable schemes are more convenient (see Chap. 4). However, such experiments seem to be of great importance, due to the following reasons: First, these schemes are radiation phase-insensitive, more simple, and are model objects for the study of phase-transition physics. Second, the practical importance of these schemes could increase with the transition from a thermal mechanism of nonlinearity to an electronic one. In this case the effects remain the same, but the temporal scale is significantly reduced. Thus, the velocity of switching waves could reach values of $v \sim 10^7$ m/s.

The effects of longitudinal and longitudinal-transverse distributivity were investigated in a number of experiments, especially in [371, 370, 372, 373, 374]. The transition to a fast electronic nonlinearity could also be practically important here, e.g., for the generation of high-frequency pulsations.

# 3. Hybrid Bistable Devices

Considered in the present chapter are the regimes of the hybrid bistable scheme, which is an electro-optic modulator in which a part of the power of the transmitted radiation is transformed into an electric voltage which is applied then through the feedback circuit to the electro-optic crystal. The transverse distributivity is not essential in this scheme, and the longitudinal one shows itself in the form of a delay. Neglecting the response (relaxation) time in the feedback circuit, the kinetics of the scheme is described by one-dimensional point mapping, which connects the system characteristics in a time interval equal to the delay time. Then, at a constant power of incident radiation and with an absence of noise, stationary, periodic (with the period being a multiple of the delay time) and dynamically chaotic regimes form depending on the parameters and initial conditions of the scheme. A finite response time leads to a shift of stability boundaries for these regimes, to an increase in the modulation period, and, for the chaotic regimes, to a breakdown of the output radiation into pulses with a duration of the order of the response time.

## 3.1 Stationary, Periodic, and Chaotic Regimes

A diagram of a mirrorless (requiring no coherence or monochromatism of the external radiation) hybrid bistable optical device, proposed and realized in [115], is shown in Fig. 3.1. After transmission of linearly polarized radiation through an electro-optic crystal, the polarization plane rotates by an angle proportional to the electrical voltage applied to the crystal. In the feedback circuit, part of the radiation transmitted through the analyzer is transformed to electrical voltage by a photodetector. The voltage is applied, in addition to the constant bias voltage, to the electro-optic crystal, which provokes rotation of the radiation polarization plane and therefore a change in the system transmission coefficient.

Let $J_{in}(t)$ be the power of radiation incident on the crystal, $J(t)$ be the light power in the feedback circuit, and $K(t)$ be the device transmission coefficient. At a finite delay time of light and electric signal in the feedback circuit, $\tau_{del}$, the transmission coefficient $K(t) = J(t)/J_{in}(t)$ depends on the light power $J$ in the preceding time moment $t - \tau_{del}$:

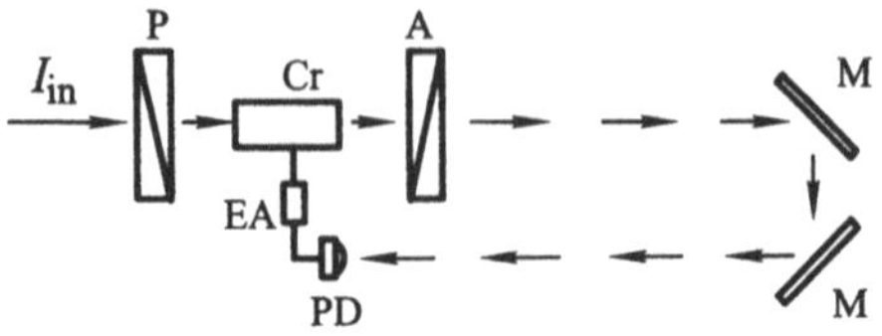

**Fig. 3.1.** Schematic of an electro-optic bistable device: P: polarizer; A: analyzer; M: mirrors; Cr: the electro-optic crystal; PD: photodetector; EA: an electric amplifier

$$K(t) = K\left(\varphi + aJ(t - \tau_{\text{del}})\right). \tag{3.1.1}$$

Here $\varphi$ is the constant phase shift proportional to the bias voltage at the electro-optic crystal, and the coefficient $a$ characterizes the sensitivity of the photodetector and amplification of the electrical signal. For discrete moments of time $t_n = n\tau_{\text{del}}$ separated by the time interval $\tau_{\text{del}}$, one can write down the recurrent equation (point map), which has the following form for dimensionless values $P = aJ_{\text{in}}$ and $Q = aJ$:

$$Q_{n+1} = P_{n+1}K(\varphi + Q_n), \quad n = 0, 1, 2, \ldots \tag{3.1.2}$$

The power of the output radiation will be $T_{\text{out}}P_{n+1}\left(1 - K(\varphi + Q_n)\right)$, where $T_{\text{out}}$ is the transmission coefficient of the exit mirror (by intensity). Under the ideal conditions of the scheme in Fig. 3.1 and at $P_n = P = \text{const.}$, the form of the point mapping and the iteration function $f$ is

$$Q_{n+1} = f(Q_n), \quad f(Q_n) = P\sin^2(\varphi + Q_n), \quad K(\Psi_n) = \sin^2\Psi_n. \tag{3.1.3}$$

Variation of the form of the transmission coefficient $K$ influences the main results only slightly.[1] Point mapping (recurrent relation) (3.1.2) or (3.1.3), together with the specification of the incident radiation power temporal dependence $P(t)$, as well as of the initial values $Q(t)$ within the time interval $t_0 < t < t_0 + \tau_{\text{del}}$, determines uniquely the kinetics of the scheme, which we will describe below, following [275]; a general outline of the point mapping theory and presentation of regular and chaotic dynamics of various nonlinear systems can be found in [22, 66, 139, 356].

Stationary regimes $Q_n = \bar{Q} = \text{const.}$, as well as the conditions of bistability and multistability, are easily found from graphic analysis of (3.1.3). The dynamics of $Q_n$ variation is conveniently represented by means of "*Lamerey's diagrams*" [22]. In Fig. 3.2 the stationary regimes $Q_n = \bar{Q} = \text{const.}$ correspond to the points of intersection of the iteration function graph $f$ (curve

---

[1] This corresponds to the structural stability of regimes of physical schemes, i.e., the character and asymptotics of the sequences $Q_n$ vary continuously with a smooth variation in the form and parameters of iteration function $f$, except for a countable number of bifurcations.

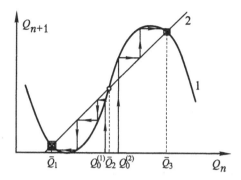

**Fig. 3.2.** Determination of stationary and transient regimes

1) with straight line 2 ($Q_{n+1} = Q_n$). Bi- and multistability correspond to
several points of intersection. The dynamics of $Q_n$ variation, with two examples of initial conditions $Q_0^{(1)}$ and $Q_0^{(2)}$, is shown in Fig. 3.2 by segments
with arrows [Lamerey's diagrams following from (3.1.3)]. The stability and
the characteristic transition time of a stationary regime are determined by
the slope of the curve $PK$ at $Q = \bar{Q}$. It can be shown by a standard linear
stability analysis, with introduction of a small perturbation, $v_n = Q_n - \bar{Q}$,
and linearization of (3.1.3):

$$v_{n+1} = qv_n , \quad q = \frac{df}{dQ}\bigg|_{\bar{Q}} . \tag{3.1.4}$$

Sequence (3.1.4) determines the geometric progression $v_n = v_0 q^n$; for continuous time description $v(t) = v_0 \exp(\gamma t)$, where $q = \exp(\gamma \tau_{del})$. The progression is converging and the stationary regime is stable under the following
condition:

$$-1 < q < 1 . \tag{3.1.5}$$

Otherwise, for $|q| > 1$, the regime is unstable. In the case of (3.1.3) condition
(3.1.5) takes the following form:

$$P|\sin[2(\varphi + \bar{Q})]| < 1 . \tag{3.1.6}$$

In the variant shown in Fig. 3.2, there are two stable ($\bar{Q}_1$ and $\bar{Q}_3$) and one
unstable ($\bar{Q}_2$) stationary regimes (bistability). By means of Lamerey's diagrams we can also determine the domain of the initial conditions at which one
or another regime is formed (domains of attraction for stationary regimes).
    With a slow variation of the incident radiation power $P(t)$, switching
between branches of the transfer function corresponds in Fig. 3.2 to tangency
of the $PK$ curve and the straight line $Q_{n+1} = Q_n$. Under the conditions close
to tangency, changes of $Q_n$ at one time step are small. Then one can replace
(3.1.2) by the following differential equation:

$$\tau_{\text{del}} \frac{dQ}{dt} \approx -Q + P(t)K(\varphi + Q) \,. \tag{3.1.7}$$

Description by means of (3.1.7) corresponds to a point scheme – but not a distributed one. Therefore, the critical slowing down demonstrated in Sect. 2.2 and its anomalously slight dependence on the scanning rate take place for the considered device too [275]. Analogous also are manifestations of noise in hybrid bistable devices [238].

Even at constant incident light power $P(t) = P = \text{const.}$, output power may not steady at a stationary level. According to (3.1.5), a stationary regime may lose its stability with control parameters ($P$ and $\varphi$) variation by one of two ways. One, the value $q - 1$ may change sign; then $\gamma$ is real and changes sign, too. The boundary of the stability domain where $q = 1$ corresponds to tangency of the $f(Q)$ curve and the straight line $Q_{n+1} = Q_n$, or to a double root of the equation for stationary regimes: $\bar{Q} = f(\bar{Q})$. This is the bifurcation of the merging of two stationary solutions (one stable and the other unstable), with subsequent disappearance of these solutions; then the trajectory $Q_n$ moves away from this region. Two, the value $q + 1$ may change sign. At the boundary of the stability domain the value $\gamma$ is purely imaginary, $\gamma = \pm i\pi/\tau_{\text{del}}$. Note that for $q = -1$ small deviations $v_n$ alternate: $v_n = v_0(-1)^n$; therefore, they coincide in time period $T = 2\tau_{\text{del}}$. Then, in the vicinity of the boundary, even small terms nonlinear in $v_n$ may result in the creation of a 2-periodic regime (with period $T = 2\tau_{\text{del}}$), when the sequence $Q_n$ has the form $\bar{Q}_1, \bar{Q}_2, \bar{Q}_1, \bar{Q}_2, \ldots$ (*bifurcation of period doubling*). To demonstrate this, let us set $q = -1 + \kappa$, $\kappa^2 \ll 1$, and present the recurrent equation for small deviations $v_n$ in the form

$$v_{n+1} = (-1 + \kappa)v_n + \alpha v_n^2 + \beta v_n^3 + \ldots , \tag{3.1.8}$$

$$\alpha = \frac{1}{2!} \left. \frac{d^2 f}{dQ^2} \right|_{\bar{Q}} , \quad \beta = \frac{1}{3!} \left. \frac{d^3 f}{dQ^3} \right|_{\bar{Q}} .$$

Then, assuming $v_n \propto |\kappa|^{1/2}$ and retaining terms up to $\kappa^2$, we have

$$v_{n+2} = (-1 + 2\kappa)v_n - \kappa\alpha v_n^2 - 2(\alpha^2 + \beta)v_n^3 \,. \tag{3.1.9}$$

These are the following fixed points of (3.1.9): $\bar{v}_1 = 0$, which represents a stationary regime, stable at $\kappa > 0$ and unstable at $\kappa < 0$, and $\bar{v}_2 = \pm\sqrt{-\kappa/(\alpha^2 + \beta)}$, which represents a 2-periodic regime. Since

$$\left. \frac{dv_{n+2}}{dv_n} \right|_{\bar{v}_2} = 1 + 4\kappa + 0(\kappa^2) \,,$$

stability of the 2-periodic regime is opposite to stability of the stationary regime. Therefore, the stable 2-periodic regime exists at $\alpha^2 + \beta > 0$ and $\kappa < 0$ (where the stationary regime is unstable), while at $\alpha^2 + \beta < 0$ the 2-periodic regime exists at $\kappa > 0$ and is unstable. In the latter case the

2-periodic regime indicates boundaries of the domain of attraction of a stable stationary regime.

Generally, periodic regimes with any period being a multiple of the delay time $T = N\tau_{\text{del}}$ ($N = 2$, 3, ...) are possible. If the cycle multiplicity $N$ is not too large, then the existence, stability, and attraction domains of $N$-periodic regimes are easily determined by means of multiple iteration functions:

$$Q^{(2)} = Q_{n+2}(Q_n) = Q_{n+2}\Big(Q_{n+1}(Q_n)\Big),\qquad (3.1.10)$$

$$Q^{(m)} = Q_{n+m}(Q_n) = Q_{n+m}\Big(Q_{n+m-1}\big(\cdots\big(Q_{n+1}(Q_n)\big)\cdots\big)\Big).\quad (3.1.11)$$

$N$-periodic regimes are stable under the following condition:

$$\left|\frac{dQ^{(N)}}{dQ_n}\right|_{\bar{Q}} = \left|\frac{dQ^{(n+N)}}{dQ_{n+N-1}}\right|_{\bar{Q}_1}\left|\frac{dQ^{(n+N-1)}}{dQ_{n+N-2}}\right|_{\bar{Q}_2}\cdots\left|\frac{dQ^{(n+1)}}{dQ_n}\right|_{\bar{Q}_n} \leq 1.\quad (3.1.12)$$

For example, the conditions of existence and stability of 2-periodic regimes ($N = 2$) for case (3.1.3) have the form

$$\bar{Q} = P\sin^2[\varphi + P\sin^2(\varphi + \bar{Q})],\qquad (3.1.13)$$

$$4P^2|\sin[2(\varphi + \bar{Q})]\sin[2(\varphi + P\sin^2(\varphi + \bar{Q}))]| \leq 1.\qquad (3.1.14)$$

An example of the multiple iteration functions ($m = 1$–4) is shown in Fig. 3.3. Under these conditions the stationary ($N = 1$) and 2-periodic ($N = 2$) regimes are unstable. The only stable regime is the 4-periodic cycle with a periodic repetition of $\bar{Q}$: $1.966 \to 0.297 \to 1.829 \to 0.131 \to 1.966 \to \cdots$ All $Q_n$ sequences approach it as $n \to \infty$ ($t \to \infty$), i.e., the cycle is globally stable. Here the method of point mapping allows us to obtain a complete description of the kinetics.

An idealized scheme corresponding to iteration function (3.1.3) is determined by only two dimensionless parameters, $P$ and $\varphi$, dependence on parameter $\varphi$ being periodic with the period $\pi$. Due to this, a fairly complete analysis of stationary and periodic regimes can be performed [237]. The results are shown in Fig. 3.4. Here solid lines represent the boundaries of the domains of stable stationary regimes. These domains are hatched (I), and bistability of the stationary regimes takes place in the zones of double hatching. The areas of the parameters corresponding to 2-periodic regimes (with the period $T = 2\tau_{\text{del}}$) are marked with circles. Figure 3.4 also presents the narrow areas of 3-periodic ($T = 3\tau_{\text{del}}$) regimes (III).

After the first bifurcation of period doubling, with further variation of a control parameter, the 2-periodic regime may also lose its stability in the same way, which results in the formation of a 4-periodic regime (with period $T = 4\tau_{\text{del}}$), and so on. If we fix the parameter $\varphi$ (bias voltage) and increase the power $P$, then successive bifurcations of *period doubling* will take place: stationary regimes will be followed by 2-periodic regimes ($T = 2\tau_{\text{del}}$),

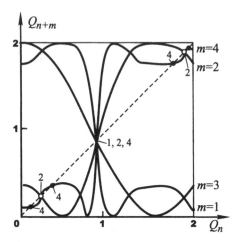

**Fig. 3.3.** Multiple iteration functions; points of intersection of corresponding curves with the straight line $Q_{n+m} = Q_n$ (*dashed line*) indicating unstable (1, 2) and stable (4) periodic regimes of the corresponding multiplicity [275]

then by 4-periodic regimes ($T = 4\tau_{\text{del}}$), 8-periodic regimes ($T = 8\tau_{\text{del}}$), etc. (Feigenbaum's scenario [356]).[2] It should be mentioned that the higher the multiplicity of the periodic regime, the narrower is its stability domain. For boundary values $P_N$ of the loss of stability of the $N$-fold periodic regime, Feigenbaum's relation holds at $N \gg 1$:

$$P_N = P_\infty - \text{const.} \times \delta^{-N}, \qquad (3.1.15)$$

where $\delta = 4.6692$ is Feigenbaum's constant [356]. Practically, only a finite number of period-doubling bifurcations is found in physical systems. It can be attributed to the effect of noise in real experiments and the finite accuracy in numerical simulations. One more reason can be connected with the way of variation of the control parameters. For example, if instead of $P$ we introduce a new parameter, $\tilde{P} = \tilde{P}(P)$, such that $P$ does not reach the value $P_\infty$ for variations of $\tilde{P}$, then the number of bifurcations will be finite and will depend on the form of the function $\tilde{P} = \tilde{P}(P)$.

---

[2] Although this scenario is universal and does not depend on the detailed form of the iteration function $f$ (3.1.3), it is realized not just for any function $f$. The necessary conditions for the scenario of period doubling are: (i) existence of a finite number of function $f$ maxima within finite intervals of argument $Q$, which leads to a many-valued (with an infinite number of branches in the case of (3.1.3)) inverse function $Q_n = f^{-1}(Q_{n+1})$; and (ii) negative value of the *Schwarzian derivative* [356],

$$S(f) = \frac{f'''}{f'} - \frac{3}{2}\left(\frac{f''}{f'}\right)^2 ;$$

in the case of (3.1.3) $S(f) = -4 - 6\cot^2(2\varphi + 2Q) \le -4$.

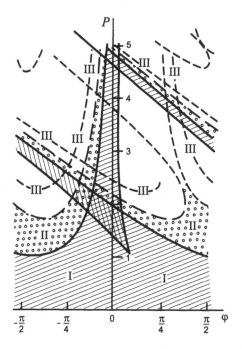

**Fig. 3.4.** Separation of parameter plane into zones, corresponding to different formed regimes: I and II are zones of stationary (*hatched*) and 2-periodic (*zones with circles*) regimes; 3-periodic regimes are formed at parameters corresponding to the narrow zones in the vicinity of the *dashed curves* III; *open areas* correspond to chaotic regimes [237]

The stability domains for stationary and periodic regimes do not fill the whole plane of the parameters $P, \varphi$. The problem of stability of aperiodic regimes is complicated and is not solved analytically even for rather simple models. However, in certain cases it is possible to prove instability of all the sequences $Q_0$, $Q_1$, $Q_2$, $Q_3, \ldots$, i.e., to show that, under a small variation of the initial condition ($Q_0' = Q_0 + \delta Q_0$), divergence between trajectories $|Q_n' - Q_n|$ increases, on the average, exponentially with $n$ (the exponents are termed Lyapunov exponents [356]). In this case, if the initial value $Q_0$ is distributed over a small interval, then, after a sufficiently large number of iterations, values $Q_n$ will be distributed over the whole space occupied by the *strange attractor*, being characterized statistically by a stationary probability density independent of the initial conditions (see below). Such regimes with quasirandom kinetics are called stochastic or dynamically chaotic, and the corresponding set of values $Q_n$ forms a strange attractor. Note that this chaos does not need any external noise.

The main features of strange attractors in physical systems are as follows [356]: (i) the sequence $Q_0$, $Q_1$, $Q_2$, $Q_3, \ldots$ is an *attractor*, i.e., it occupies a part of the phase space, and all other sequences $Q_n$ from the domain of its

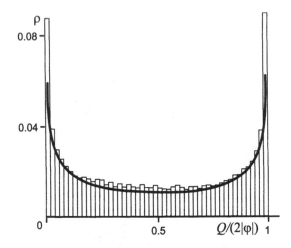

**Fig. 3.5.** Histogram and probability density (3.1.16) [275]

attraction tend to it with increases in $n$; (ii) one can say that the attractor consists of a single trajectory; this means that the trajectory approaches any point of the attractor for sufficiently large $n$; (iii) stable globally, a strange attractor is locally unstable, and an arbitrary small variation of the initial value $Q_0$ results in a finite variation of $Q_n$ at sufficiently large $n$; due to this extreme sensitivity to initial conditions the system has practically unpredictable behaviour (more precisely, prediction is possible for only a finite time interval, and the length of this interval depends on the precision of the initial values assignment only logarithmically); (iv) the attractor is structurally stable and typical: after a small variation of the form of the iteration function $f$, the "new" attractor is topologically equivalent to the "old" one, and the set of control parameters, for which the attractor exists, should not have zero measure. The main means for analysing regimes of "deterministic (dynamic) chaos" is computer simulations.

Stochastic regimes are present in the considered hybrid device as well. In the cases when the whole trajectory $Q_0$, $Q_1$, $Q_2$,... is localized in the neighbourhood of the only extreme value of the iteration function, one can reduce (3.1.2) or (3.1.3) to the quadratic ("logistic") mapping. Using the results of the analysis of this mapping, we can confirm that the regime is stochastic on some lines in the $P, \varphi$ plane. Figure 3.5 presents the histogram (probability density) obtained for iteration function (3.1.3) for the statistics of $10^5$ tests (iterations). The histogram practically does not depend on the initial value $Q_0$ in the interval $0 < Q_0 < 2|\varphi|$ and is well approximated by the theoretical distribution [356] (the curve in Fig. 3.5):

$$\rho(q) = \pi^{-1}[q(1-q)]^{-1/2}, \quad q = Q/2|\varphi| . \tag{3.1.16}$$

With a relatively small power decrease, the character of the regime does not change. Along with stochastic regimes, stationary and periodic regimes can be formed at the initial value $Q_0 > 2|\varphi|$.

It follows from the calculations that there are whole domains in the $P, \varphi$ plane which correspond to stochastic behaviour. They are located between the zones of stability of the stationary and periodic regimes and can partly overlap them.

Thus, the $P, \varphi$ plane has a fairly complex zonal structure. Along with simple attractors (stationary and periodic regimes), there are stochastic ones. The formation of one or another regime depends on, apart from power $P$ and bias voltage $\varphi$, the initial conditions. If we specify a constant value of $Q(t) = Q_0 = $ const. within the initial interval with duration $\tau_{\text{del}}$, the series of pulses of transmitted radiation will have the form of steps with duration $\tau_{\text{del}}$.

To conclude the section, note that deterministic chaos and strange attractors are known not only in systems described by point mapping, but also in a wide variety of nonlinear dynamical systems. Feigenbaum's period doubling is only one of the possible ways to deterministic chaos. Other important scenarios are the formation of a strange attractor after bifurcation from a quasiperiodic regime [349] and so called *intermittency*, when long periods of regular dynamics interchange with fairly short periods of irregular dynamics [219].

## 3.2 Effect of Finite Feedback Bandwidth

The description of hybrid device kinetics in terms of point mapping is based on the assumption of the dependence of the output signal at time $t$ on the input signal at only one moment $t - \tau_{\text{del}}$. This corresponds to the infinite feedback bandwidth (zero relaxation time). In this case, there is no connection between signal values at close time moments, and even discontinuous functions are solutions of governing equations. This assumption can be justified for stationary or slowly varying regimes, otherwise we need a consistent description of continuous (not discrete, as in Sect. 3.1) temporal dependence of the signals, taking into consideration the finite response time $\mu$ or the finite bandwidth (of the order of $\mu^{-1}$) in the feedback circuit. A more complete model of the scheme corresponds to substitution of (3.1.2) with the following equations:

$$Q(t) = PK\left(\varphi + \Phi(t)\right), \tag{3.2.1}$$

$$\mu\frac{\mathrm{d}\Phi}{\mathrm{d}t} + \Phi(t) = Q(t - \tau_{\text{del}}), \tag{3.2.2}$$

where $Q(t)$ is the angular displacement of the polarization plane of light transmitted through the crystal, the angle being proportional to the electrical voltage across the crystal. Excluding the value $Q$ from (3.2.1) and (3.2.2), we obtain

$$\mu\frac{d\Phi}{dt} + \Phi(t) = PK\left(\varphi + \Phi(t - \tau_{\text{del}})\right) . \qquad (3.2.3)$$

In contrast to the recurrent relations (3.1.2) or (3.1.3), (3.2.3) is a differential-difference equation. As in Sect. 3.1, initial conditions for $\Phi$ are to be given within the time interval $t_0 - \tau_{\text{del}} < t < t_0$.

In the limit of a small delay, it follows from (3.2.3) that

$$\mu\frac{d\Phi}{dt} + \Phi = PK(\varphi + \Phi) . \qquad (3.2.4)$$

Equation (3.2.4) describes a point scheme, and inequality $\tau_{\text{del}} \ll \mu$ serves as the condition for neglecting delay (i.e., longitudinal distributivity). As we have seen in Sect. 2.2, only stationary regimes are formed with time in the point (one-component) model at constant power $P = \text{const}$. At $\mu = 0$, (3.2.1) and (3.2.2) pass into recurrent relation (3.1.2).

In the general case (3.2.3) is solved by the method of steps [82]. To do this, the right-hand sides of (3.2.2) and (3.2.3) can be treated as known functions of time.[3] Therefore the solution of (3.2.2) has the form

$$\Phi(t) = \Phi_0 \exp[-(t - t_0)/\mu] + \frac{1}{\mu}\int_{t_0}^{t} \exp[-(t - t')/\mu]Q(t' - \tau_{\text{del}})\, dt . \qquad (3.2.5)$$

Relation (3.2.5) together with (3.2.1) allows us to find $\Phi(t)$ and $Q(t)$ sequentially at any time moment. For formed processes, assuming $t_0 = -\infty$ $(t - t_0 \gg \mu)$, we obtain

$$Q(t) = PK\left(\varphi + \int_{0}^{\infty} \exp(-\theta)Q(t - \tau_{\text{del}} - \mu\theta)\, d\theta\right) . \qquad (3.2.6)$$

The characteristics of stationary regimes do not depend on $\mu$, as is seen from (3.2.1) and (3.2.2). However, the quantity $\mu$ enters the condition for stability of these regimes, giving rise to a shift of the boundaries of the corresponding parameter domains, as is shown in Fig. 3.6 [here and below the results are given for iteration function (3.1.3)].

For the periodic regimes, calculations according to (3.2.6) show smoothing of pulse fronts, which is more substantial at higher values of $\mu$. In this case the pulse duration increases by a value of the order of $\mu$. The shape of the formed pulses is shown in Fig. 3.7 at $\mu/\tau_{\text{del}} = 0.01$ and 0.1 (for other parameters corresponding to the 2-periodic regime at $\mu = 0$). As $\mu \to 0$ the formed pulses approach a rectangular shape.

Especially important is the role of the finite response time for stochastic regimes. Calculations show sharply pronounced (at $\mu \ll \tau_{\text{del}}$) fragmentation

---

[3] At the first step it is given by the initial conditions.

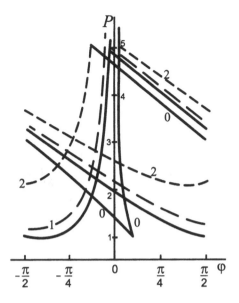

**Fig. 3.6.** Shift of boundaries of domains of stability of stationary regimes upon variations of $\mu$: $\mu/\tau_{\mathrm{del}} = 0.3$ (1) and 1 (2); boundaries labelled with a 0 are not shifted upon variations of $\mu$ [237]

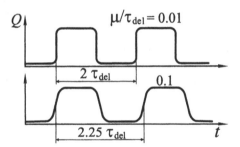

**Fig. 3.7.** Temporal dependence of the output radiation intensity in the case of a 2-periodic regime at $\mu = 0.01\tau_{\mathrm{del}}$ and $\mu = 0.1\tau_{\mathrm{del}}$ [237]

of transmitted radiation to pulses with duration of the order of $\mu$. This is characterized by the autocorrelation function introduced by the relation

$$R(\Delta t) = \frac{1}{N_0} \lim_{T \to \infty} \frac{1}{T} \int_0^T \tilde{Q}(t + \Delta t)\tilde{Q}(t)\, dt \,,$$

$$\tilde{Q}(t) = Q(t) - \lim_{T \to \infty} \frac{1}{T} \int_0^T Q(t)\, dt \,, \tag{3.2.7}$$

where the factor $N_0$ provides normalization $R(0) = 1$. The form of $R(\Delta t)$ in the stochasticity domain is shown in Fig. 3.8. Successive reduction of main

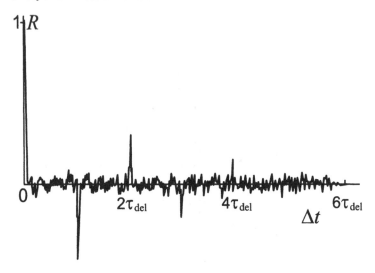

**Fig. 3.8.** Autocorrelation function of output radiation intensity in the case of a chaotic regime [237]

maxima spaced by the delay time $\tau_{\mathrm{del}}$ indicates uncoupling of the correlations with $\Delta t$, i.e., stochasticity. When the parameters $P$ and $\varphi$ change, reduction of the main maxima of $R(\Delta t)$ can be slowed down, which reflects the influence of unstable periodic trajectories.

Note that even at $\mu \ll \tau_{\mathrm{del}}$ the influence of the finite response time for stochastic regimes is not reduced. On the contrary, it is especially striking at $\mu \to 0$, causing more and more fine-scale transmitted radiation fragmentation (at a constant power of input radiation). In this case, the number of positive Lyapunov exponents increases proportionally to $\tau_{\mathrm{del}}/\tau$, and the chaotic signal becomes more and more complex (hyperchaos). Therefore the description of stochastic regimes in the framework of recurrent relations for discrete time moments (see Sect. 3.1) is rather qualitative.

Finiteness of the response time is also essential in another problem. Let it be recalled that in Sect. 2.6 we considered the kinetics of local perturbations over the background of a homogeneous stationary state in a transversely distributed scheme of increasing absorption bistability. A similar problem can be solved for a hybrid scheme (with considerable delay), where temporal analogues of stable [193, 313] and unstable [313] switching waves and of critical nuclei [313] arise. However, it is necessary to note the following distinction: In the basic model of Chap. 2, the rod length can be assumed to be infinite; one of the stationary regimes is stable (as large perturbation over the entire aperture is needed for its switching), and one of two others is metastable. However, it is impossible to speak about an infinite delay time $\tau_{\mathrm{del}}$ in a hybrid scheme. Here, with fixed parameters of the scheme and radiation intensity, each of the steady states $\bar{Q}_1$ and $\bar{Q}_3$ switches to the other under the influence of a sufficiently large perturbation. Therefore both of these states are metastable, although their lifetimes can differ considerably. This leads to an essential difference in the switching kinetics.

In conclusion it should be recalled that excitation of periodic and stochastic regimes caused by longitudinal distributivity (by delay effects) in hybrid schemes was first considered, following Ikeda's paper on nonlinear cavities [146], in [275, 147], and later on in a large number of theoretical and experimental works, references to which are given in [117, 241]. In [193] similarity was stressed between the kinetics of local perturbations in a hybrid scheme with a large delay time and in the one-dimensional systems with a phase transition of the first kind, such as the scheme of increasing absorption bistability (Chap. 2). Though such an analogy is useful, it is not complete (see above). To restore the analogy, it is necessary to limit the range of perturbation parameters. In this way, temporal analogues of stable [193, 313] and unstable [313] switching waves and of critical nuclei [313] were found; see also some experimental investigations in [148].

A recent rebirth of interest in chaotic regimes in optical systems with delayed feedback was caused by the idea of hiding a message signal in chaos [254, 394, 257, 258]. To ensure a high degree of communication security, it is necessary to produce well-developed chaotic regimes (with a large number of positive Lyapunov exponents) with sufficiently wide bandwidth, and, to extract the message information, chaotic synchronization at the receiver has to be allowed. An example of successful implementation of such chaos generation can be found in [189]. Another promising application of hybrid devices connected with transverse pattern formation is based on the use of large-scale arrays of opto-electronic feedback circuits [407].

# 4. Driven Nonlinear Interferometers

Coherent radiation in an interferometer is characterized by the spatial distributions of the amplitude and phase of the field; additional degrees of freedom are connected with the state of radiation polarization and the existence of counterpropagating beams. Diffraction and scattering of radiation cause transverse redistribution of the field. The specific time scales are the time of light round-trip in the interferometer and the field relaxation time in the interferometer. The medium inside the interferometer has frequency dispersion, nonlinearity of different kinds with several specific times of relaxation that differ essentially by magnitude; diffusion processes in the medium also lead to transverse redistribution of the fields. Additional linear optical elements (amplitude and phase transparencies, spatial filters, etc.) can be present inside the interferometer. All these factors predetermine extreme variety of nonlinear optical phenomena and of spatio-temporal structures in such systems.

In the case of fast (as compared with relaxation times in the medium) field relaxation and the predominance of diffusive transverse interaction over a diffractive one, the basic equations are similar to those analyzed in Chap. 2 for schemes of increasing absorption bistability. Therefore we can also observe here the "diffusive" switching waves, the spatial hysteresis, the nontrivial kinetics of local perturbations and other already-known phenomena of spatial distributivity. Besides, if the interferometer round-trip time exceeds the time of medium relaxation (i.e., longitudinal distributivity is essential), then stationary regimes can become unstable with parameter variation, being substituted by periodic and chaotic regimes, as it takes place in hybrid schemes of bistability (Chap. 3).

New features specific to media with fast nonlinearity are the development of absolute instabilities of field transverse structure, the high sensitivity of the hysteresis phenomena to the angle of radiation incidence, the new types of switching waves which correspond to the spatial transition between the stationary and nonstationary states, and the spatial field oscillations on the switching wave front with a diffractive mechanism of transverse interaction. The latter circumstance leads to fundamental differences in features of diffusive and diffractive patterns. In particular, with homogeneous illumination of wide-aperture interferometers the formation of localized (particle-like) field structures – "diffractive autosolitons", or "dissipative optical solitons" – be-

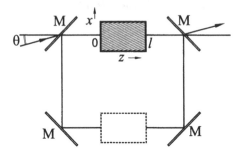

**Fig. 4.1.** Schematic of a nonlinear ring interferometer. M: mirrors; the layer of the nonlinear medium is *hatched*; additional linear elements can be added in the *dashed box*

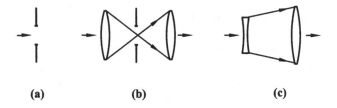

**Fig. 4.2.** Examples of additional linear elements: diaphragm (**a**), spatial filter (**b**), telescope system (**c**)

comes possible. With fixed parameters in a system there is a discrete set of states of single dissipative solitons, and multi-soliton structures can form with a discrete set of distances between them. The presence of dissipative solitons changes the kinetics of interferometer switching down (with a slow decrease in external beam power).

## 4.1 Models of Nonlinear Interferometers

### 4.1.1 Driven Ring Interferometers

Let us consider a ring interferometer (a cavity) driven by external coherent radiation and filled partly with a nonlinear medium (Fig. 4.1); additional linear elements can be added inside the interferometer (Fig. 4.2). Incident radiation has a fixed (for definiteness, linear) polarization. A nonzero component of the electric field of incident radiation $\tilde{E}_{\text{in}}$ is presented in the form:

$$\tilde{E}_{\text{in}} = \text{Re}\left\{ E_{\text{in}}(\boldsymbol{r}_\perp, z, t) \exp[\text{i}(kz - \omega_0 t)]\right\} . \tag{4.1.1}$$

Here $z$ is the coordinate along the interferometer axis; $\boldsymbol{r}_\perp$ is the vector of the transverse coordinates $x, y$; $\omega_0$ is the radiation carrier frequency; $k = \omega_0 \varepsilon_1^{1/2}/c$ is the wave number in the initial (linear) medium with dielectric permittivity

$\varepsilon_1$; and $c$ is the speed of light in vacuum. The envelope (complex amplitude) $E_{in}$ is assumed to slowly vary during a period of optical oscillations ($2\pi/\omega_0$) and at the light wavelength ($\lambda = 2\pi/k$).

We divide the entire interferometer extension $0 < z < L$ along the longitudinal coordinate $z$ into three regions: the layer of the nonlinear medium ($0 < z < d$), the linear interval ($d < z < L$) and the region adjacent to the entrance mirror, where external radiation enters. In the layer of the nonlinear medium with polarization $P = P_0 + \delta P$, where $P_0$ is the component of medium polarization linear with respect to the field (see Appendix A), we present the field $\tilde{E}$ in a form analogous to (4.1.1):

$$\tilde{E} = \text{Re}\left\{E(\boldsymbol{r}_\perp, z, t) \exp[i(k_0 z - \omega_0 t)]\right\} . \tag{4.1.2}$$

Here $k_0 = \omega_0 \varepsilon_0^{1/2}/c$, $\varepsilon_0 = \varepsilon_0(\omega_0)$ is the linear part of the dielectric permittivity of the medium at the frequency $\omega_0$. The envelope $E(\boldsymbol{r}_\perp, z, t)$ is also assumed to be a slowly varying function of its arguments. Its transformation at the extension of the nonlinear layer can be written in the form

$$E(\boldsymbol{r}_\perp, d, t) = \hat{M} E(\boldsymbol{r}_\perp, 0, t - \tau_d) , \quad \tau_d = d/v_{gr} , \tag{4.1.3}$$

where $v_{gr}$ is the group velocity (see Appendix A). The operator $\hat{M}$ is nonlinear and nonlocal (integral), in the general case with respect to both the transverse coordinates (because of radiation diffraction) and time (because of frequency dispersion and relaxation in the medium). Its form is determined by a paraxial equation (A.21) for the field envelope $E$:

$$2ik_0 \left(\frac{\partial E}{\partial z} + \frac{1}{v_{gr}} \frac{\partial E}{\partial t}\right) + \Delta_\perp E - D_2 \frac{\partial^2 E}{\partial t^2} + 4\pi \frac{\omega_0^2}{c^2} \delta P = 0 . \tag{4.1.4}$$

Here $\Delta_\perp = \partial^2/\partial x^2 + \partial^2/\partial y^2$ is the transverse Laplace operator and $D_2$ is the parameter of quadratic dispersion (see Appendix A). For the present, we do not specify the form of the constituent equation relating $\delta P$ to $E$ (see Appendix B).

Transformation of the field outside the layer of the nonlinear medium also takes place in the linear (air) regions of the interferometer, with reflection of radiation from the mirrors and when passing through additional linear optical elements (Fig. 4.2). We will write the resulting linear transformation of the field envelope in a form analogous to (4.1.3):

$$E(\boldsymbol{r}_\perp, L, t) = \hat{L} E(\boldsymbol{r}_\perp, d, t - \tau_L) , \tag{4.1.5}$$

where $\tau_L$ is the corresponding time delay, and the linear operator $\hat{L}$ also can be integral with respect to the transverse coordinates and time. In the linear intervals the paraxial equation for nondispersing media,

$$2ik_0 \left(\frac{\partial E}{\partial z} + \frac{1}{v_0} \frac{\partial E}{\partial t}\right) + \Delta_\perp E = 0 , \tag{4.1.6}$$

where $v_0 = c/\varepsilon_1^{1/2}$, describes the diffraction transformation of the field

$$E(\boldsymbol{r}_\perp, z_2, t) = \hat{\boldsymbol{D}} E(\boldsymbol{r}_\perp, z_1, t - (z_2 - z_1)/v_0) . \qquad (4.1.7)$$

The form of the integral (with respect to the transverse coordinates) diffraction operator $\hat{\boldsymbol{D}}$ follows from solution of (4.1.6):

$$E(\boldsymbol{r}_\perp, z_2, t) = [\lambda_0(z_2 - z_1)]^{-m/2} \exp\left(-\frac{i\pi}{2m}\right)$$
$$\times \int \exp\left(\frac{ik_0(\boldsymbol{r}_\perp - \boldsymbol{r}_\perp'')^2}{2(z_2 - z_1)}\right) E\left(\boldsymbol{r}_\perp'', z_1, t - \frac{z_2 - z_1}{v_0}\right) \mathrm{d}\boldsymbol{r}_\perp'' , \qquad (4.1.8)$$

where $m = 1$ for a one-dimensional geometry ($\boldsymbol{r}_\perp = x$, slit beams) and $m = 2$ for a two-dimensional ($\boldsymbol{r}_\perp = x, y$) geometry.

In actual schemes, there are aperture limitations to the field by diaphragms and because of the finite sizes of cavity mirrors. The action of the operator of the aperture limitation $\hat{\boldsymbol{A}} = \hat{\boldsymbol{A}}(\boldsymbol{r}_\perp)$,

$$E'(\boldsymbol{r}_\perp, z, t) = \hat{\boldsymbol{A}} E(\boldsymbol{r}_\perp, z, t) , \qquad (4.1.9)$$

is reduced to vanishing the field amplitude $E'$ directly behind the diaphragm in the area out of the aperture. For a mirror the amplitude of reflected radiation inside the aperture $E'$ is obtained by multiplying the amplitude of radiation incident on the mirror $E$ by the amplitude (complex) reflection coefficient $R$:

$$E(\boldsymbol{r}_\perp, z, t) = R\hat{\boldsymbol{A}} E(\boldsymbol{r}_\perp, z, t) . \qquad (4.1.10)$$

Possible curvature of the mirror is taken into account by corresponding coordinate dependence of the phase of the coefficient $R$. Dependence of module $|R|$ on the coordinates allows us to smooth oscillations caused by diffraction on mirror edges [21].

Inside the interferometer, such linear elements as a telescope and the spatial filters can also be present. Various spatial field structures in nonlinear cavities with different additional linear elements including those providing field rotation are considered in [7, 8, 404, 406].

Figure 4.2b shows a spatial filter – a pair of confocal lenses with a diaphragm in their common focal plane. The field in the focal plane of a lens is a Fourier transform of the complex amplitude of the input field $E_{\mathrm{in}}$ [127]. The corresponding density of the field amplitude for the "spatial frequency" $\boldsymbol{k}_\perp$ is (we omit the arguments $z$ and $t$)

$$G(\boldsymbol{k}_\perp) = (2\pi)^{-m} \int_{-\infty}^{\infty} E_{\mathrm{in}}(\boldsymbol{r}_\perp) \exp(i\boldsymbol{k}_\perp \boldsymbol{r}_\perp) \mathrm{d}\boldsymbol{r}_\perp . \qquad (4.1.11)$$

The diaphragm in the focal plane cuts off the field components with spatial frequencies exceeding the boundary frequency of the filter $q_{\mathrm{f}}$. After the inverse

Fourier transform performed by the second lens of the filter, we obtain the field as it exits:

$$E_{\text{out}}(\boldsymbol{r}_\perp) = \hat{\boldsymbol{F}} E_{\text{in}}(\boldsymbol{r}_\perp) = \int_{k<q_{\text{f}}} d\boldsymbol{k}_\perp \, G(\boldsymbol{k}_\perp) \exp(-i\boldsymbol{k}_\perp \boldsymbol{r}_\perp)$$

$$= (2\pi)^{-m} \int_{k<q_{\text{f}}} d\boldsymbol{k}_\perp \, \exp(-i\boldsymbol{k}_\perp \boldsymbol{r}_\perp) \int d\boldsymbol{r'}_\perp \, E_{\text{in}}(\boldsymbol{r'}_\perp) \exp(i\boldsymbol{k}_\perp \boldsymbol{r'}_\perp). \quad (4.1.12)$$

In the absence of the diaphragm ($q_{\text{f}} \to \infty$), the operator of spatial filtration is unitary ($\hat{\boldsymbol{F}} = 1$). Besides, a spatial filter or a telescope system changes the effective optical length of the linear interval. In a telescope system with different focal distances of the lenses, the transverse scale also changes:

$$E_{\text{out}}(\boldsymbol{r}_\perp) = \hat{\boldsymbol{T}} E_{\text{in}}(\boldsymbol{r}_\perp) = E_{\text{in}}(\eta \boldsymbol{r}_\perp), \quad (4.1.13)$$

where $\eta$ is the coefficient of magnification. The operator of the transformation of the field amplitude in the linear interval $\hat{\boldsymbol{L}}$ is the product of the described operators $\hat{\boldsymbol{D}}$, $\hat{\boldsymbol{A}}$, $\hat{\boldsymbol{F}}$, and $\hat{\boldsymbol{T}}$ (and of reflection coefficient $R$), and the effective time delay is the sum of the corresponding values of the partial time delays.

Finally, coherent superposition of the fields of external radiation and the one circulating inside the interferometer takes place on the linear section in the vicinity of the input mirror. In view of the importance of phase relations, it should be described by summation of their full (quickly varying) amplitudes:

$$\tilde{E}(\boldsymbol{r}_\perp, 0, t) = \tilde{E}''_{\text{in}}(\boldsymbol{r}_\perp, 0, t) + R_{\text{in}} \tilde{E}(\boldsymbol{r}_\perp, L, t - \tau_{\text{in}}). \quad (4.1.14)$$

Here $\tilde{E}''_{\text{in}} = T_{\text{in}} \tilde{E}_{\text{in}}$, $T_{\text{in}}$ is the amplitude coefficient of transmission of the input mirror, $\tau_{\text{in}}$ is the time delay on the considered interval (small compared with the diffraction length, see Appendix A). If in (4.1.14) we introduce the slowly varying amplitudes (4.1.2) and denote by $R$ the product of the amplitude reflection coefficients of all interferometer mirrors, by $\tau_{\text{del}}$ — the total time delay — and by $\Delta_{\text{ph}}$ — the phase shift related to this delay — we obtain

$$E(\boldsymbol{r}_\perp, 0, t) = E'_{\text{in}}(\boldsymbol{r}_\perp, 0, t) + R \exp(i\Delta_{\text{ph}}) \hat{\boldsymbol{L}} \hat{\boldsymbol{M}} E(\boldsymbol{r}_\perp, 0, t - \tau_{\text{del}}). \quad (4.1.15)$$

Without loss of generality, it is possible to consider $|\Delta_{\text{ph}}| \le \pi$, and the value $R$ as real and positive, removing the possible phase factor by redenoting the phase detuning $\Delta_{\text{ph}}$.

The form of the operator of the nonlinear transform of the field $\hat{\boldsymbol{M}}$ is determined by the form of the constituent equation for $\delta P$ or $\delta\varepsilon$ ($\delta P = \delta\varepsilon E/4\pi$, see Appendix B). In the cases considered in the present chapter, it is sufficient to use the relaxation equation accounting for diffusion in the medium,

$$\frac{\partial \delta\varepsilon}{\partial t} = -\frac{1}{\tau_{\text{rel}}}[\delta\varepsilon - \delta\varepsilon_{\text{nl}}(|E|^2)] + D\,\Delta\delta\varepsilon, \quad (4.1.16)$$

where $\tau_{\text{rel}}$ is the relaxation time of nonlinearity, $D$ is the diffusion coefficient, and $\Delta$ is the Laplace operator. Neglecting diffusion corresponds to $D = 0$:

$$\frac{\partial \delta \varepsilon}{\partial t} = -\frac{1}{\tau_{\text{rel}}} [\delta \varepsilon - \delta \varepsilon_{\text{nl}}(|E|^2)] . \qquad (4.1.17)$$

At $\tau_{\text{rel}} \ll \tau_E$, nonlinearity can be considered fast and completely determined by the instant value of field intensity:

$$\delta \varepsilon = \delta \varepsilon_{\text{nl}}(|E|^2) . \qquad (4.1.18)$$

The time of field relaxation in the interferometer $\tau_E$ depends on the cavity characteristics (time of light round-trip $\tau_{\text{del}}$ and losses $1-R$) and nonlinearity; it can be estimated by the rate of decay of small field deviations from the steady state (see Sect. 4.2).

Equations (4.1.15) and (4.1.16) give a general closed description of the spatio-temporal field structures in a nonlinear interferometer. With some additional assumptions this description can be essentially simplified. In particular, only one equation (4.1.15) remains under the conditions of applicability of (4.1.18). As above, it is useful to separate the effects of transverse and longitudinal distributivity.

### 4.1.2 Model of Transverse Distributivity

This model is often referred to as the mean-field approximation [204]. Neglect of longitudinal extension is justified if longitudinal changes of the field in one round-trip in the interferometer are small:

$$|E - \langle E \rangle| \ll |\langle E \rangle| . \qquad (4.1.19)$$

Here, as in Sect. 2.1, the angular brackets mean averaging in the longitudinal direction:

$$\langle E \rangle = \frac{1}{d} \int_0^d E(\mathbf{r}_\perp, z, t) \, dz . \qquad (4.1.20)$$

For (4.1.19) to be true, extensions of the linear and nonlinear regions ought to be small compared with the typical lengths of diffraction and nonlinear distortions (see Appendix A). Besides, phase detuning and losses in the interferometer should be small:

$$\Delta_{\text{ph}}^2 \ll 1 , \quad 1 - R \ll 1 . \qquad (4.1.21)$$

Any additional linear elements inside the interferometer are assumed to be absent.

The derivation of the approximate equation is close to the one described in Sect. 2.1 (see also [204]). We average paraxial equation (4.1.5) over the thickness of the nonlinear layer $d$:

$$2ik_0\left(\frac{1}{d}[E(\boldsymbol{r}_\perp, d, t) - E(\boldsymbol{r}_\perp, 0, t)] + \frac{1}{v_{gr}}\frac{\partial\langle E\rangle}{\partial t}\right)$$

$$+ \Delta_\perp\langle E\rangle - D_2\frac{\partial^2\langle E\rangle}{\partial t^2} + 2\pi\frac{\omega_0^2}{c^2}\langle\delta P\rangle = 0 . \qquad (4.1.22)$$

For simplicity we assume the interferometer to be completely full of a non-linear medium. Using boundary condition (4.1.15), neglecting the terms quadratic in small values [in view of (4.1.19) and (4.1.21)], and replacing $E'_{in} \to E_{in}$, we obtain an equation which describes the effects of purely transverse distributivity:

$$\frac{1}{v_{gr}}\frac{\partial\langle E\rangle}{\partial t} - \frac{i}{2k_0}\Delta_\perp\langle E\rangle + \frac{iD_2}{2k_0}\frac{\partial^2\langle E\rangle}{\partial t^2}$$

$$-i\frac{2\pi k_0}{\varepsilon_0}\langle\delta P\rangle + \frac{1}{d}\{[1 - R\exp(i\Delta_{ph})]\langle E\rangle - E_{in}\} = 0 . \qquad (4.1.23)$$

According to (4.1.23), temporal variation of the field inside an interferometer is caused by diffraction (term $\propto \Delta_\perp E$), frequency dispersion ($\propto D_2$), instant nonlinearity of the medium ($\propto \langle\delta P\rangle$), multiple reflection from the mirrors and input of external radiation [the latter terms on the left-hand side of (4.1.23)]. In contrast to the "one-component" model in Chap. 2, the system described by (4.1.23) is at least a two-component one, because the amplitude $\langle E\rangle$ is complex. Under the approximations noted, (4.1.23) describes the field dynamics both in ring (Fig. 4.1) and Fabry–Perot (Fig. 1.5) interferometers.

If smooth spatial filtering with characteristic boundary frequency $q_f$ is present in the scheme, i.e., losses for waves with oblique incidence grow quadratically with (small) angle of incidence, the second term on the right-hand side of (4.1.23) changes in the following way [298]:

$$-\frac{i}{2k_0}\Delta_\perp\langle E\rangle \to \left(-\frac{i}{2k_0} + \frac{1-R}{q_f^2 d}\right)\Delta_\perp\langle E\rangle . \qquad (4.1.24)$$

It is important to consider the following feature of (4.1.23) true for a non-dispersive medium ($D_2 = 0$) with fast nonlinearity (4.1.18) in the case when incident radiation amplitude does not depend on the coordinate $x$, i.e., $E_{in}^{(0)} = $ const. Suppose we know the corresponding solution of (4.1.23) of the form $E^{(0)} = E^{(0)}(x, y, t)$. Then one can see that for

$$E_{in} = E_{in}^{(0)}(y, t)\exp(ik_0\theta x)\exp(-i\nu t) \qquad (4.1.25)$$

the solution of (4.1.23) will be

$$E = E^{(0)}(x - vt, y, t)\exp(ik_0\theta x)\exp(-i\nu t) ,$$

$$v = v_{gr}\theta , \quad \nu = \frac{v_{gr}k_0}{2}\theta^2 . \qquad (4.1.26)$$

The physical meaning of this "Galilean transform" is the following: At oblique incidence of external radiation at a small angle $\theta$ with respect to the interferometer axis, the field transverse structure is subjected to motion due to the geometric shift of rays. The speed of shift of structure as a whole, $v$, is proportional to the angle of incidence, $\theta$. Variation of the optical path at oblique incidence $\delta L = L\theta^2/2$ is compensated by the frequency shift $\nu = \omega_0 \delta L/L$. This feature is violated, and the influence of angle of incidence weakens upon consideration of dispersion ($D_2 \neq 0$) or a nonlinear medium finite relaxation time $\tau_{rel} \neq 0$.

Formally the Galilean transform corresponds to the transition to the co-ordinate frame moving in the transverse direction $x$ with a constant velocity $v$: $\xi = x - vt$. In this frame, for one-dimensional steady-state field structures and nondispersing medium ($D_2 = 0$), (4.1.23) takes the form

$$\frac{d^2 E}{d\xi^2} - 2ik_0 \frac{v}{v_{gr}} \frac{dE}{d\xi} + \left(\frac{\omega_0}{c}\right)^2 \delta\varepsilon E$$
$$+ i\frac{2k_0}{d}\{[1 - R\exp(i\Delta_{ph})]E - E_{in}\} = 0 . \tag{4.1.27}$$

The field transversely homogeneous stationary states $E = E_{st} = E_{st1}$, $E_{st2}, \ldots$ are found from

$$\left(1 - R\exp(i\Delta_{ph}) - i\frac{k_0 d}{2\varepsilon_0}\delta\varepsilon(|E_{st}|^2)\right) E_{st} = E_{in} . \tag{4.1.28}$$

Besides them, the ordinary differential equation (4.1.27) can have transversely inhomogeneous solutions with asymptotic $dE/d\xi \to 0$ at $\xi \to \pm\infty$. More precisely, the asymptotic is determined by introduction of the field's small deviations from the stationary transversely homogeneous states, $E = E_{st}(1 + \delta E)$, and by linearization of (4.1.27) with respect to these deviations:

$$\frac{d^2 \delta E}{d\xi^2} - 2ik_0 \frac{v}{v_{gr}} \frac{d\delta E}{d\xi} + \alpha\delta E + \beta\delta E^* = 0 , \tag{4.1.29}$$

$$\alpha = \left(\frac{\omega_0}{c}\right)^2 (\delta\varepsilon_{st} + \delta\varepsilon' I_{st}) + i\frac{2k_0}{d}[1 - R\exp(i\Delta_{ph})] ,$$

$$\beta = \left(\frac{\omega_0}{c}\right)^2 \delta\varepsilon' I_{st} , \quad \delta\varepsilon' = \left(\frac{d\delta\varepsilon}{dI}\right)_{I=I_{st}} , \quad I = |E|^2 . \tag{4.1.30}$$

We seek the solution of (4.1.29) in the form

$$\delta E = a\exp(p\xi) + b^*\exp(p^*\xi) . \tag{4.1.31}$$

The condition of solvability of the system of linear algebraic equations with respect to $a$ and $b$ (vanishing of the system determinant) gives the algebraic quartic (fourth-order) equation for the determination of the value $p$. It is reduced to a quadratic one for motionless structures ($v = 0$) when

$$p^2 = -\text{Re}\,\alpha \pm [|\beta|^2 - (\text{Im}\,\alpha)^2]^{1/2} . \tag{4.1.32}$$

It follows from (4.1.32) that at $|\beta| < |\operatorname{Im}\alpha|$ there are field diffraction oscillations distinct from aperiodic diminishing, which is only possible in the case of diffusion transverse coupling. As will be shown in Sect. 4.4, this new feature radically changes the character of spatio-temporal structures. Fulfillment of this condition depends on the nonlinearity type and intensity $I_{\text{st}}$.

### 4.1.3 Model of a Nonlinear Screen

We come to some other model in the case of a thin nonlinear layer, in which we can neglect diffraction. However, diffraction is considered in the more lengthy linear region inside the interferometer. Dispersion is considered inessential; this is true for the regimes with sufficiently smooth temporal variation. In this case the operator of the nonlinear field transformation $\hat{M}$ in (4.1.15) is local:

$$\hat{M} = \exp\left(\frac{i\omega_0^2}{2k_0 c^2}\int_0^d \delta\varepsilon\,\mathrm{d}z\right). \tag{4.1.33}$$

If nonlinearity is fast, then $\hat{M}$ represents a complex amplitude coefficient of transmission of the nonlinear layer:

$$\hat{M} = K(|E(\boldsymbol{r}_\perp,0,t)|^2) = \frac{E(\boldsymbol{r}_\perp,d,t)}{E(\boldsymbol{r}_\perp,0,t-\tau_{\mathrm{d}})}. \tag{4.1.34}$$

For the quasistationary response of the two-level medium with saturation intensity $I_{\text{sat}}$ and the linear coefficient of (unsaturated) absorption $\alpha_0$, the transmission coefficient is determined from

$$\frac{\mathrm{d}E}{\mathrm{d}z} = -\frac{\alpha_0}{2}(1-iw_0)\frac{E}{1+|E|^2/I_{\text{sat}}}. \tag{4.1.35}$$

The value $w_0$ represents frequency detuning (in the units of the spectral line width). From (4.1.35) we obtain the following relation between field intensity at the input into the nonlinear layer $I_{z=0} = |E_{z=0}|^2$ and the intensity transmission coefficient of the layer $K_{\text{i}} = |K|^2$:

$$I_{z=0} = \frac{\alpha_0 d + \ln K_{\text{i}}}{1-K_{\text{i}}}. \tag{4.1.36}$$

The amplitude coefficient of transmission of the layer is expressed in terms of $K_{\text{i}}$:

$$K = K_{\text{i}}^{(1-iw_0)/2}. \tag{4.1.37}$$

It is possible to extend the capabilities of such a model by replacing the layer by several nonlinear screens separated by linear regions. Actually it corresponds to the scheme of the numerical solution of paraxial equation (4.1.4) by the method of splitting, where, in the case of wide-aperture interferometers, the algorithm of the fast Fourier transform is quite effective [323]. Note that in the variant proposed in [323], the paraxial approximation is not actually used. Instead, it is enough to limit ourselves by the weaker approximation of unidirectional propagation of radiation.

### 4.1.4 Transversely Confined and Point Schemes

It is possible to come to such a model either in the case of a single-mode cavity or strong discrimination of the losses of transverse modes, or for the regimes of plane waves of a wide-aperture interferometer, for which the transverse variation of field is neglected. If we also consider stationary states $E = E_{st}$, (4.1.15) is reduced to the form specific to point schemes [compare with (4.1.28)]:

$$E_{in} = E_{st}[1 - R\exp(i\Delta_{ph})K(|E_{st}|^2)] .\tag{4.1.38}$$

From this we obtain the equation determining the stationary values of intensity inside the interferometer $I_{st} = |E_{st}|^2$:

$$I_{in} = I_{st}|1 - R\exp(i\Delta_{ph})K(I_{st})|^2 .\tag{4.1.39}$$

Under conditions of bi- (multi-)stability, (4.1.39) has several solutions for $I_{st}$, for each value of intensity of incident radiation in some ranges of $I_{in}$ variation. Knowing the intensity $I_{st}$, we determine the field phase unambiguously, because it follows from (4.1.38) that

$$E_{st} = E_{in}/[1 - R\exp(i\Delta_{ph})K(I_{st})] .\tag{4.1.40}$$

For example, in the case of a fast Kerr medium with the dielectric permittivity

$$\varepsilon = \varepsilon_0 + \varepsilon_2|E|^2 ,\tag{4.1.41}$$

the amplitude coefficient of transmission of a thin nonlinear layer is

$$K = \exp(i\varphi_2 I_{st}) ,\tag{4.1.42}$$

where $\varphi_2 = \varepsilon_2 k_0 d/(2\varepsilon_0)$. Then (4.1.39) takes the form

$$I_{in} = I_{st}\left\{(1 - R)^2 + 4R\sin^2[(\Delta_{ph} + \varphi_2 I_{st})/2]\right\} .\tag{4.1.43}$$

Not presenting a demonstration of multistability by graphic analysis of (4.1.43), let us consider an important case of low-threshold bistability. In addition to (4.1.21), we assume the nonlinear shift to be small: $|\varphi_2|I_{st} \ll 1$. Then (4.1.43) is reduced to the cubic equation for $I_{st}$:

$$I_{in} = I_{st}[(1 - R)^2 + (\Delta_{ph} + \varphi_2 I_{st})^2] .\tag{4.1.44}$$

Bistability takes place under the conditions

$$|\Delta_{ph}| > \sqrt{3}(1 - R) , \quad \varphi_2\Delta_{ph} < 0\tag{4.1.45}$$

in the range of intensities

$$I_{min} < I_{in} < I_{max} ,\tag{4.1.46}$$

where

$$I_{\max(\min)} = \frac{1}{3}\{2|\Delta_{ph}| \pm [\Delta_{ph}^2 - 3(1-R)^2]^{1/2}\} \ . \qquad (4.1.47)$$

Within the framework of the present model, the kinetics is easily described for the case of fast nonlinearity and in the absence of frequency dispersion. Then (4.1.15) connects the values of the field amplitude separated by the time interval $\tau_{del}$ – the light round-trip time in the cavity. Introducing discrete time $t_n = n\tau_{del}$, we obtain the recurrent relation

$$E_{n+1} = E_{in} + R\exp(i\Delta_{ph})K(|E_n|^2)E_n \ , \quad n = 0, 1, 2, \ldots \qquad (4.1.48)$$

In contrast to the case considered in Chap. 3, the map described by (4.1.48) is two-component (the field amplitude is complex).

The following substitution is possible for the regimes with a weak change in the field in the cavity during one round-trip:

$$E(t + \tau_{del}) \approx E(t) + \tau_{del}\frac{dE}{dt} \ . \qquad (4.1.49)$$

It results in the transformation of (4.1.48) to

$$\tau_{del}\frac{dE}{dt} = E_{in} - [1 - R\exp(i\Delta_{ph})K(|E|^2)]E \ . \qquad (4.1.50)$$

Stationary states for the systems described (4.1.48) and (4.1.50) coincide [relation (4.1.38)]. The precision of the kinetics description in terms of (4.1.50) is determined by the accuracy of approximation (4.1.49).

### 4.1.5 Model of Slow Nonlinearity

In the limit opposite to the case of fast nonlinearity

$$\tau_{rel} \gg \tau_E \ , \qquad (4.1.51)$$

the field in the cavity follows adiabatically the slowly varying state of the medium. Therefore, having found stationary solution of (4.1.15) at the fixed instant distribution $\delta\varepsilon$, we can write

$$E = E\Big(\delta\varepsilon(\boldsymbol{r}_\perp, z, t)\Big). \qquad (4.1.52)$$

Practically it is convenient to use the model of a thin screen (4.1.33) and the variant of purely transverse distributivity with averaging over the layer thickness:

$$\langle\delta\varepsilon\rangle = \frac{1}{d}\int_0^d \delta\varepsilon(\boldsymbol{r}_\perp, z, t)\,dz \ . \qquad (4.1.53)$$

In the case when the interferometer is filled completely with a nonlinear medium, radiation incidence is normal ($\theta = 0$) and diffraction is negligible, we find for intensity $I = |\langle E\rangle|^2$ from (4.1.15)

$$I = I(\langle\delta\varepsilon\rangle) = I_{\rm in}|1 - R\exp\{{\rm i}[\Delta_{\rm ph} + \omega_0^2 d\langle\delta\varepsilon\rangle/(2k_0c^2)]\}|^{-2}\ . \qquad (4.1.54)$$

If we now average (4.1.16) in the longitudinal direction, we obtain the closed equation describing the kinetics of the transverse variation of $\langle\delta\varepsilon\rangle$ (and, according to (4.1.54), of intensity $I$):

$$\frac{\partial\langle\delta\varepsilon\rangle}{\partial t} = -\frac{1}{\tau_{\rm rel}}\{\langle\delta\varepsilon\rangle - \delta\varepsilon_{\rm nl}[I(\langle\delta\varepsilon\rangle)]\} + D\Delta_\perp\langle\delta\varepsilon\rangle\ . \qquad (4.1.55)$$

Such an approach is valid if diffusion (but not radiation diffraction or spatial filtering) is the prevailing mechanism of the transverse coupling in the interferometer. For purely dispersive (${\rm Im}\,\delta\varepsilon = 0$) or absorptive (${\rm Re}\,\delta\varepsilon = 0$) types of nonlinearity, (4.1.55) corresponds to the one-component system considered in Chap. 2. Therefore, the approach in Chap. 2 with the introduction of the "potential curves" can be completely transferred to the case of these nonlinear interferometers. Correspondingly, the results of analysis of the spatial hysteresis and switching with slow time variation of the intensity of incident radiation beam are true; the features of the switching waves are also the same. Thus we will not repeat these conclusions, referring for details to [277]. Below we will mainly consider new (as compared with the case of bistability at increasing absorption) factors and peculiar spatio-temporal structures of the field in nonlinear interferometers.

## 4.2 Stability of Stationary Regimes

The instability of some of stationary states is an integral part of bistability. Already in a point (spatially nondistributed) scheme the intermediate branches correspond to unstable states; other instabilities are absent for one-component point schemes (Sect. 2.2). Additional types of instability appear in two-component schemes. As McCall has shown [223], the conditions of instability are reached in the cavities, in which nonlinearity of the medium is caused by two competitive mechanisms with different relaxation times. As a result, the regimes of regenerative pulsations form. This and a number of other types of instability are discussed in [117].

In distributed systems the variety of types of instability greatly increases (see Sect. 4.9). A general scheme of analysis of stability of stationary and homogeneous states with respect to small perturbations is standard. Linearized kinetic equations with constant coefficients are true for such perturbations. Therefore their solution is represented in the form of expansion in eigen-types of perturbations, which change in time exponentially. The sign of the real part of these Lyapunov exponents determines the nature of the state stability. As before, it is convenient to separate manifestation of the longitudinal (this section) and transverse (Sect. 4.3) distributivity for stability analysis.

### 4.2.1 Self-pulsations

Let us start our discussion with one of the simplest two-component models determined by difference [(4.1.48), further relations will be marked with label (a)] and differential [(4.1.50), label (b)] equations [286]. The stationary state, whose stability we are interested in, is characterized by the field amplitude upon entrance of the nonlinear element $E_{st} = E_{z=0}$ (Fig. 4.1); this state is the same for both variants of the description and is determined by (4.1.39-4.1.40).

We introduce a small perturbation of the amplitude $\delta E$:

$$E = E_{z=0} = E_{st}(1 + \delta E). \tag{4.2.1}$$

The kinetics equations linearized with respect to $\delta E$ have the form

$$\delta E_{n+1} = R \exp(i\Delta_{ph})[(K_{st} + K' I_{st})\delta E_n + (K' I_{st})\delta E_n^*] \text{ (a)},$$

$$\tau_{del}\frac{d\delta E}{dt} = -\delta E + R \exp(i\Delta_{ph})[(K_{st} + K' I_{st})\delta E + (K' I_{st})\delta E^*] \text{ (b)}, \tag{4.2.2}$$

where

$$I_{st} = |E_{st}|^2, \quad K_{st} = K(I_{st}), \quad K' = (dK/dI)_{I=I_{st}}. \tag{4.2.3}$$

We seek the solution of (4.2.2) in the form

$$\delta E_n = a\Lambda^n + b^* \Lambda^{*n} \qquad \text{(a)},$$
$$\delta E(t) = a \exp(\gamma t) + b^* \exp(\gamma^* t) \quad \text{(b)}. \tag{4.2.4}$$

The dispersion relation for the determination of $\gamma$ and $\Lambda \propto \exp(\gamma\tau_{del})$ has the form of the following quadratic equation:

$$\Lambda^2 + (p - 2)\Lambda + (1 - p + q) = 0 \quad \text{(a)},$$
$$(\gamma\tau_{del})^2 + p(\gamma\tau_{del}) + q = 0 \qquad \text{(b)}. \tag{4.2.5}$$

It is convenient to express the coefficients $p$ and $q$ through two types of differential gain: the differential gain of the nonlinear cavity $D_c$ and the differential gain of the nonlinear element $D_e$, which are calculated at the stationary value of intensity $I = I_{st}$:

$$q = D_c^{-1}, \quad p = q + 1 - |R|^2 D_e, \tag{4.2.6}$$

$$D_c = \left(\frac{dI_{in}}{dI}\right)^{-1}_{I=I_{st}}, \quad D_e = \left(\frac{dI_d}{dI_{in}}\right)_{I=I_{st}}, \tag{4.2.7}$$

where $I_d = |E_{z=d}|^2$ is the intensity at the nonlinear element output. For the differential gain of the element we have

$$D_e = \left[\frac{d}{dI}(|K|^2 I)\right]_{I=I_{st}}. \tag{4.2.8}$$

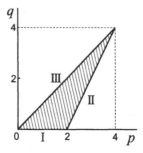

**Fig. 4.3.** Domain of stability of a stationary regime (*hatched*); the loss of stability is possible upon leaving the domain across the segments I, II, and III [286]

The explicit expression for the differential gain of the cavity $D_c$, through its parameters and characteristics of nonlinearity, follows from (4.1.39):

$$D_c^{-1} = 1 - 2R\operatorname{Re}\left[\exp(i\Delta_{\mathrm{ph}})\left(\frac{\mathrm{d}(KI)}{\mathrm{d}I}\right)_{I=I_{\mathrm{st}}}\right] + R^2 D_e \,. \qquad (4.2.9)$$

The condition of stability of the stationary regime,

$$\begin{aligned}\max|\Lambda|^2 < 1 \qquad &\text{(a)}\,,\\ \max\operatorname{Re}\gamma < 0 \qquad &\text{(b)}\,,\end{aligned} \qquad (4.2.10)$$

is written in the form

$$\begin{aligned}0 < q < p\,, \quad q > 2(p-2) \quad &\text{(a)}\,,\\ p > 0\,, \quad q > 0 \qquad\qquad\quad &\text{(b)}\,.\end{aligned} \qquad (4.2.11)$$

It is convenient to represent this result of the description of the kinetics in terms of (4.1.48) in the plane of the parameters $p, q$ (Fig. 4.3). The stability of the stationary regime can be lost in three different ways:

(I) In view of (4.2.6) and (4.2.7), changing the sign of $q$ ($q < 0$) will mean negative differential gain of the cavity, i.e., the unstable states corresponding to intermediate branches of the transfer function.

(II) At the boundary of the stability region $q = 2(p-2)$, $2 < p < 4$, the value $\Lambda = -1$; therefore in the vicinity of the boundary the field oscillates with the period $T = 2\tau_{\mathrm{del}}$. Such instability (Ikeda instability [146]) is specific to nonlinear systems with a delay and has already been considered in Chap. 3. For this type of instability, approximation (4.1.49) is inappropriate, and it cannot be described in the framework of (4.1.50).

(III) The boundary

$$0 < p = q < 4 \quad \text{(a)} \qquad\qquad (4.2.12)$$

corresponds to the oscillatory loss of stability of the stationary state (the Andronov–Hopf bifurcation). At small deviations $\delta$ of the sign due from the bifurcation values of the parameters, the field fluctuations are close to the harmonic ones with the modulation amplitude proportional to $\sqrt{|\delta|}$ and the period

$$
\begin{aligned}
T &= 2\pi\tau_{\text{del}}/\arccos(1 - q/2) \quad &\text{(a)}, \\
T &= 2\pi\tau_{\text{del}}/\sqrt{q} \quad &\text{(b)}.
\end{aligned} \tag{4.2.13}
$$

Note that (4.2.13a) is transformed into (4.2.13b) at $q \ll 1$. Approximation (4.1.49) is true in the case of considerable periods, $T \gg \tau_{\text{del}}$, for which $q \ll 1$ is necessary as well.

In view of (4.2.6) and (4.2.7), (4.2.12) can be presented in the form

$$
(|R|^2 D_{\text{e}})_{I=I_{\text{st}}} = \begin{cases} 1 & \text{(a)}, \\ 1 + q, \ q > 0 & \text{(b)}. \end{cases} \tag{4.2.14}
$$

The criterion of instability (4.2.14a) has a clear physical meaning: the differential gain of small perturbations (deviations from the stationary level) in the nonlinear element should balance their attenuation due to partial radiation transmission through the cavity mirrors. In this case the intensity and phase of the field (more exactly, the phase difference of the field in the cavity and of the external radiation) oscillate with time in the vicinity of the unstable position of equilibrium. In such a form the condition of the self-pulsation instability is also applicable to Fabry–Perot nonlinear cavities. Note that in the framework of the differential equation (4.1.50) the boundaries of the stability region are determined with considerable inaccuracy, and one should assume that $p \ll q \ll 1$ to justify approximation (4.1.49).

The presence of differential gain in the nonlinear element of the cavity $(dI_{\text{d}}/dI_{\text{in}} > 1)$ is the necessary condition to achieve the oscillatory instability. It is absent for the media usually considered in optical bistability (the Kerr nonlinearity or a medium of two-level particles). Oscillatory instability is also absent if nonlinearity is purely amplitudinal (the transmission coefficient $K$ is real) and detuning $\Delta_{\text{ph}} = 0$, but it can take place with amplitudinal nonlinearity $(K = K^*)$ if $\Delta_{\text{ph}} \neq 0$. An example of this type of instability has been presented using radiation scattering in a nonlinear heterogeneous medium [20, 19, 286].

As (4.1.50) corresponds to a real differential equation of the second order, only periodic steady-state regimes (the limiting cycles) are possible within its framework (besides the stationary ones). The kinetics described by (4.1.48) is more diverse. In the general case, the formation of almost-periodic oscillations and of strange attractors of Henon type [356] is possible here.

### 4.2.2 Modulation Instability

Another type of instability arises in cavities with media that have frequency dispersion. Similar regimes can arise in nonlinear cavities where frequency

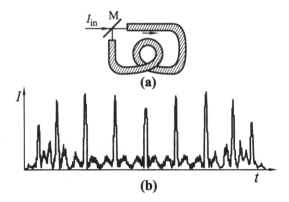

**Fig. 4.4.** Diagram of a nonlinear optical fiber cavity (**a**) and an example of the regime of self-pulsations (**b**) [412]

dispersion is attributed not to the properties of the homogeneous medium, but to the existence of media interfaces (for example, additional cavities). It is similar to low-threshold modulation instability in one-mode optical fibers [9, 13]. The cavity scheme considered differs from ordinary optical fibers by the presence of feedback (mirror M in Fig. 4.4a). It has been studied in [239, 240]; our consideration is based on [287].

We will write the dielectric permittivity of the medium in a fiber cavity as $\varepsilon = \varepsilon_0 + \delta\varepsilon$. The medium is assumed to be transparent (Im $\varepsilon = 0$) with linear dispersion [$\varepsilon_0 = \varepsilon_0(\omega)$] and fast nonlinearity [$\delta\varepsilon = \delta\varepsilon_{\mathrm{nl}}(|E|^2)$]. Radiation propagation in the medium is described by paraxial equation (4.1.4) neglecting transverse phenomena:

$$2ik_0\left(\frac{\partial E}{\partial z} + \frac{1}{v_{\mathrm{gr}}}\frac{\partial E}{\partial t}\right) - D_2\frac{\partial^2 E}{\partial t^2} + k_0^2\frac{\delta\varepsilon_{\mathrm{nl}}}{\varepsilon_0}E = 0 . \qquad (4.2.15)$$

The boundary condition has the form (4.1.14), and we assume the cavity to be completely full with the nonlinear medium ($d = L$), the additional linear elements to be absent and $\tau_{\mathrm{del}} \to 0$.

In the unperturbed stationary regime with frequency $\omega_0$ we have

$$E_{\mathrm{st}}(z) = E_{\mathrm{st}0}\exp[ik_0(\delta\varepsilon_{\mathrm{st}}/2\varepsilon_0)z] , \qquad (4.2.16)$$

where

$$\varepsilon_0 = \varepsilon_0(\omega_0) , \quad \delta\varepsilon_{\mathrm{st}} = \delta\varepsilon_{\mathrm{nl}}(I_{\mathrm{st}}) . \qquad (4.2.17)$$

The relations between the radiation holding intensity $I_{\mathrm{in}}$ and $I_{\mathrm{st}} = |E_{\mathrm{st}}|^2$ have the form [compare with (4.1.43)]

$$I_{\mathrm{in}} = I_{\mathrm{st}}[(1-R)^2 + 4R\sin^2(\tilde{\Delta}/2)] , \quad \tilde{\Delta} = \Delta_{\mathrm{ph}} + (k_0L/2)(\delta\varepsilon_{\mathrm{st}}/\varepsilon_0) . \quad (4.2.18)$$

For small deviations of the field amplitude from the stationary value $\delta E$, we obtain, after passing on to the running system of the coordinates $z$, $\eta = t - z/v_{\mathrm{gr}}$,

$$2ik_0 \frac{\partial \delta E}{\partial z} - D_2 \frac{\partial^2 \delta E}{\partial \eta^2} + J(\delta E + \delta E^*) = 0 , \qquad (4.2.19)$$

$$J = k_0^2 I_{st}(\delta\varepsilon'/\varepsilon_0) , \qquad \delta\varepsilon' = (d\delta\varepsilon/dI)_{I=I_{st}} . \qquad (4.2.20)$$

Boundary condition (4.1.14) takes the form

$$\delta E(0,\eta) = R \exp(i\tilde{\Delta})\delta E(L, \eta - L/v_{gr}) . \qquad (4.2.21)$$

To solve (4.2.19) we assume

$$\delta E(z,\eta) = a(z) \exp(i\Omega\eta) + b^*(z) \exp(-i\Omega^*\eta) , \qquad (4.2.22)$$

where $\Omega = \Omega' + i\Omega''$ is the (complex) frequency of perturbations. Substituting (4.2.22) into (4.2.19), one finds transformation of perturbation on the length of the nonlinear medium (in the matrix form):

$$\begin{pmatrix} a(z) \\ b(z) \end{pmatrix} = \begin{pmatrix} m_{11} & m_{12} \\ m_{12}^* & m_{11}^* \end{pmatrix} \begin{pmatrix} a(0) \\ b(0) \end{pmatrix} , \qquad (4.2.23)$$

$$m_{11} = (1 + iG)\cosh(\kappa z) , \qquad m_{12} = i[J/(2k_0\kappa)]\sinh(\kappa z) ,$$

$$(2k_0\kappa)^2 = -D_2\Omega^2(2J + D_2\Omega^2) , \quad G = (J + D_2\Omega^2)/(2k_0\kappa) . \quad (4.2.24)$$

Using boundary condition (4.2.21) leads to the following dispersion equation for the determination of $\Omega$:

$$Z^2 - 2p(\Omega)Z + 1 = 0 , \quad Z = \exp(i\Omega L/v_{gr})/R , \qquad (4.2.25)$$

$$p(\Omega) = \cosh(\kappa L)\cos\tilde{\Delta} - G\sinh(\kappa L)\sin\tilde{\Delta} . \qquad (4.2.26)$$

If the values $\kappa$ determined by (4.2.24) are purely imaginary, then the hyperbolic functions in (4.2.26) are transformed into trigonometric functions. Instability of the stationary regime is achieved at $\Omega'' < 0$, when according to (4.2.22), initially small perturbations increase exponentially with time. The boundary of the region of stability of the stationary regime we are interested in corresponds to $\Omega'' = 0$. Then the value $p = p(\Omega')$ is real, and it follows from (4.2.25) that

$$\exp(i\Omega'L/v_{gr}) = \pm 1 , \quad \Omega' = \pi n v_{gr}/L , \quad n = 0, 1, 2, \ldots \qquad (4.2.27)$$

The value $n = 0$ in (4.2.27) corresponds to nonoscillatory instability of the states representing the intermediate branches of the transfer function (with the negative slope); $n = 1$ represents the Ikeda instability with the period $2\tau_{del}$, where $\tau_{del} = L/v_{gr}$ is the round-trip time for the cavity. The differential gain for a layer of a transparent nonlinear medium $dI(z = d)/dI(z = 0) = 1$; so the self-pulsations discussed above do not exist. At $n \geq 1$ the period of the oscillations at the threshold of their onset is

$$T_n = 2\pi/\Omega' = 2\tau_{del}/n , \qquad (4.2.28)$$

and the threshold coefficient of reflection

$$R = |p| - \sqrt{p^2 - 1} \,. \qquad (4.2.29)$$

This threshold value corresponds to balancing of the losses in the cavity by amplification of perturbations due to power transfer from basic (unperturbed) radiation to them.

Using (4.2.18) and (4.2.26–4.2.29), one can easily construct the dependence of the threshold value of $R$ on the intensity of incident radiation $I_{in}$ and on other problem parameters. The results have a more obvious form in the case of small values of nonlinearity and detuning:

$$|\kappa|L \ll 1 \,, \quad (k_0 L/2)|\delta\varepsilon_{st}|/\varepsilon_0 \ll 1 \,, \quad |\tilde{\Delta}| \ll 1 \,. \qquad (4.2.30)$$

In this limit the Ikeda instability ($n = 1$) does not exist, $n = 2N$, $N = 1, 2, \ldots$, and

$$p = p_{max} - q^2/2 \,, \quad p_{max} = 1 + (k_0 L \delta\varepsilon'/2\varepsilon_0)^2/2 \,,$$
$$q = k_0 L(\delta\varepsilon_{st} + \delta\varepsilon' I_{st})/(2\varepsilon_0) + 2\pi^2 D_2 v_{gr}^2 N^2/(k_0 L) + \Delta_{ph} \,. \qquad (4.2.31)$$

The lowest threshold value of the reflection coefficient

$$R_{min} = 1 - (k_0 L/2)|\delta\varepsilon'|I_{st}/\varepsilon_0 \qquad (4.2.32)$$

is achieved at $q = 0$, i.e., at

$$N^2 = -\frac{k_0 L}{2\pi^2 D_2 v_{gr}^2} \left( \frac{k_0 L}{2} \frac{\delta\varepsilon_{st} + \delta\varepsilon' I_{st}}{\varepsilon_0} + \Delta_{ph} \right) \,. \qquad (4.2.33)$$

Under the conditions (4.2.30) the relation (4.2.18) is also simplified:

$$I_{in} = I_{st}[(1 - R)^2 + \tilde{\Delta}^2] \,. \qquad (4.2.34)$$

Note that in the absence of dispersion ($D_2 = 0$), it follows from (4.2.31) and (4.2.34) that the type of instability considered is achieved only for the states which correspond to the intermediate branches of the transfer function (with negative slope, $dI_{st}/dI_{in} < 0$). However, the instability behaviour changes in the presence of dispersion. For the self-focusing Kerr nonlinearity ($\delta\varepsilon' = \varepsilon_2 > 0$) in the region of anomalous dispersion, $D_2 < 0$, the most rapidly growing perturbations have the period $T_N = \tau_{del}/N \ll \tau_{del}$ ($k_0 L \gg 1$), i.e., at each moment the field inside the cavity consists of a large number of soliton-like structures.

Let us present a numerical example, assuming $\Delta_{ph} = 0$, $\varepsilon_2 I_{st}/\varepsilon_0 = 10^{-5}$, $D_2 c^2 = -10^{-2}$, $k_0 L/2 = 10^{-4}$. Then the minimum threshold value of the reflection coefficient $R_{min} = 0.9$ is achieved at $N = 140$ (pulse period $\tau_{del}/140$). In view of the condition $k_0 L \gg 1$ the perturbation growth rate for the neighbouring values of $N$ decreases only slightly.

It is also fairly easy to derive the conditions of modulation instability in a nonlinear cavity in the case of higher-order dispersion, nonlinear dependence of the group velocity, etc. For example, to take into account the finite time of relaxation of the nonlinear medium $\tau_{\rm rel}$, (4.1.17) is used in addition to (4.2.15). Dispersion equation (4.2.25) in this case preserves its form upon substitution in (4.2.26) and (4.2.33):

$$J \to J/(1 + i\Omega\tau_{\rm rel}) . \tag{4.2.35}$$

If the frequencies of the most rapidly growing perturbations determined by (4.2.25) and (4.2.33) are small enough,

$$\Omega' \ll \tau_{\rm rel}^{-1} , \tag{4.2.36}$$

the relaxation time of nonlinearity is inessential.

Exponential temporal increase in the modulation depth is terminated at the nonlinear stage [9, 13]. Its consistent analysis can be based on the numerical solution of (4.2.15) with boundary condition (4.1.14). The results of calculations demonstrating a pulsed-periodic regime (at constant input intensity $I_{\rm in} = $ const.) are presented in Fig. 4.4b. The greatest contrast is achieved under conditions of multistability.

# 4.3 Instabilities of Transverse Field Structure

## 4.3.1 Matrix Representation

Instability of transverse field structure is similar to modulation instability. As Bespalov and Talanov showed [42], when a plane wave with initially weak perturbations of transverse structure propagates in a medium with the refraction index increasing with growth of the field intensity, the perturbations are amplified, finally causing the decay of the radiation beam into separate intensive filaments. This effect of "small-scale self-focusing" is the main obstacle for increasing the brightness of high-power solid-state laser systems, and in this regard the methods of its suppression have been thoroughly studied [211]. As applied to the nonlinear interferometers driven by external radiation, instabilities of the transverse radiation structure were found in [323] and actively studied in subsequent publications. Note that, in contrast to classical small-scale self-focusing [42], instabilities in the nonlinear interferometer are absolute (perturbations grow with time because of the presence of feedback).

Considering these instabilities, one can neglect medium dispersion and use the quasistationary description.[1] For a transparent medium with fast nonlin-

---

[1] Generally speaking, in the presence of growing perturbations the total field is not stationary, and in the initial equation (4.1.4) the temporal derivative should be preserved, as in Sect. 4.2. But this refinement does not influence the position of the boundaries of the stability regions and the values of the perturbation growth rate, if dispersion is weak and the nonlinear response is fast [a condition such as (4.2.36)].

earity of the refraction index (4.1.18), paraxial equation (4.1.4) becomes

$$2ik_0\frac{\partial E}{\partial z} + \Delta_\perp E + k_0^2\frac{\delta\varepsilon_{nl}(|E|^2)}{\varepsilon_0}E = 0 . \tag{4.3.1}$$

Propagation of the plane wave with the phase velocity depending on its intensity $I_0 = |E_0|^2$ corresponds to the unperturbed solution of (4.3.1) [compare with (4.2.16)]:

$$E_0(z) = E_0(0)\exp\left(\frac{ik_0\delta\varepsilon_{st}}{2\varepsilon_0}z\right) , \quad \delta\varepsilon_{st} = \delta\varepsilon_{nl}(I_0) . \tag{4.3.2}$$

To describe evolution of weak perturbations in the nonlinear medium, we assume

$$E(\mathbf{r}_\perp, z) = E_0(z)[1 + \delta E(\mathbf{r}_\perp, z)] , \quad |\delta E|^2 \ll 1 . \tag{4.3.3}$$

Linearization of (4.3.1) with respect to $\delta E$ gives

$$2ik_0\frac{\partial\delta E}{\partial z} + \Delta_\perp\delta E + J(\delta E + \delta E^*) = 0 , \tag{4.3.4}$$

$$J = \frac{k_0^2\delta\varepsilon' I_0}{\varepsilon_0} , \quad \delta\varepsilon' = \left(\frac{d\delta\varepsilon}{dI}\right)_{I=I_0} . \tag{4.3.5}$$

At each $z$ the perturbation $\delta E$ can be expanded into a spectrum with respect to spatial frequencies $\mathbf{q}_\perp = (q_x, q_y)$:

$$\delta E(\mathbf{r}_\perp, z) = \int \delta E_q(z)\exp(i\mathbf{q}_\perp\mathbf{r}_\perp)\,d\mathbf{q}_\perp . \tag{4.3.6}$$

The linearized equation (4.3.4) connects only the perturbation components with the opposite spatial frequencies $\mathbf{q}_\perp$ and $-\mathbf{q}_\perp$. For harmonics with fixed value of $q^2$ we have

$$2ik_0\frac{d(\delta E_q)}{dz} - q^2\delta E_q + J(\delta E_q + \delta E_q^*) = 0 . \tag{4.3.7}$$

Introducing the two-component vector (T denotes transposition)

$$\delta\mathbf{E}_q(z) = \begin{pmatrix} \mathrm{Re}\,\delta E_q(z) \\ \mathrm{Im}\,\delta E_q(z) \end{pmatrix} = \left(\mathrm{Re}\,\delta E_q(z),\ \mathrm{Im}\,\delta E_q(z)\right)^{\mathrm{T}} , \tag{4.3.8}$$

the solution of (4.3.7) can be presented in matrix form:

$$\delta\mathbf{E}_q(z) = \mathbf{U}_q\delta\mathbf{E}_q(0) . \tag{4.3.9}$$

The explicit form of the matrix $\mathbf{U}_q$ is (see [327])

$$U_q = \begin{pmatrix} \cosh(BX) & \frac{q^2}{JX}\sinh(BX) \\ \frac{2J-q^2}{JX}\sinh(BX) & \cosh(BX) \end{pmatrix} \quad (q^2 < 2J) ,$$

$$U_q = \begin{pmatrix} \cos(BX) & \frac{q^2}{JX}\sin(BX) \\ \frac{JX}{q^2}\sin(BX) & \cos(BX) \end{pmatrix} \quad (q^2 > 2J) , \quad (4.3.10)$$

$$X = \frac{q}{J}\sqrt{|2J - q^2|} , \quad B = \frac{Jd}{2k_0} = \varphi_2 I_0 . \quad (4.3.11)$$

The coefficient $\varphi_2$ was determined in Sect. 4.1. The value $B$ coincides with the nonlinear phase shift of the unperturbed plane wave over the extent of the layer of the nonlinear medium; in the theory of self-focusing it is called the "break-up integral" (see Appendix A). In the limit of low spatial frequencies

$$q^2 \ll |J| , \quad q\sqrt{|J|}d/k_0 \ll 1 , \quad (4.3.12)$$

from (4.3.10) we obtain the following:

$$U_0 = \begin{pmatrix} 1 & 0 \\ 2B & 1 \end{pmatrix} . \quad (4.3.13)$$

Transformation of the perturbation amplitude in the linear (air) interval, which can include lens elements and is specified by the effective length $l_0$ and magnification (beam widening) coefficient $M$, is described by

$$V_q = \begin{pmatrix} \cos\alpha & \sin\alpha \\ -\sin\alpha & \cos\alpha \end{pmatrix} , \quad (4.3.14)$$

where

$$\alpha = \frac{q^2 l_0}{2k^{(0)}} - \Delta_0 - B , \quad k^{(0)} = \frac{\omega_0}{c} ,$$
$$\exp(i\Delta_0) = \exp(ik^{(0)}l_0) , \quad |\Delta_0| \leq \pi . \quad (4.3.15)$$

In the case of a beam with widening ($M \neq 1$) in addition to (4.3.14) it is necessary to multiply the spatial frequency and the perturbation amplitude by the inverse magnification $\eta = |M|^{-1}$. Note that

$$\det U_q = \det V_q = 1 . \quad (4.3.16)$$

The matrix describing the consequent passage of a perturbation through nonlinear and linear elements is the product of the matrices corresponding to these elements. Taking into account that in each round-trip in the interferometer the perturbation amplitude is also multiplied by the value $R$, where $R$ is the product of the amplitude coefficients of refraction of the interferometer mirrors, we can write the perturbation transformed during $n$ such passages as

$$\begin{pmatrix} \mathrm{Re}\, \delta E_{q_{n-1}}^{(n)} \\ \mathrm{Im}\, \delta E_{q_{n-1}}^{(n)} \end{pmatrix} = (R\eta)^n \boldsymbol{W}^{(n)} \begin{pmatrix} \mathrm{Re}\, \delta E_q \\ \mathrm{Im}\, \delta E_q \end{pmatrix} , \qquad (4.3.17)$$

where

$$\boldsymbol{W}^{(n)} = \boldsymbol{V}_{q_{n-1}} \boldsymbol{U}_{q_{n-1}} \boldsymbol{V}_{q_{n-2}} \boldsymbol{U}_{q_{n-2}} \cdots \boldsymbol{V}_q \boldsymbol{U}_q , \qquad q_m = \eta^m q . \qquad (4.3.18)$$

Besides, in the values $\alpha$ (4.3.15) determining the matrix $\boldsymbol{V}_q$, it is necessary to make the substitution $\Delta_0 \to \Delta_{\mathrm{ph}}$, where $\Delta_{\mathrm{ph}}$ is the total phase detuning of the unperturbed plane wave after one round-trip in the interferometer. In view of (4.3.16)

$$\det \boldsymbol{W}^{(n)} = 1 . \qquad (4.3.19)$$

Under the condition $BX \ll 1$, i.e.,

$$\frac{qd}{2k_0} \sqrt{|2J - q^2|} \ll 1 , \qquad (4.3.20)$$

matrices $\boldsymbol{U}_q$ can be presented as the products of the matrices of the "air interval" $\boldsymbol{V}_q$ and the thin phase screen $\boldsymbol{U}_0$:

$$\boldsymbol{U}_q = \begin{pmatrix} 1 & \frac{q^2 d}{2k_0} \\ -\frac{q^2 d}{2k_0} & 1 \end{pmatrix} \times \begin{pmatrix} 1 & 0 \\ 2B & 1 \end{pmatrix} . \qquad (4.3.21)$$

These two matrices commute with an accuracy up to the square terms in the small non-diagonal elements. Multiplying the first of these matrices by the matrix $\boldsymbol{V}_q$ (4.3.14) leads also to the matrix of the air interval $\boldsymbol{V}_q$, but with modified length $l_0$. Therefore it is possible to use an approximation of the thin screen and relation (4.3.13) without limitation of generality, if (4.3.20) is true.

### 4.3.2 Cavities Without and with Magnification

**Cavity Without Magnification.** In this case in which $\eta = M = 1$, the spatial frequency of perturbations is conserved for any number of the round-trips in the cavity: $q_n = q$. Therefore the transformation matrix $\boldsymbol{W}^{(n)}$ is the $n$th power of the transformation matrix for one round-trip:

$$\boldsymbol{W}_q^{(n)} = (\boldsymbol{W}_q^{(1)})^n = (\boldsymbol{V}_q \times \boldsymbol{U}_q)^n . \qquad (4.3.22)$$

The analysis of the eigenvalues of the matrix $\boldsymbol{W}_q^{(1)}$ was performed initially in the same way as that applied to periodic structures of the nonlinear layers and air intervals [400]. Eigenvalues $\Lambda_{1,2}$ of the matrix $\boldsymbol{W}_q^{(1)}$ are found from the quadratic equation

$$\Lambda^2 - 2C\Lambda + 1 = 0 , \qquad (4.3.23)$$

where

$$C = \frac{1}{2}(W_{11}^{(1)} + W_{22}^{(1)}) =$$

$$\begin{cases} \cosh(BX)\cos\alpha + \frac{J-q^2}{JX}\sinh(BX)\sin\alpha & (q^2 < 2J) , \\ \cos(BX)\cos\alpha + \frac{1}{2}[\frac{JX}{q^2} - \frac{q^2}{JX}]\sin(BX)\sin\alpha & (q^2 > 2J) . \end{cases} \quad (4.3.24)$$

In the approximation of a thin nonlinear screen [(4.3.12) and (4.3.20)] the form of $C$ is simplified considerably:

$$C = \cos\alpha + B\sin\alpha = \sqrt{1+B^2}\cos(\alpha - \arctan B) . \quad (4.3.25)$$

The eigenvalues are

$$\Lambda_{1,2} = C \pm \sqrt{C^2 - 1} . \quad (4.3.26)$$

At $C^2 < 1$ their modulus $|\Lambda_{1,2}| = 1$; as $R < 1$, the perturbations will vanish with an increase in the number of the round-trips $n$. At $C^2 > 1$ both eigenvalues $\Lambda_1$ and $\Lambda_2$ are real and their product $\Lambda_1\Lambda_2 = 1$. We will denote by $\Lambda_m$ the bigger (by modulus) value ($|\Lambda_m| \geq 1$). Then the perturbations will grow with an increase in $n$ unrestrictedly (in the approximations accepted), if $R|\Lambda_m| > 1$. The spatial frequency range corresponding to instability

$$q_{min} < q < q_{max} \quad (4.3.27)$$

is determined by the relation

$$|\Lambda_m(q_{min})| = |\Lambda_m(q_{max})| = 1/R . \quad (4.3.28)$$

In the general case there can be several such instability zones at fixed $B$.

It should be noted that among the instabilities considered in the present section, there are also ones not connected with perturbations of the field's transverse structure. Actually, at $q = 0$ we come to the instabilities studied in Sect. 4.2 for the intermediate branch of the transfer function and to the Ikeda instability. The third type of instability (self-pulsation) shown in Fig. 4.3 is not realized in a transparent nonlinear medium because the differential gain of the nonlinear element is equal to unity there. If the rate of growth of the transverse structure perturbations ($q \neq 0$) is higher than, e.g., the growth rate of Ikeda instabilities in approximation of the plane waves ($q = 0$), then it will be difficult to observe the latter instabilities because of the decay of the radiation beam inside (and at the exit of) the interferometer into separate intensive filaments.

The possibility of the development of transverse perturbations in the media not only with self-focusing ($\delta\varepsilon' > 0$), but with self-defocusing ($\delta\varepsilon' < 0$) as well, is a peculiarity of the discussed instabilities in the nonlinear interferometers, distinguishing them from small-scale self-focusing in homogeneous or layered nonlinear media without feedback. This we will illustrate by applying [323] for the case of low-threshold instability.

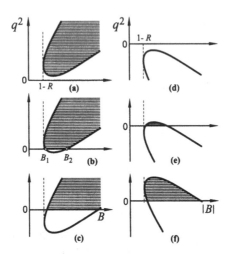

**Fig. 4.5a–f.** Domains of instability of stationary regimes in the nonlinear cavity (*hatched*) [323]

Under conditions (4.1.21) and (4.3.12), the value $C$ is close to unity:

$$C = 1 + \delta C , \quad 2\delta C \approx B^2 - \left( \frac{q^2 l_0}{2k^{(0)}} - \Delta_{\text{ph}} - 2B \right)^2 . \tag{4.3.29}$$

Then, after making $\Lambda_{\text{m}} \approx 1 + \sqrt{2\delta C}$ equivalent to $1/R \approx 1 + (1 - R)$, we obtain the relation determining the boundaries of the instability region:

$$\frac{q^2 l_0}{2k^{(0)}} = \Delta_{\text{ph}} + 2B \pm \sqrt{B^2 - (1 - R)^2} . \tag{4.3.30}$$

In the plane of the parameters $(q^2, B)$ these boundaries are hyperbolas with the vertical tangent at $B^2 = (1 - R)^2$ and the horizontal tangent at $B^2 = (4/3)(1-R)^2$. At $B^2 < (1-R)^2$ instabilities do not exist. Analyzing (4.3.30), we can distinguish three variants for self-focusing nonlinearity ($B > 0$):

(a) $\Delta_{\text{ph}} > -\sqrt{3}(1 - R)$ (Fig. 4.5a). In this case bistability is absent, and transversely homogeneous states of the single branch of the transfer function are stable at $B < 1 - R$ and unstable at $B > 1 - R$ in the limited interval of spatial frequencies (4.3.27) with $q_{\text{min}} > 0$.

(b) $\sqrt{3}(1 - R) < -\Delta_{\text{ph}} < 2(1 - R)$ (Fig. 4.5b). Bistability is present, and the interval $B_1 < B < B_2$ corresponds to the intermediate branch of the transfer function (the states are unstable even at $q = 0$). A part of the lower branch $(1 - R < B < B_1)$ and the upper branch are unstable with respect to the perturbations with spatial frequencies in range (4.3.27) where $q_{\text{min}} > 0$.

(c) $-\Delta_{\text{ph}} > 2(1 - R)$ (Fig. 4.5c). The interval $B_1 < B < B_2$ again corresponds to the intermediate branch and unstable states. The lower branch represents stable states, and the upper one is related to states unstable with respect to perturbations with spatial frequencies in the interval (4.3.27); for them $q_{\text{min}} > 0$ as well.

For self-defocusing nonlinearity $(B < 0)$ we can also distinguish three variants:

(d) $\Delta_{ph} < \sqrt{3}(1 - R)$ (Fig. 4.5d). There is neither bistability nor instability.

(e) $\sqrt{3}(1 - R) < \Delta_{ph} < 2(1 - R)$ (Fig. 4.5e). Bistability exists, and the lower and upper branches of the transfer function correspond to stable states.

(f) $\Delta_{ph} > 2(1 - R)$ (Fig. 4.5f). Bistability takes place. Instabilities are absent for the upper branch and are present for a part of the lower branch $(1 - R < |B_1| < |B_2|)$ in the range of the spatial frequencies (4.3.27) with $q_{min} > 0$.

It is in the latter case that instabilities of the transverse field structure develop in the interferometer with self-defocusing nonlinearity. Simultaneous stability of the states of the lower and upper branches of the transfer function is possible only in variant (e).

Low-threshold Ikeda instabilities are present under the same conditions except in the range of phase detunings $\Delta_{ph}$. Let us assume $\Delta_{ph} = \pi + \Delta'$ and $\Delta'^2 \ll 1$. Then for the unperturbed stationary solutions we obtain, instead of (4.1.44), the regimes close to the linear one:

$$I_{in} = I_{st}[(1 + R)^2 + (\Delta' + \varphi_2 I_{st})^2] \approx 4I_{st} . \tag{4.3.31}$$

Bistability does not exist. At the same time, (4.3.30) and all conclusions for instabilities in the variants (a) – (f) are valid with the substitution $\Delta_{ph} \to \Delta'$.

The approach considered corresponds to the idealized scheme. In particular, the analytical investigation of instabilities in the case of limited radiation beams requires applying a considerably more complex technique [325], so numerical calculations [323] become less labour-intensive. Suppression of instabilities of the transverse structure, for which the lower boundary of spatial frequencies $q_{min} > 0$, can be reached by introducing a spatial filter inside the interferometer (see Sect. 4.1) with the cut-off frequency of the filter $q_f < q_{min}$ [323]. Though a method like this is convenient for carrying out numerical calculations, in practice the use of supplementary elements is possible only with a large base (longitudinal extension) of the interferometer, and it demands higher requirements for the accuracy of the scheme alignment. Instead, the following scheme can be used for suppression of instabilities of the transverse field structure [289].

**Cavity with Magnification.** After a complete round-trip of the radiation beam in the cavity (interferometer), the beam's transverse dimensions increase by $|M|$ times $(|M| > 1)$ and the spatial frequency of perturbation decreases by $|M|$ times. Therefore, in the nonlinear interferometer a perturbation with any initial spatial frequency will experience only a limited number of round-trips, $n < n_{max}$, in the range of "dangerous" spatial frequencies (4.3.27), and

$$n_{max} = \ln(q_{max}/q_{min})/\ln|M| . \tag{4.3.32}$$

Thereby perturbation growth is restricted. At the same time, the influence of cavity magnification on the main (unperturbed) beam is only a decrease in the radiation intensity by $|M|^2$ times, which is equivalent to a decrease in the reflection coefficient of one of the cavity mirrors. Suppression of instabilities of the transverse structure occurs at any coefficients of magnification $|M| > 1$. However, if the coefficient $|M|$ is close to unity, then initial perturbations can increase during a considerable number of round-trips in the cavity, reaching high values.

Purely temporal (Sect. 4.2) and purely spatial (this section) instabilities are particular cases of spatio-temporal instabilities, as instability threshold could be lower for perturbations modulated both in time and space. Perturbation growth can be restricted by nonlinearity. Then different structures can form in the vicinity of the instability threshold: spatially homogeneous, periodic in time (purely temporal instabilities); stationary, modulated in space (purely spatial instabilities); and, in the general case, spatio-temporal. For example, if there is not only transverse diffraction interaction of radiation but also diffusion mixing in the medium, pulsing periodic spatial structures arise in the interferometer [34]. General techniques of analysis of the nonlinear stage of instabilities are presented in the review [68]. Stable field patterns of different types (including "rolls" and hexagonal structures) are demonstrated in [100].

## 4.4 Model of Threshold Nonlinearity

### 4.4.1 General Solution

In this section we consider the basic structures in driven nonlinear interferometers analytically in the framework of the model of transverse distributivity (4.1.27) for the case of threshold-type fast nonlinearity [290]:

$$\delta\varepsilon = \delta\varepsilon(I) = \begin{cases} \delta\varepsilon_0, & I < I_{\text{sat}} , \\ 0, & I > I_{\text{sat}} \end{cases} \qquad (4.4.1)$$

(sharp saturation of nonlinearity at intensity $I = |E|^2 = I_{\text{sat}}$). A similar form of nonlinearity was used in the analysis of increasing absorption bistability [142, 184], see also [161]. Although the form of (4.4.1) can be justified, e.g., for the case of absorption saturation in semiconductors, more important is the fact that the general features of the structures we are interested in are not sensitive to the specific type of nonlinearity, as it follows from the calculation results presented in the following sections.

For threshold nonlinearity (4.4.1), the transfer function for transversely homogeneous states has a zigzag form (Fig. 4.6) with the bistability range

$$E_{\text{min}} < E_{\text{in}} < E_{\text{max}} . \qquad (4.4.2)$$

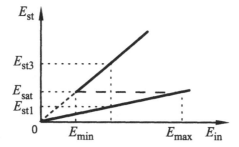

**Fig. 4.6.** Transfer function of an interferometer with threshold nonlinearity

The lower ($E = E_{st1}$) and upper ($E = E_{st3}$) branches of the transfer function correspond to the regimes stable with respect to small perturbations. In the absence of interferometer detuning and for purely absorptive nonlinearity, i.e.,

$$\Delta_{ph} = 0 , \quad \delta\varepsilon_0 = i\delta\varepsilon_0'' , \quad \mathrm{Re}\,\delta\varepsilon_0 = 0 , \quad \mathrm{Im}\,\delta\varepsilon_0 = \delta\varepsilon_0'' > 0 , \qquad (4.4.3)$$

we find

$$E_{min} = (1 - R)E_{sat} = \eta E_0 , \quad E_{max} = (1 - R + \mu)E_{sat} = \frac{1}{\eta}E_0 ,$$

$$E_{st1} = \frac{E_{in}}{1 - R + \mu} = \eta\frac{E_{sat}E_{in}}{E_0} ,$$

$$E_{st3} = \frac{E_{in}}{1 - R} = \frac{1}{\eta}\frac{E_{sat}E_{in}}{E_0} , \qquad (4.4.4)$$

$$\mu = k_0 d\frac{\delta\varepsilon_0''}{2\varepsilon_0} , \quad \eta = \sqrt{(1 - R)/(1 - R + \mu)} ,$$

$$E_0 = \sqrt{E_{min}E_{max}} , \quad E_{sat} = \sqrt{I_{sat}} . \qquad (4.4.5)$$

Transversely inhomogeneous field distributions are found as follows: For nonlinearity (4.4.1), the entire interval $-\infty < \xi < +\infty$ is divided into a number of regions where intensity $I < I_{sat}$ or $I > I_{sat}$. Inside every such region (4.1.27) has the form of a linear ordinary differential equation with constant coefficients; therefore it is easy to find its general solution. At the boundaries of regions where $I = I_{sat}$, conditions for continuity of values $E$ and $dE/d\xi$ are set down, and at $\xi \to \pm\infty$ the field should pass into the corresponding transversely homogeneous state with the amplitude $E = E_{st}$.

### 4.4.2 Switching Waves

For definiteness, we consider the switching wave with asymptotic $E \to E_{st1}$ at $\xi \to -\infty$ and $E \to E_{st3}$ at $\xi \to +\infty$ (Fig. 4.7).[2] Because, moving along

---

[2] This is the *left* switching wave; the *right* one has the opposite asymptotic (Fig. 4.7**a,b**).

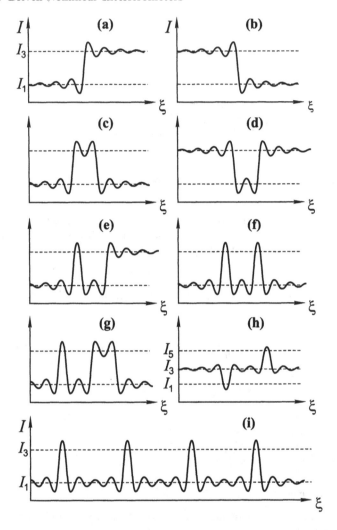

**Fig. 4.7.** Transverse profiles of intensity: "left" (**a**) and "right" (**b**) switching waves; positive (**c**) and negative (**d**) single dissipative optical solitons; switching wave coupled with soliton (**e**); symmetric (**f**) and asymmetric (**g**) two-soliton structures; "dipole" – a bound state of positive and negative solitons in the case of tristability (**h**); periodic chain of solitons (**i**)

the wave front, we come from one stationary point in the system phase space ($E = E_{st1}$) to another ($E = E_{st3}$), a switching wave represents a *heteroclinic trajectory*, as in Chap. 2. One can choose the coordinate origin at the switching wave front, i.e., $|E_{sw}(x = 0)|^2 = I_{sat}$. In this case the field has the form

$$
E_{sw} = \begin{cases} E_{st1} + A \exp(p\xi) , & I < I_{sat} , \\ E_{st3} + B \exp(-q\xi) , & I > I_{sat} , \end{cases}
\qquad (4.4.6)
$$

$$p = ik_0 \frac{v}{v_{gr}}$$

$$+ \left\{ -k_0^2 \left( \frac{v}{v_{gr}} \right)^2 - \left[ \left( \frac{\omega_0}{c} \right)^2 \delta\varepsilon_0 + i\frac{2k_0}{d}[1 - R\exp(i\Delta_{ph})] \right] \right\}^{1/2} . \quad (4.4.7)$$

The value $q$ is obtained from $p$ (4.4.7) by the substitution $v \to -v$, $\delta\varepsilon_0 \to 0$. The root branch is chosen with a non-negative real part. This ensures the right asymptotic at $\xi \to \pm\infty$; note that the complexity of $p$ and $q$ (4.4.7) manifests itself in the oscillatory type of the field approaching the stationary regimes with amplitudes $E_{st1,3}$. From the continuity conditions we find

$$A = \frac{E_{st3} - E_{st1}}{1 + p/q} , \quad B = \frac{E_{st3} - E_{st1}}{1 + q/p} . \quad (4.4.8)$$

The velocity of the switching wave is determined from the condition at the regions' boundary $E_{sw}(0) = I_{sat}$, i.e.,

$$\left| E_{st1} + \frac{E_{st3} - E_{st1}}{1 + p/q} \right|^2 = I_{sat} . \quad (4.4.9)$$

We assume below that (4.4.2) and (4.4.3) are satisfied and take into account (4.4.4) and (4.4.5). Then instead of (4.4.9) we obtain

$$I_{in} = \frac{I_0}{\eta^2 \left| 1 + \frac{q}{p+q} \left( \frac{1}{\eta^2} - 1 \right) \right|^2} , \quad (4.4.10)$$

$$I_0 = |E_0|^2 = E_{min} E_{max} . \quad (4.4.11)$$

It is convenient to assign some value to the switching wave velocity $v$ and then find the values $p$ and $q$ from (4.4.7); thereafter (4.4.10) gives a unique value of intensity $I_{in}$. Varying $v$, we obtain a single-valued dependence $I_{in}(v)$, from which one can find the inverse dependence $v(I_{in})$. The latter is of an alternating-sign nature. It follows from (4.4.10) that, at normal incidence, the switching wave is motionless only at $I_{in} = I_0$, so the value $I_0$ determined by (4.4.11) is Maxwell's value of the incident radiation intensity. At small intensity deviations $\delta I_{in} = I_{in} - I_0$,

$$v = -\frac{v_{gr}}{1 - \eta} \sqrt{\frac{1 - R}{k_0 d}} \frac{\delta I_{in}}{I_0} . \quad (4.4.12)$$

### 4.4.3 Stability of Switching Waves

For simplicity we consider stability of the motionless switching wave (4.4.6) ($v = 0$, $\xi = x$, $I_{in} = I_0$) with respect to the field small perturbations $\delta E$:

$$E(x,t) = E_{sw}(x) + \delta E(x,t) ,$$
$$\delta E(x,t) = f(x)\exp(\gamma t) + g^*(x)\exp(\gamma^* t) . \quad (4.4.13)$$

Linearizing the initial kinetic equation (4.1.23) with respect to $\delta E$, we obtain a set of equations for the determination of the perturbation eigenfunctions $f(x)$, $g(x)$ and the eigenvalues $\Gamma = \gamma d/v_{\mathrm{gr}}$ $[\mu(x) = \mu$ at $x < 0$ and $0$ at $x > 0$, $\sigma = \sigma(x) = (\mathrm{d}\delta\varepsilon_0''/\mathrm{d}I)_{I=I_{\mathrm{sw}}}]$:

$$-\mathrm{i}\frac{d}{2k_0}\frac{\mathrm{d}^2 f}{\mathrm{d}x^2} + [\Gamma + 1 - R + \mu(x)]f + \frac{k_0 d}{2\varepsilon_0}\sigma I_{\mathrm{sat}}(f+g) = 0 ,$$

$$\mathrm{i}\frac{d}{2k_0}\frac{\mathrm{d}^2 g}{\mathrm{d}x^2} + [\Gamma + 1 - R + \mu(x)]g + \frac{k_0 d}{2\varepsilon_0}\sigma I_{\mathrm{sat}}(f+g) = 0 . \qquad (4.4.14)$$

A switching wave will be stable if all the eigenvalues have a nonpositive real part: $\mathrm{Re}\,\Gamma \leq 0$. If $\Gamma$ is real, it is sufficient to hold only the first equation in (4.4.14), assuming $g = f^*$. The existence of the zero eigenvalue $\Gamma = 0$ ("a neutral mode") follows from the translational symmetry (with respect to the transverse coordinate shift) of the kinetic equation (4.1.23) for any type of nonlinearity.

In the intervals $x < 0$ ($I < I_{\mathrm{sat}}$) and $x > 0$ ($I > I_{\mathrm{sat}}$), set (4.4.14) decomposes into separate equations for $f$ and $g$. Thus we find

$$f_{x<0} = f_0 \exp(\tilde{p}_1 x) , \quad f_{x>0} = f_0 \exp(-\tilde{q}_1 x) ,$$

$$g_{x<0} = g_0 \exp(\tilde{p}_2 x) , \quad g_{x>0} = g_0 \exp(-\tilde{q}_2 x) , \qquad (4.4.15)$$

$$\pm\mathrm{i}\frac{d}{2k_0}\tilde{p}_{1,2}^2 = \Gamma + 1 - R + \mu , \quad \mathrm{Re}\,\tilde{p}_{1,2} > 0 ,$$

$$\pm\mathrm{i}\frac{d}{2k_0}\tilde{q}_{1,2}^2 = \Gamma + 1 - R , \quad \mathrm{Re}\,\tilde{q}_{1,2} > 0 . \qquad (4.4.16)$$

Functions $f(x)$ and $g(x)$ are continuous, and their derivatives $\mathrm{d}f/\mathrm{d}x$ and $\mathrm{d}g/\mathrm{d}x$ have a jump at $x = 0$. Integrating (4.4.14) over a small interval including the value $x = 0$ (see also [184]), we come to the following set of equations that are linear in $f_0$ and $g_0$:

$$\mathrm{i}\frac{d}{k_0}(\tilde{p}_1 + \tilde{q}_1)f_0 - \frac{\mu(f_0 + g_0)}{(1-\eta)\sqrt{(1-R+\mu)k_0/d}} = 0 ,$$

$$\mathrm{i}\frac{d}{k_0}(\tilde{p}_2 + \tilde{q}_2)g_0 + \frac{\mu(f_0 + g_0)}{(1-\eta)\sqrt{(1-R+\mu)k_0/d}} = 0 . \qquad (4.4.17)$$

Taking the set's determinant to be equal to zero and assuming $\mathrm{Re}\,\Gamma > -(1-R)$ (otherwise the perturbation decays with time), we obtain the following equation for $\Gamma$:

$$\sqrt{\Gamma + 1 - R + \mu} + \sqrt{\Gamma + 1 - R} = \nu , \qquad (4.4.18)$$

$$\nu = \sqrt{1 - R + \mu} + \sqrt{1 - R} . \qquad (4.4.19)$$

This equation can be shown to have the same form for real $\Gamma$. Equation (4.4.18) has only a trivial solution, $\Gamma = 0$, which proves the switching wave stability with respect to weak spatial modulation.

This result can be generalized, including perturbations with additional spatial modulation in the $y$-direction:

$$\delta E(x, y, t) = f(x) \exp(\gamma t + i\kappa_y y) + g^*(x) \exp(\gamma^* t - i\kappa_y y) . \qquad (4.4.20)$$

The dispersion relation for the determination of $\Gamma$ will take the form

$$[(1 + i)X(Q) - \nu][(1 - i)X(-Q) - \nu] = \nu^2 , \qquad (4.4.21)$$
$$X(Q) = \sqrt{\Gamma + 1 - R + \mu + iQ} + \sqrt{\Gamma + 1 - R + iQ} ,$$
$$Q = \frac{\kappa_y^2 d}{2k_0} . \qquad (4.4.22)$$

If $\kappa_y = 0$, then $Q = 0$, and (4.4.21) passes into (4.4.18). At zero spatial frequency $\kappa_y = 0$ the maximum eigenvalue is $\Gamma = 0$. One can show that at small spatial frequencies ($Q \ll 1 - R$) this value is shifted: $\Gamma \approx -Q < 0$. Therefore, perturbations modulated in the $y$-direction decay faster. Thus, waves of switching between stable transversely homogeneous regimes are stable.

### 4.4.4 Switching Waves: Influence of Inhomogeneities

In the case of normal incidence of external plane-wave radiation considered above, and for an arbitrary type of nonlinearity, there are right and left switching waves with velocities $v^{(0)}$ and $-v^{(0)}$, correspondingly. At oblique incidence, the degeneracy of the velocity modules is removed; for the case of fast nonlinearity, in accordance with the Galilean transform (4.1.26), the velocities of the right $v_{\rm r}$ and left $v_{\rm l}$ switching waves are

$$v_{\rm r} = v^{(0)} + v_{\rm gr}\theta , \quad v_{\rm l} = -v^{(0)} + v_{\rm gr}\theta . \qquad (4.4.23)$$

The Galilean transform is not valid if one takes into account some additional factors, e.g., finite relaxation time of the medium nonlinearity or spatial filtering of radiation inside the interferometer. Then we have instead of (4.4.23)

$$v_{\rm r} = v^{(0)} + d_{\rm g}v_{\rm gr}\theta , \quad v_{\rm l} = -v^{(0)} + d_{\rm g}v_{\rm gr}\theta . \qquad (4.4.24)$$

The value of the drag coefficient $d_{\rm g}$ can be found by the perturbation approach (see below). In the case of threshold nonlinearity and spatial filtering, the drag coefficient can be found analytically [305].

If the intensity $I_{\rm in}$ and phase $\Phi_{\rm in}$ of external radiation or the interferometer's parameters $p$ (phase detuning $\Delta_{\rm ph}$, reflection coefficient $R$, etc.) vary with the transverse coordinate $x$ slowly (as compared with the width of the switching wave front), one can speak of the switching wave as having slowly varying characteristics (see Chap. 2). Its velocity will be determined by local (at the wave's front) values of the holding radiation intensity and interferometer parameters and by the local gradient of the radiation phase. Let us introduce the coordinates of the right $x_{\rm r}(t)$ and left $x_{\rm l}(t)$ switching wave fronts by using the relation $I(x_r) = I(x_1) = I_{sat}$. Then a natural generalization of (4.4.24) is

$$\dot{x}_r = v\left(I_{in}(x_r), p(x_r)\right) + d_g \frac{v_{gr}}{k_0} \frac{d\Phi_{in}}{dx} ,$$
$$\dot{x}_1 = -v\left(I_{in}(x_1), p(x_1)\right) + d_g \frac{v_{gr}}{k_0} \frac{d\Phi_{in}}{dx} . \qquad (4.4.25)$$

With interferometer excitation by a radiation beam, e.g., a Gaussian one, the switching wave front can stop at two positions, where the local intensity is equal to Maxwell's value $I_0$. Linearizing (4.4.25) in small deviations from these positions, we find that only one of them is stable (where $dI_{in}/dx < 0$ for the right switching wave and $dI_{in}/dx > 0$ for the left one).

### 4.4.5 Single Localized Structures

Now we consider the field localized structures of the type presented in Fig. 4.7c,d, i.e., those with the same asymptotics at $\xi \to -\infty$ and at $\xi \to +\infty$. Correspondingly, these structures represent homoclinic trajectories in the system phase space. It should be recalled that in one-component systems such regimes – the critical nuclei – exist, but are unstable (Chap. 2). As we will show, they can be stable in driven nonlinear interferometers and in many other nonlinear optical systems. We term these localized structures the *dissipative optical solitons* (DOSs).

The construction of DOSs resembles the solution of the quantum-mechanical problem of the spectrum of a particle in a rectangular potential well [186]; however, in our case the "potential" is complex and the width of the well, $2a$, is an unknown eigenvalue of the problem. A single DOS is motionless ($v = 0$, $\xi = x$). We choose the coordinate origin to be at the DOS centre ($dE/dx = 0$ at $x = 0$). For a positive, or bright DOS ($E \to E_{st1}$ at $\xi \to \pm\infty$, see Fig. 4.7c), the field has the form

$$E_{dos} = \begin{cases} E_{st1} + A \exp(-px) , & |x| > a , I < I_{sat} , \\ E_{st3} + B \cosh(qx) , & |x| < a , I > I_{sat} , \end{cases} \qquad (4.4.26)$$

$$p = (1 - i)\kappa/\eta , \quad q = (1 - i)\kappa , \quad \kappa = \sqrt{(1 - R)k_0/d} . \qquad (4.4.27)$$

Joining the solutions at $|x| = a$ leads to the following set of equations:

$$E_{st1} + A \exp(-pa) = E_{st3} + B \cosh(qa) ,$$
$$-pA \exp(-pa) = qB \sinh(qa) ,$$
$$|E_{st1} + A \exp(-pa)|^2 = I_{sat} . \qquad (4.4.28)$$

Under conditions of monostability, when $E_{st1} = E_{st3}$, there are no solutions of (4.4.28). Inside the bistability range (4.4.2), we find from (4.4.28) the following equation for the determination of a DOS half-width, $a$:

$$\left| E_{st3} - \frac{E_{st3} - E_{st1}}{1 + \eta Z} \right|^2 = I_{sat} , \tag{4.4.29}$$

$$Z = \tanh[(1 - i)\kappa a] . \tag{4.4.30}$$

Using (4.4.4-4.4.5), one can present (4.4.29) in the form

$$I_{in} = I_0 \eta^2 \left| 1 - \frac{1 - \eta^2}{1 + \eta Z} \right|^{-2} . \tag{4.4.31}$$

Specifying and varying DOS width $a$, we determine in terms of (4.4.31) a single-valued and nonmonotone dependence $I_{in}(a)$. The inverse dependence $a(I_{in})$, which is of paramount interest, is multivalued (infinite number of branches, Fig. 4.8b). The number of branch $n$ serves as a DOS "quantum number" and coincides with a number of intensity oscillations in the DOS central region (see Fig. 4.9). Taking into account the character of stability of DOSs (see below), we find that the most narrow DOS (in the "ground" state) exists in the widest range of external radiation intensity: $I_1 < I_{in} < I_2$. In the case considered, this range is narrower than the bistability range (4.4.2), and includes Maxwell's value of intensity $I_0$ (4.4.11); this is also the case for wider DOSs (in "excited" states). For greater-width DOSs ($n \gg 1$, $\kappa a \gg 1$), which exist in the vicinity of Maxwell's value of intensity, we have

$$\exp(-2\kappa a) \cos(2\kappa a) \approx \frac{1}{4} \frac{1 + \eta}{1 - \eta} \left( 1 - \frac{I_0}{I_{in}} \right) , \tag{4.4.32}$$

$$|\delta I_{in}| \ll I_0 , \quad \delta I_{in} = I_{in} - I_0. \tag{4.4.33}$$

The right-hand side of (4.4.32) is determined by the intensity of the incident radiation and vanishes at its Maxwell's value $I_{in} = I_0$. At this value of intensity, there is an infinite number of possible widths $a$ of DOSs with the difference between the adjacent values $\delta a \approx \pi/\kappa$ ($n \gg 1$).

At the holding intensity deviation from its Maxwell's value, only a finite number of DOSs persists, the latter having different widths and different numbers of field oscillations in the DOS central part (Fig. 4.9). When $I_{in}$ approaches the Maxwell's value $I_0$, the number of solutions increases unrestrictedly.

Negative, or dark DOSs ($E \to E_{st1}$ at $\xi \to \pm\infty$, see Fig. 4.7d) are found similarly. The same equations (4.4.29) and (4.4.31) are applied for them with the substitution

$$Z = \coth(pa) = \coth[(1 - i)/\eta\kappa a)] . \tag{4.4.34}$$

The character of the solutions is evident from Fig. 4.8e.

### 4.4.6 Stability of Dissipative Optical Solitons

The procedure of the analysis is similar to the one for switching waves (see above). Some distinction is connected with uncoupling of the dispersion relation for $\Gamma$ for perturbations even and odd in $x$. For even perturbations of

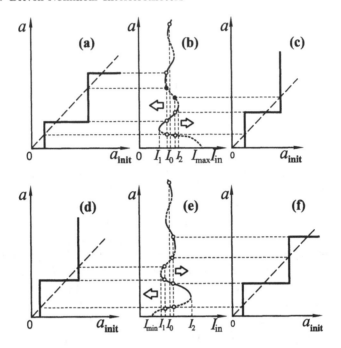

**Fig. 4.8.** Dependence of half-width of positive (**a–c**) and negative (**d–f**) dissipative optical solitons on holding intensity $I_{in}$ [(**b,e**), unstable states are shown by *dotted lines*] and on the initial distance between fronts of two switching waves $2a_{init}$ at $I_{min} < I_{in} < I_0$ (**a,d**) and $I_0 < I_{in} < I_{max}$ (**c,f**)

a positive DOS, the dispersion relation is

$$
\left[\sqrt{\Gamma+1-R+\mu} + \sqrt{\Gamma+1-R}\tanh(\tilde{q}_1 a)\right]
$$
$$
\times\left[\sqrt{\Gamma+1-R+\mu} + \sqrt{\Gamma+1-R}\tanh(\tilde{q}_2 a)\right]
$$
$$
+(1+i)\zeta\{(1-i)\sqrt{\Gamma+1-R+\mu}
$$
$$
+\sqrt{\Gamma+1-R}[\tanh(\tilde{q}_1 a) - i\tanh(\tilde{q}_2 a)]\} = 0 , \qquad (4.4.35)
$$

$$
\zeta = \frac{\mu\sqrt{k_0/dI_{sat}}}{(dI_{dos}/dx)_{x=a}} . \qquad (4.4.36)
$$

For odd perturbations one should substitute $\tanh \rightarrow \coth$ in (4.4.35).

An analytical solution of (4.4.35) can be found for the case of wide DOSs ($\kappa a \gg 1$). At $I_{in} = I_0$ the most "dangerous" eigenvalues $\Gamma$ appear real. For the odd perturbations the maximum root $\Gamma = 0$; so they do not increase. Stable solutions alternate with unstable ones (with respect to even perturbations), for which

$$
da/dI_{in} < 0 . \qquad (4.4.37)
$$

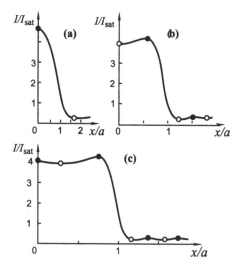

**Fig. 4.9.** Intensity profiles for dissipative optical solitons in the ground (**a**) and first two excited states (**b,c**): $\eta = 0.5$, $\kappa a = 2$ (**a**), 5 (**b**) and 8 (**c**); *solid* and *open circles* indicate local maxima and minima [290]

The numerical solution of (4.4.35) confirms the stability of DOSs for which $da/dI_{\mathrm{in}} > 0$.

Wide DOSs ($n \gg 1$, $\kappa a \gg 1$) can be interpreted as bound states of switching waves (see below). Under conditions (4.4.33), neglecting distortions of switching waves' fronts at their interaction, one can derive the following equations for the coordinates of the left and right fronts ($d_{\mathrm{g}} = 1$):

$$
\dot{x}_{\mathrm{r}} = \frac{v_{\mathrm{gr}}}{k_0} \left[ \frac{d\Phi_{\mathrm{in}}}{dx} - \frac{\kappa}{1-\eta} \left( -\frac{\delta I_{\mathrm{in}}}{I_{\mathrm{in}}} \right. \right.
$$
$$
\left. \left. + 4\frac{1-\eta}{1+\eta} \exp[-\kappa(x_{\mathrm{r}} - x_{\mathrm{l}})] \cos[\kappa(x_{\mathrm{r}} - x_{\mathrm{l}})] \right) \right] ,
$$
$$
\dot{x}_{\mathrm{l}} = \frac{v_{\mathrm{gr}}}{k_0} \left[ \frac{d\Phi_{\mathrm{in}}}{dx} + \frac{\kappa}{1-\eta} \left( -\frac{\delta I_{\mathrm{in}}}{I_{\mathrm{in}}} \right. \right.
$$
$$
\left. \left. + 4\frac{1-\eta}{1+\eta} \exp[-\kappa(x_{\mathrm{r}} - x_{\mathrm{l}})] \cos[\kappa(x_{\mathrm{r}} - x_{\mathrm{l}})] \right) \right] . \tag{4.4.38}
$$

At large distances between the fronts $\kappa(x_{\mathrm{r}} - x_{\mathrm{l}}) \gg 1$, (4.4.38) gives expressions for the velocities of left and right switching waves conforming to (4.4.25).

Let us introduce a DOS's centre coordinate $x_0$ and half-width $a$:

$$
x_0 = (x_{\mathrm{r}} + x_{\mathrm{l}})/2 , \quad a = (x_{\mathrm{r}} - x_{\mathrm{l}})/2 . \tag{4.4.39}
$$

Equations for these values follow from (4.4.38):

$$
\dot{x}_0 = \frac{v_{\mathrm{gr}}}{k_0} \frac{d\Phi_{\mathrm{in}}}{dx} = v_{\mathrm{gr}}\theta , \tag{4.4.40}
$$

$$\dot{a} = \frac{v_{gr}}{k_0} \frac{\kappa}{1 - \eta} \left( \frac{\delta I_{in}}{I_{in}} - 4\frac{1 - \eta}{1 + \eta} \exp(-2\kappa a) \cos(2\kappa a) \right) . \qquad (4.4.41)$$

We find again from (4.4.40) that the DOS's centre moves in the transverse direction with a velocity proportional to the angle of incidence $\theta$. Equation (4.4.41) describing the kinetics of the DOS's width, gives the same steady-state values of $a$ as (4.4.32). It also gives the condition of these states' instability in the form of (4.4.37). Therefore stable and unstable states of DOS alternate, as shown in Fig. 4.8b. Close to Maxwell's intensity value (4.4.33) the time of transition to a stationary DOS increases sharply (exponentially) with the increase in the stationary value of the DOS's width (i.e., for highly excited states of DOS).

Unstable states of DOS serve as boundaries for domains of attraction of stable states (see Fig. 4.8a,c). Because transversely homogeneous regimes with amplitudes $E = E_{st1}$ and $E = E_{st3}$ are stable with respect to weak perturbations, DOS excitation is "hard". It means that a DOS forms under sufficiently large initial perturbation on the background of transversely homogeneous holding radiation, whereas initially small or narrow perturbations dissolve with time when the system transfers to the homogeneous regime. Within the range of the initial perturbation width $a_{init}$ shown in Fig. 4.8a,c, a DOS in the ground state forms ($n = 0$). With an increase in $a_{init}$, a DOS in the first excited state forms ($n = 1$, Fig. 4.8a). At $I_{in} > I_0$ and large $a_{init}$, when interaction of the switching wave fronts is reduced, these fronts move in the opposite directions, switching consecutively the entire interferometer to the transversely homogeneous state (see Fig. 4.8c).

Equations similar to (4.4.40-4.4.41) are also valid for negative DOSs. Results for them are illustrated in Fig. 4.8d–f.

### 4.4.7 Dissipative Solitons: Effect of Inhomogeneities

Similarly to the case of switching waves (Sect. 4.4.4), we can consider the effect of inhomogeneities as applied to DOSs. It is natural to distinguish two aspects of the effect: (i) distortions of DOSs' shape (the case of "small-scale inhomogeneities"), and (ii) motion of a DOS as a whole under the effect of smooth, or "large-scale" inhomogeneities, a natural scale being the DOS width.

Let us consider a particular case of stationary one-dimensional variations of the holding radiation

$$E_{in}(x) = E_{in}^{(0)} + \delta E_{in}(x) , \qquad (4.4.42)$$

where $E_{in}^{(0)}$ corresponds to an ideal scheme without inhomogeneities, the latter being assumed to be small, $|\delta E_{in}| \ll |E_{in}^{(0)}|$. For stationary (motionless) distributions we can rewrite (4.1.27) in the following form:

$$\frac{d^2E}{dx^2} + \frac{\omega_0^2}{c^2}\delta\varepsilon E + 2i\frac{k_0}{d}\{[1 - R\exp(i\Delta_{ph})]E - E_{in}^{(0)}\}$$
$$= 2i\frac{k_0}{d}\delta E_{in}(x) . \tag{4.4.43}$$

The general solution of (4.4.43) consists of the general solution of the corresponding homogeneous equation $E^{(0)}$ and a particular solution of inhomogeneous equation (4.4.43). Since for the threshold nonlinearity considered $\delta\varepsilon$ is a step function, it is easy to find the general solution of the homogeneous equation; then we can find the general solution of (4.4.43) analytically for any form of inhomogeneity $\delta E_{in}(x)$. If, for instance, $\delta E_{in}(x) = a_m\cos(k_m x)$, then a particular solution of (4.4.43) is

$$\delta E(x) = \frac{2i\frac{k_0}{d}a_m\cos(k_m x)}{(\frac{\omega_0}{c})^2\delta\varepsilon + 2i\frac{k_0}{d}[1 - R\exp(i\Delta_{ph})] - k_m^2} . \tag{4.4.44}$$

According to (4.4.44), modulation depth changes step-like. For high modulation spatial frequencies $k_m$ the depth is small ($\sim k_m^{-2}$), i.e., high-frequency modulation produces only slight distortions. Resonance effects could be possible for dispersion nonlinearity (with real $\delta\varepsilon$) or for substantial phase detunings $\Delta_{ph}$; however, in the latter case the bistability necessary for soliton existence is not low-threshold.

Soliton shape distortions apart, let us consider the effect of soliton motion induced by inhomogeneities. Since in an ideal scheme (no inhomogeneities) a single soliton is motionless, for smooth inhomogeneities the velocity of the soliton "centre-of-mass" with coordinate $x_0$ is proportional to the inhomogeneities gradients [293, 294]:

$$\dot{x}_0 = \alpha\frac{d\Phi_{in}}{dx}(x_0) + \beta\frac{dI_{in}}{dx}(x_0) , \tag{4.4.45}$$

where $\Phi_{in}(x)$ and $I_{in}(x)$ are the holding radiation phase and intensity, respectively. Coefficients $\alpha$ and $\beta$ can be found separately. The term with the coefficient $\alpha$ corresponds to the locally oblique driving radiation incidence. Since we consider the scheme with the "Galilean transformation" (4.1.25-4.1.26), we have $\alpha = v_{gr}/k_0$. The term with the coefficient $\beta$ reflects the motion of a DOS composed of two switching wave fronts with differing velocities of fronts, since a local value of intensity $I_{in}$ depends on the front position $\beta \sim \pm w(dv_{sw}/dI_{in})$ ($w$ is the DOS width, $v_{sw}$ is the switching wave velocity). More precisely, the value $\beta$ should be found by a multi-scale perturbation approach [209, 90]. To this end, let us go in the governing equation (4.1.23) to the frame of reference, moving jointly with the DOS $\xi = x - \int^t v\,dt$, and linearize this equation with respect to perturbations of external $\delta E_{in}$ and internal $\delta E$ fields, and other inhomogeneities. Then, neglecting, in the first-order approximation, distortions of the DOS shape, we obtain (for simplicity, we take $\Delta_{ph}^{(0)} = 0$)

$$-L\delta E = \frac{v}{v_{\mathrm{gr}}}\frac{\mathrm{d}E^{(0)}}{\mathrm{d}\xi} + \delta P\,, \tag{4.4.46}$$

$$Lf = L_1\frac{\partial^2 f}{\partial\xi^2} + L_2 f + L_3 f^*\,,\quad L_1 = \frac{\mathrm{i}}{2k_0}\,,$$

$$L_2 = \frac{\mathrm{i}k_0}{2\varepsilon_0}(\delta\varepsilon + \delta\varepsilon''|E^{(0)}|^2) - \frac{1}{d}(1-R)\,,\quad L_3 = \frac{\mathrm{i}k_0}{2\varepsilon_0}\delta\varepsilon'' E^{(0)2}\,, \tag{4.4.47}$$

$$\delta P = \frac{1}{d}(\delta E_{\mathrm{in}} + \mathrm{i}RE^{(0)}\delta\Delta_{\mathrm{ph}} + \delta R)\,. \tag{4.4.48}$$

To solve the linear inhomogeneous equation (4.4.46), let us introduce the scalar product

$$\langle f, g\rangle = \int \mathrm{Re}[f^*(\xi)g(\xi)]\,\mathrm{d}\xi \tag{4.4.49}$$

and the adjoint operator $L^\dagger$; it follows from the definition $\langle Lf, g\rangle = \langle f, L^\dagger g\rangle$ that $L^\dagger g = L_1^*\frac{\mathrm{d}^2 g}{\mathrm{d}\xi^2} + L_2^* g + L_3 g^*$. Eigenfunctions with zero eigenvalues of operators $L$ ("neutral mode") and $L^\dagger$ corresponding to translation invariance have the form [see (4.4.26)]

$$Lu_0 = 0 \Rightarrow u_0 = \frac{\mathrm{d}E^{(0)}}{\mathrm{d}\xi} = \begin{cases} -\mathrm{sign}(\xi)pA\exp(-p|\xi|)\,, & |\xi| > a\,, \\ qB\sinh(q\xi)\,, & |\xi| < a\,, \end{cases} \tag{4.4.50}$$

$$L^\dagger u_0^\dagger = 0 \Rightarrow u_0^\dagger = \begin{cases} -\mathrm{sign}(\xi)p^*A^*\exp(-p^*|\xi|)\,, & |\xi| > a\,, \\ q^*B^*\sinh(q^*\xi)\,, & |\xi| < a\,. \end{cases} \tag{4.4.51}$$

The solvability condition for (4.4.46) is orthogonality of its right-hand side to $u_0^\dagger$. Therefore

$$v = \dot{x}_0 = \frac{\langle u_0^\dagger, P\rangle}{\langle u_0^\dagger, u_0\rangle}\,. \tag{4.4.52}$$

Calculating the scalar products in (4.4.52) for linear coordinate dependences of inhomogeneities, we obtain again $\alpha = c/k_0$ in the case of an inhomogeneity of the driven radiation phase. Considering the inhomogeneity of the driven radiation intensity, we obtain

$$\beta = -\frac{c}{d\sqrt{I_{\mathrm{in}}}}\frac{\eta}{\eta^2-1}\frac{\mathrm{Re}\,r_1}{\mathrm{Re}\,r_2}\,, \tag{4.4.53}$$

$$r_1 = \frac{(\eta^2 + \eta qa - 1)\sinh(qa) + qa\cosh(qa)}{q[\eta\sinh(qa) + \cosh(qa)]}\,,$$

$$r_2 = q\frac{\eta[\cosh^2(qa) - 1] + \cosh(qa)\sinh(qa) - qa}{[\eta\sinh(qa) + \cosh(qa)]^2}\,. \tag{4.4.54}$$

Similarly, it is easy to find a response of soliton motion to smooth inhomogeneities of the cavity losses ($\sim \delta R$) and optical path ($\sim \delta\Delta_{\mathrm{ph}}$). Generalization and analysis of the equation of DOS motion will be given in Sect. 4.6.4.

### 4.4.8 Cylindrically Symmetric Solitons

The model of threshold nonlinearity also allows us to construct cylindrically symmetric, at normal incidence, localized structures, for which the field amplitude depends only on the transverse radial coordinate $\varrho$. In this case (4.1.27) reads

$$-\mathrm{i}\frac{d}{2k_0}\frac{1}{\varrho}\frac{d}{d\varrho}\left(\varrho\frac{dE}{d\varrho}\right) + \left((1-R) + \frac{k_0 d}{2}\frac{\delta\varepsilon''(\varrho)}{\varepsilon_0}\right)E = E_{\mathrm{in}}\,. \qquad (4.4.55)$$

In regions with constant $\delta\varepsilon$, a general solution of (4.4.55) is expressed in cylindrical functions. For positive DOSs we have (in the notation of [417])

$$E_{\mathrm{dos}} = \begin{cases} E_{\mathrm{st}1} + AH_0^{(1)}(\mathrm{i}p\varrho)\,, & \varrho > a\,,\ I < I_{\mathrm{sat}}\,, \\ E_{\mathrm{st}3} + BJ_0(\mathrm{i}q\varrho)\,, & \rho < a\,,\ I > I_{\mathrm{sat}}\,, \end{cases} \qquad (4.4.56)$$

where coefficients $p$ and $q$ are determined by (4.4.27).

Joining solutions (4.4.56) at $\varrho = a$ leads to the same relations (4.4.29, 4.4.31) between the DOS half-width and the intensity $I_{\mathrm{in}}$, where now

$$Z = -\frac{J_1(\mathrm{i}qa)H_0^{(1)}(\mathrm{i}pa)}{J_0(\mathrm{i}qa)H_1^{(1)}(\mathrm{i}pa)}\,. \qquad (4.4.57)$$

For a negative DOS with the field distribution

$$E_{\mathrm{dos}} = \begin{cases} E_{\mathrm{st}1} + AJ_0(\mathrm{i}p\varrho)\,, & \varrho < a\,, \\ E_{\mathrm{st}3} + BH_0^{(1)}(\mathrm{i}q\varrho)\,, & \varrho > a\,, \end{cases} \qquad (4.4.58)$$

the same equations are valid with the substitution

$$Z = -\frac{J_0(\mathrm{i}pa)H_1^{(1)}(\mathrm{i}qa)}{J_1(\mathrm{i}pa)H_0^{(1)}(\mathrm{i}qa)}\,. \qquad (4.4.59)$$

The functions entering (4.4.56–4.4.59) are tabulated in [417]. For wide DOSs [$\kappa a \gg 1$, (4.4.33)] the value $Z \approx 1$. Then distinctions between the cylindrically symmetric and one-dimensional geometries become insignificant.

Apart from the structures indicated, there are other cylindrically symmetric structures, including DOSs in the form of bright or dark rings. At sufficiently large ring radii, these structures approach positive and negative one-dimensional DOSs, respectively [290].

### 4.4.9 Combined Switching Waves

Now we return to a transversely one-dimensional geometry and consider a more complex object – the switching wave connected with the DOS by field diffraction oscillations [292]. Such a stationary field structure, $E = E(\xi)$, moving in the transverse direction $x$ with unified velocity $v$ is presented in Fig. 4.7e. Dispersion relations for the DOS width $\xi_{21}$ and for the distance from the DOS and the switching wave front $\xi_{32}$ ($\xi_{mn} = \xi_m - \xi_n$, $\xi_n$ are front coordinates, $I(\xi_n) = I_{\text{sat}}$) are found again by joining the solutions of (4.1.27) of type (4.4.26). Solution of the dispersion relations gives a discrete spectrum of values $\xi_{mn}$ and $v$. Velocity $v$ differs from the velocity of an "ordinary" (in the absence of DOS) switching wave and depends on the DOS type (on its "quantum number" or width). Combined switching waves can form when an "ordinary" switching wave approaches an initially motionless DOS.

### 4.4.10 Bound Dissipative Optical Solitons

A two-soliton complex (Fig. 4.7f,g) is characterized by the three "quantum numbers": the numbers of the field oscillations at each of the two DOSs (regions $\xi_1 < \xi < \xi_2$ and $\xi_3 < \xi < \xi_4$) and the number of oscillations between the DOSs ($\xi_2 < \xi < \xi_3$). The procedure of their construction is the same, but the dispersion relation has a more cumbersome form [290, 292]. As compared with the case of single DOSs, the spectrum of their widths is split due to interaction of the constituent DOSs. Important is the separation of the complexes into two types: symmetric (coinciding "quantum numbers" of the first and second DOSs) and asymmetric ("quantum numbers" of DOSs differ, see Fig. 4.7f,g). At normal incidence of the holding radiation, the symmetric complexes are motionless ($v = 0$), and the asymmetric complexes move in the transverse direction with velocity determined by the holding radiation intensity and the three "quantum numbers" indicated above. Note that in the approximate analysis with equations of type (4.4.38, 4.4.40), the motion of asymmetric complexes is not described, because for these two-particle interactions the balance between the effective forces of action and counteraction is violated. In the vicinity of Maxwell's intensity value equilibrium distances between two wide positive DOSs approach a set of widths of a negative DOS. As compared with single DOSs, a spectrum of characteristics of two-soliton complexes is split, and they exist and are stable in a wider range of parameters (i.e., for certain ranges of parameters there are stable two-soliton complexes, whereas single DOSs do not exist).

In the general case one can construct bound $n$-soliton complexes (homoclinic trajectories, the same field asymptotic at $\xi \to -\infty$ and $\xi \to +\infty$), and combined switching waves, where an "ordinary" switching wave is coupled with a $n$-soliton structure (heteroclinic trajectories). Both types of structures have a discrete spectrum of parameters; "quantum numbers" are represented by the numbers of field oscillations between the fronts of "ordinary" switching waves. The $n$-soliton complexes are naturally separated into symmetric (transversely motionless at normal incidence) and asymmetric (moving with a constant velocity) complexes. With an increase in the number of solitons in the complex, there is a multiple split and condensation of spectra of the complex parameters. In the limit of an infinite periodic chain of DOSs (Fig. 4.7i), these spectra become continuous (zonal). Stable infinite (periodic) structures exist in a wider range of parameters than two-soliton complexes and especially as single DOSs. Since there is a set of equilibrium distances between individual DOSs, it is possible to construct almost arbitrary infinite (not necessarily periodic) configurations, including "quasicrystals" of dissipative solitons.

### 4.4.11 The Case of Multistability

This case can be realized under a more complicated form of nonlinearity with two characteristic values of intensity of saturation:

$$
\delta\varepsilon = \delta\varepsilon(I) = \begin{cases} \delta\varepsilon_1, & I < I_{\text{sat1}} , \\ \delta\varepsilon_2, & I_{\text{sat1}} < I < I_{\text{sat2}} , \\ \delta\varepsilon_3, & I > I_{\text{sat2}} . \end{cases} \tag{4.4.60}
$$

Now there are three stable transversely homogeneous field distributions with amplitudes $E_1$, $E_2$, and $E_3$ in a certain range of the holding radiation intensity $I_{\text{in}}$ ("tristability"). Correspondingly, the following are possible: diffraction switching waves of different types, bound states of switching waves and DOSs, and asymmetric moving DOSs. It is not surprising that spectra of these structures' parameters are discrete again. Not discussing these questions more, note here only the possibility of the analytical construction of a "dipole" – a bound state of positive and negative DOSs with a unified field asymptotic corresponding to the stable intermediate branch of the transfer function ($E \to E_2$ at $x \to \pm\infty$, Fig. 4.7h).

We conclude this section with an instructive analogy between spectra of dissipative optical soliton parameters and those of quantum objects. Single DOSs can be associated with atoms with a discrete spectrum of states. Bound DOSs are classical analogues of molecules with a discrete spectrum of distances between atoms (solitons). In addition, an infinite number of bound DOSs is similar to a solid state with a continuous (zonal) spectrum. Note also the following: the fact that two-soliton and multisoliton configurations survive even under conditions when single solitons cannot exist points, in a sense, to the "social" nature of dissipative optical solitons.

## 4.5 Structures for Other Nonlinearities

The threshold nonlinearity considered in Sect. 4.4 seems to be the only type that allows us to analytically determine spatio-temporal structures in nonlinear interferometers. For other types of nonlinearity, numerical calculations are necessary.

### 4.5.1 Switching Waves

Originally diffractive switching waves in driven nonlinear interferometers were found numerically in [323, 324, 343] for a number of different types of nonlinearities in the model of a nonlinear screen (see Sect. 4.1). This model allows us to check whether instabilities are really absent, and to take into account the scheme's longitudinal distributivity. The following results should be noted.

The waves of switching between stable stationary states still exist in the entire range of bistability. A typical width of the switching wave's front is close to Fresnel's zone $\sqrt{\lambda L}$, where $\lambda$ is the radiation wavelength and $L$ is the interferometer length. The front is shifted to this distance during one roundtrip through the interferometer. Thus, a rough estimation of the switching wave's velocity is

$$v \approx v_{\mathrm{gr}}\sqrt{\lambda/L} \ . \tag{4.5.1}$$

When substantial spatial filtering occurs [filter's boundary spatial frequency $q_{\mathrm{f}} < (\lambda L)^{-1/2}$], the front width of the switching wave is of the order of $q_{\mathrm{f}}^{-1}$ and its typical velocity is

$$v \approx \frac{v_{\mathrm{gr}}}{q_{\mathrm{f}} L} \ . \tag{4.5.2}$$

The specific feature of switching waves in the case of the diffractive transverse coupling considered here is that the spatial transition along their front from one transversely homogeneous state to another is accompanied by field oscillations (see Sect. 4.1). These oscillations are similar to those in the classical problem of light diffraction on a half-plane; the role of the edge here is played by the variation of the nonlinear component of the medium dielectric permittivity induced by a sharp variation of the radiation intensity on the front of the switching wave. Recall that the spatial transfer between the states on the front of the switching waves is smooth (nonoscillatory) in the case of the diffusive mechanism of the transverse coupling. The importance of this distinction will be demonstrated below.

The dependence of the switching wave velocity $v$ on the intensity of radiation (plane wave) incident on the interferometer once again has an alternating sign: at Maxwell's value of the incident radiation, $I_{\mathrm{in}} = I_0$, the velocity $v = 0$ (Fig. 4.10). For simultaneous propagation of several switching waves in the nonlinear interferometer, they can be considered to be independent if the distance between their fronts exceeds the front width. The switching waves moving apart (the intensity $I_{\mathrm{in}} > I_0$) are shown in Fig. 4.11a, and the colliding and annihilating waves $(I_{\mathrm{in}} > I_0)$ are presented in Fig. 4.11b.

In the case of multistability, waves of switching between different states are possible; Fig. 4.12a represents a "snap shot" of such waves. An interesting peculiarity of diffractive switching waves in a multistable interferometer, e.g., with the stable homogeneous states $E_{\mathrm{st1}}$, $E_{\mathrm{st3}}$, and $E_{\mathrm{st5}}$, is the velocity "locking" for the partial switching waves between adjacent states ($E_{\mathrm{st1}} \to E_{\mathrm{st3}}$ and $E_{\mathrm{st3}} \to E_{\mathrm{st5}}$) due to diffractive field oscillations in the intermediate (stable)

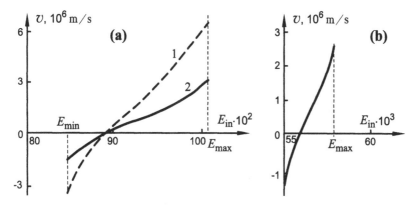

**Fig. 4.10.** Dependence of the switching wave velocity on the amplitude of incident radiation for two types of nonlinearity – (**a**) a saturable absorption and (**b**) the striction nonlinearity in nondissipative plasma – for two cavity lengths (1: $L = 250\,\lambda$; 2: $L = 1000\,\lambda$) [324, 343]

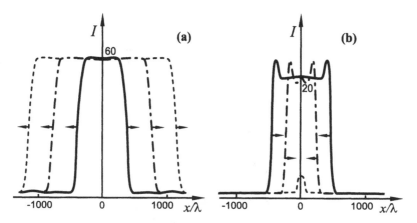

**Fig. 4.11.** Moving apart (**a**) and colliding (**b**) switching waves. Intensity profiles are given at intervals of 200 round-trips in the interferometer; the initial distribution is step-like [324, 343]

state $E_{st3}$. The velocity spectrum is discrete at fixed intensity of incident radiation (in correspondence with the number of oscillations mentioned above, see Sect. 4.4). In Fig. 4.12b,c one can see the "cylindrical switching waves" which form and are close to one-dimensional waves if the radius of the switching wave front exceeds the front width.

## 4.5.2 Metastability and Kinetics of Local Perturbations

Maxwell's intensity value $I_0$ discriminates again sections of metastability on the lower ($I_0 < I_{in} < I_{max}$) and upper ($I_{min} < I_{in} < I_0$) branches of the

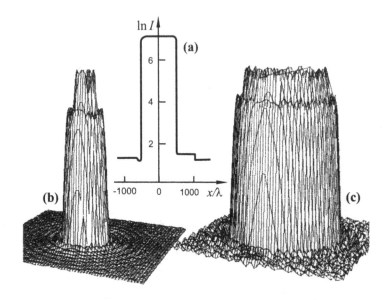

**Fig. 4.12.** Different types of switching waves: (**a**) in a multistable interferometer [343]; a cylindrical switching wave at the initial time moment (**b**) and after 200 round-trips (**c**)

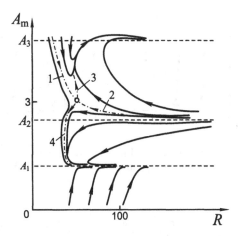

**Fig. 4.13.** The kinetics of field local perturbations, with width $R$ and maximum amplitude $A_m$, over the background of metastable states, $I_0 < I_{in} < I_{max}$; lines 1–4 represent separatrices [316]

transfer function. One can also obtain the picture of the kinetics of local perturbations in a form similar to that presented in Sect. 2.6 (see [316]); it is illustrated in Fig. 4.13.

### 4.5.3 Dissipative Optical Solitons

In Sect. 4.4 we demonstrated the field localized structures in driven wide-aperture interferometers with a threshold nonlinearity. These structures were first found in 1988 by Rosanov, Khodova and Fedorov [319, 332] and termed "diffractive autosolitons"; other terms used for them are "cavity solitons", "dissipative optical solitons", or DOSs, and "dissitons" (an abbreviation of the previous term). Here we will continue the study of DOSs for various types of optical nonlinearity. It is remarkable that modulation instability of transversely homogeneous regimes is not a *sine qua non* for their existence; moreover, the variety of DOSs is richer under conditions of instability absence. Because of this, and taking into account recent experiments [380], we begin this section with a simple interpretation of a one-dimensional DOS as a coupled state of switching waves [321].

**Interaction of Switching Waves and Single Dissipative Solitons.** Note that the motion of switching waves' fronts can be stopped by an essential inhomogeneity of external radiation amplitude or phase (see Sects. 4.4 and 4.6). Let two switching waves be excited in the interferometer, the distance between the fronts exceeding their width. Then interaction of the fronts is weak, and they, for instance, approach one another with a velocity close to double the velocity of single switching waves, $2v$. However, as the distance decreases, the field diffraction oscillations in the vicinity of the fronts (see Sect. 4.1) manifest themselves. The left front is thus affected by, apart from the homogeneous external radiation, the tail of the right front overlapping with that of the left front and including the field oscillations. Therefore the left front can be stopped by these oscillations if they are strong enough. The right front can be stopped due to the same reason. Thus a coupled state of switching waves representing a field localized structure – a DOS – arises in a nonlinear interferometer.

Moving from the DOS's centre to its periphery, we will come, at $x \to -\infty$ and $x \to +\infty$, to the same transversely homogeneous state (corresponding to the lower branch of the transfer function $E = E_{st1}$ for a "positive" DOS and to the upper branch $E = E_{st3}$ for a "negative" DOS). Therefore a DOS presents a homoclinic trajectory and resembles the "critical nuclei" found in Sect. 2.3. However, the latter are unstable with respect to small perturbations in a one-component bistable scheme (see Sect. 2.4). The conclusion with regard to the instability of these structures cannot be extended to a nonlinear interferometer characterized by the field amplitude and phase (a two-component system at least).

It is easier to "stop" switching waves with a small velocity which corresponds to the closeness of the intensity of the incident wave to Maxwell's value. Therefore DOSs exist usually in a range of intensities $I_{in}$ more narrow than the region of bistability, but including Maxwell's value $I_{in} = I_0$. At $I_{in} \neq I_0$, only a finite number of oscillations is "overcritical" (large enough to stop a switching wave), whereas the rest of the oscillations are subcritical.

Therefore, with fixed parameters of the nonlinear interferometer and at a fixed intensity of incident radiation $I_{in} \neq I_0$, only a finite number of DOSs exist, characterized by different widths $w_{dos}$. When $I_{in}$ approaches $I_0$, more and more oscillations become overcritical, and the set of DOSs with different widths increases. Note that these conclusions are in full agreement with the analytical consideration of the case of threshold nonlinearity (Sect. 4.4).

We will now consider a ring nonlinear interferometer (see Fig. 4.1). External radiation is a monochromatic plane wave propagating along the interferometer axis $z$, unless otherwise indicated. Nonlinearity is either fast (4.1.18) or corresponds to the relaxation equation (4.1.17). The field kinetics is described in the approximation of a nonlinear screen or of purely transverse distributivity (the mean-field model, see Sect. 4.1). The transmission coefficient of the layer of two-level particles is determined by relations (4.1.36, 4.1.37). Frequency dispersion of the medium is neglected. Transverse coupling is caused by radiation diffraction, and diffusion is negligibly small in the medium. First we will discuss transversely one-dimensional distributions of the field (dependence only on one transverse coordinate $x$).

Inside the interferometer, a spatial filter can be placed which is designed primarily for suppression of instabilities of the field transverse structure (see Sect. 4.3). Development of instabilities of the small-scale self-focusing type is specific for media with nonlinearity of the refraction index. In the case of two-level media, when the radiation frequency coincides with the frequency of working transition (frequency detuning $w_0 = 0$), nonlinearity is reduced to absorption saturation. Then instabilities of the transverse structure usually are absent, and there is no further need for spatial filtration of radiation. The intensity transverse profile of a single DOS under such conditions ($w_0 = 0$) is shown in Fig. 4.14d. The DOS width is approximately the width of the switching wave front ($\sqrt{\lambda L}$).

For nonzero frequency detuning, a spatial filter inside the interferometer with the boundary frequency $q_f < q_{min}$ prevents the development of instabilities of the transverse field structure. One more consequence of spatial filtration is an increase in diffractive oscillations, as the spatial frequencies close to $q_f$ are emphasized in radiation transmitted through the filter. At $q_f^{-1} \gg \sqrt{\lambda L}$, the width of the DOSs in the nonlinear interferometer with filtration increases noticeably, and the conditions for their existence become more favorable.

Figure 4.14 also shows different types of single DOSs in the interferometer with nonresonant nonlinearity ($w_0 = 9.3$) and spatial filtration with the boundary frequency $q_f = 5 \times 10^{-3} k_0$, the incidence of external radiation being normal ($\theta = 0$). "Positive", or "bright" (intensity in the DOS centre is above background) and "negative", or "dark" (the reverse relation of intensities) DOSs of different width are presented. The regions of existence of these DOSs differ: a DOS with a greater width ("excited state" of DOS) exists in a more narrow range of intensities $I_{in}$ than the "narrow" DOS (in the "ground"

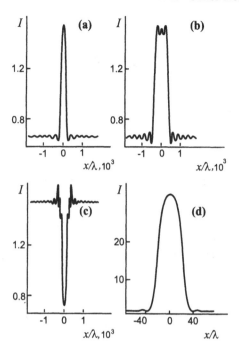

**Fig. 4.14.** Types of dissipative optical solitons: positive (**a,b,d**) and negative (**c**) solitons for absorptive-dispersive (**a–c**) and purely absorptive (**d**) nonlinearities [321]

state). Figure 4.15 shows the bistability range in an approximation of plane waves and the more narrow range of existence of DOSs in the "ground" state: $I_1 < I_{in} < I_2$. Boundary values of intensity for DOSs of other types [positive $(+)$ and negative $(-)$, in the $n$th excited state] change:

$$I_{1n}^{(\pm)} < I_{in} < I_{2n}^{(\pm)} . \qquad (4.5.3)$$

All these regions include Maxwell's intensity value $I_0$.

DOSs can be formed at the collision of two switching waves or by a hard excitation by means of an initial local perturbation of the field additional to the holding plane wave. In view of the stability of transversely homogeneous states with respect to small perturbations, the local perturbations with small amplitudes $a_p$ or the widths $w_p$ dissolve with time (Fig. 4.16a). For large $a_p$ and $w_p$ and for the proper relation of intensities $I_{in}$ and $I_0$, the perturbation can be transformed into two switching waves running apart from one another (Fig. 4.16c). These waves will switch, with time, the entire nonlinear interferometer to a transversely homogeneous state (different from the original). Therefore, DOSs will be excited only in some intermediate region of the perturbation parameters $a_p$ and $w_p$.

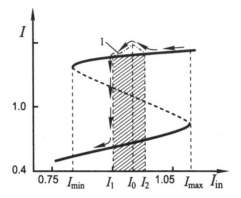

**Fig. 4.15.** Transfer function and the ranges of bistability ($I_{min}$–$I_{max}$) and of the existence of a DOS in the ground state ($I_1$–$I_2$, *hatched*). *Dash–dotted line* 1 shows the radiation axial intensity during "switching down" of the interferometer driven by a wide beam [319]

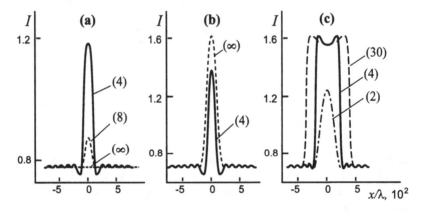

**Fig. 4.16a–c.** The kinetics of local field perturbations in the interferometer under conditions of DOS existence (time in units of $\tau_{del}$ is shown in brackets for each of the intensity profiles) [319]

An example of settling of the regime of a single DOS, confirming its stability, is presented in Fig. 4.16b. The dependence of the type of stationary DOS on the width of initial local field perturbation in the interferometer $w_p$ (for its fixed amplitude $a_p$) is given in Fig. 4.17. In the conditions of Fig. 4.17a, depending on the filtration level, there are 4 (line 2) or 8 (line 1) variants of the DOS width; in the case of Fig. 4.17b, there are two variants of the width, and, at considerable widths of initial perturbations, switching waves running apart form ($I_{in} > I_0$). With an increase in the boundary frequency of filtration $q_f$, the number of different DOSs and their widths decrease.

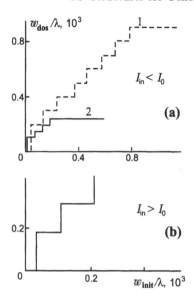

**Fig. 4.17.** Dependence of DOS width $w_{\mathrm{DOS}}$ on the initial width $w_{\mathrm{init}}$ of the field local perturbation: (a) $I_{\mathrm{in}} < I_0$, (b) $I_{\mathrm{in}} > I_0$ [333]

As the DOSs discussed are stationary, the relaxation time does not influence their characteristics, but it can affect the stability and kinetics of DOS formation. The calculations confirm the stability of DOSs in the wide range of the ratio of relaxation time $\tau_{\mathrm{rel}}$ and interferometer round-trip time $\tau_{\mathrm{del}}$, including the cases of fast nonlinearity, $\tau_{\mathrm{rel}} \ll \tau_{\mathrm{del}}$, and with comparable times, $\tau_{\mathrm{rel}} \approx \tau_{\mathrm{del}}$ [332]. "Pulsing distributions" of the soliton type (see Sect. 4.7) are also possible.

The interpretation of a DOS as a coupled state of switching waves stopped by field diffraction oscillations is asymptotic and is justified mainly for wide ("highly excited") DOSs whose switching fronts are distant from each other. The width of the narrowest DOSs (the ground state) is comparable with the width of the switching wave front. Therefore the concept of weakly interacting switching waves provides only a qualitative description in this case. Moreover, the range of the existence of DOSs can appear even wider than the range of bistability, so they can exist in the absence of both switching waves and bistability [321]; see also Sects. 4.7 and 4.9.

**Interaction of Dissipative Solitons.** If initially we form two DOSs separated in the transverse direction by the distance $d_0$, they will interact because of the overlapping of their tails. At greater distances $d_0$, the field oscillations on the tails are weak (subcritical), and one can consider that each of the DOSs is influenced by the mean gradient of radiation intensity corresponding to the tail of another DOS. Therefore, at self-focusing nonlinearity, two distant DOSs usually attract each other. But as they approach, the oscilla-

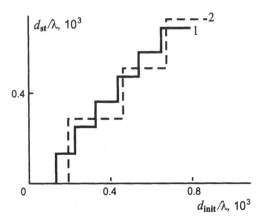

**Fig. 4.18.** Equilibrium distance between two solitons $d_{st}$ versus the initial distance between them: $q_f/k_0 = 0.005$ (1) and 0.01 (2) [321]

tions become essential and, depending on the value of $d_0$, the DOSs' mutual attraction can be replaced by their repulsion. Correspondingly, there is a limited set of the equilibrium distances $d_{DOS}$ between DOSs.

The settling of one or the other variant of the distance $d_{DOS}$ depends on the initial distance $d_0$ between the DOSs, as shown in Fig. 4.18. For sufficiently small initial distances $d_0$, two DOSs merge. Under certain conditions the equilibrium distances between two positive DOSs can be associated with a set of the widths of single negative DOSs (see Sect. 4.4). The distance $d_{DOS}$ also depends on the type ("quantum number") of the interacting DOSs. Each of them can be in the "ground" or "excited" state; both DOSs can be "positive" or "negative". At self-defocusing nonlinearity distant DOSs are usually repulsive; however, there is also a set of equilibrium distances between them. Therefore DOSs can form bound "multi-particle" structures (Fig. 4.19). It is essential that the main characteristics of a DOS – its width and maximum intensity – change only slightly in bound systems, as compared with the case of a single DOS. With the addition of one more DOSs to an equilibrium two-particle system, the equilibrium distance between the DOSs also changes only slightly. Therefore, by knowing the set of the equilibrium distances between two DOSs of different types, one can construct an unlimited number of multi-soliton structures (with an unlimited aperture of the interferometer).

**Two-Dimensional Dissipative Solitons.** The analysis of the two-dimensional transverse variation of the field (along the coordinates $x$ and $y$) is necessary to justify the one-dimensional consideration given above, to confirm the DOSs' stability, and to seek new types of spatial structures.

Calculations confirm the stability of diffractive switching waves and DOSs of different types. A single axially symmetric DOS is presented in Fig. 4.20, and two- and three-soliton structures are given in Fig. 4.21.

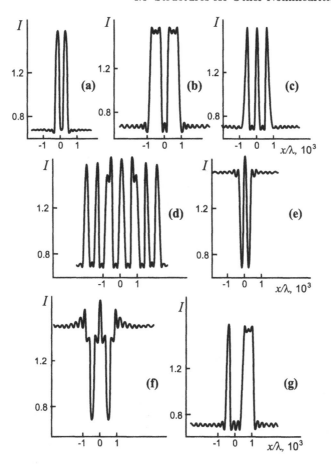

**Fig. 4.19.** Examples of two-soliton (**a,b,e,f,g**) and multi-soliton (**c,d**) structures [321]

The features of single DOSs and the equilibrium distances between them in bound structures are close to those found in one-dimensional calculations. This allows us to give a simplified geometrical recipe for the construction of two-dimensional stationary multisoliton structures. If, e.g., only one value of the equilibrium distance $d_1$ between two DOSs is possible, then the structures can be constructed by arranging DOSs in nodes of the net composed of equilateral triangles with side length $d_1$ (Fig. 4.22). If several equilibrium distances are possible $d_{\text{dos}} = d_1, d_2, \ldots$, then the structures are more diverse. For example, for two variants of the distance $d_1$ and $d_2$ ($d_1 < d_2$), at $d_2 < 2d_1$ four equilibrium three-soliton configurations can be constructed (Fig. 4.23).

Asymmetric structures move and, generally, rotate. For example, multi-soliton structures presented in Fig. 4.22 and in the two right-hand figures of Fig. 4.23 (isosceles, but not equilateral triangles) have one (and only one) axis of symmetry. Such stationary structures move along this axis with some

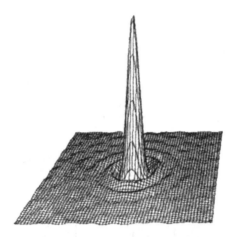

**Fig. 4.20.** Transverse distribution of intensity for a two-dimensional single DOS [321]

constant velocity. The existence of motion can be proved by consideration of the two-dimensional variant of (4.1.27), its discretization and comparison of the number of equations to that of the unknowns. For stationary structures without axes of symmetry (e.g., three-soliton complexes in the form of scalene triangles) it is necessary to use a more cumbersome equation in the reference frame moving and rotating with some angular velocity. Similar comparison of the number of equations and that of the unknowns shows rotation of these structures with a constant angular velocity. Numerical values of the velocity of translational motion and of angular velocity are fairly small under typical conditions.

Using DOSs of different types, one can obtain more complex stationary structures. It is also possible to derive approximate "mechanical equations" governing the dynamics of interacting DOSs (see the analysis in [293, 294] and Sect. 4.6.4). Most diverse are collision regimes of moving (asymmetric) DOSs. The result is determined by their approach velocity $v = v_2 - v_1$, where $v_{1,2}$ are the initial velocities of colliding DOSs [300]. If this speed is great,

$$v \gg c|\delta\varepsilon|w_n^{(\pm)}/\lambda \,, \qquad (4.5.4)$$

where $w_n^{(\pm)}$ is a characteristic width of DOSs of the type considered, and $\delta\varepsilon$ is a typical value of the nonlinear dielectric permittivity, the time of their interaction being small. As a result of this weak interaction, the DOSs pass through one another without noticeable distortions, and their trajectories vary only slightly as compared with the initial trajectories. At smaller approach speeds, DOSs may not only be scattered one by another, but can also be captured. Structures rotating in the transverse direction arise more naturally in bistable laser schemes (see Chap. 6).

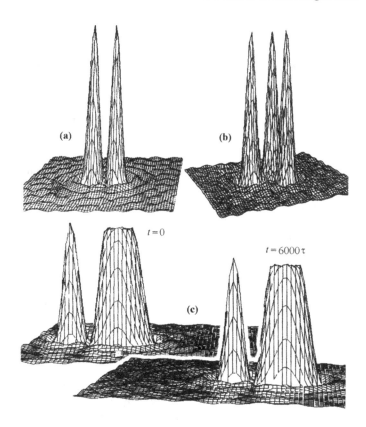

**Fig. 4.21.** Transverse distribution of intensity for two- and three-soliton structures: (**a,b**) symmetric motionless configurations; (**c**) asymmetric, moving in the transverse direction structure [321]

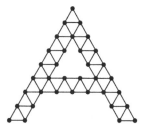

**Fig. 4.22.** Diagram showing the construction of a multisoliton configuration with only one possible equilibrium distance between two neibouring solitons

## 4.6 Effect of Inhomogeneities

Different types of inhomogeneities are of great importance in wide-aperture nonlinear interferometers. They include inhomogeneities of the refraction in-

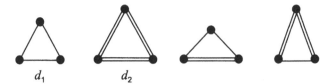

**Fig. 4.23.** Three-soliton structures with two possible equilibrium distances, $d_1$ and $d_2$

dex and absorption coefficient of the medium, irregularities of the interferometer mirrors, and stationary inhomogeneities of the amplitude and of the wave front of external radiation, including its oblique incidence. The effect of smooth (large-scale) inhomogeneities was considered in Sect. 4.4 for interferometers with threshold nonlinearity. Numerical calculations are necessary for other types of nonlinearity and for small-scale inhomogeneities. We will begin this section with an analysis of the effect of inhomogeneities in situations when dissipative solitons are absent (Sects. 4.6.1–4.6.3) and will then proceed to the treatment of similar effects for dissipative solitons (Sects. 4.6.4 and 4.6.5).

### 4.6.1 Switching Waves

The role of inhomogeneities in switching processes in interferometers is the same as for schemes of bistability with increasing absorption (see Chap. 3). Let, for example, the amplitude of the incident radiation include a stationary local perturbation characterized by width $d_{\mathrm{inh}}$ and amplitude $\Delta_{\mathrm{inh}}$:

$$E_{\mathrm{in}} = E_{\mathrm{hom}} + \Delta_{\mathrm{inh}} \exp(-x^2/d_{\mathrm{inh}}^2) \,. \qquad (4.6.1)$$

The front of the switching wave moving from the periphery ($x = -\infty$) to the centre of the local perturbation ($x = 0$) can either be captured by it or pass through it and move further ($x \to +\infty$), depending on the local perturbation parameters. These two variants are separated by the critical value of the amplitude $\Delta_{\mathrm{cr}} = \Delta_{\mathrm{cr}}(d_{\mathrm{inh}})$ (Fig. 4.24). In the first case ($\Delta_{\mathrm{inh}} > \Delta_{\mathrm{cr}}$), when the switching wave front approaches the local perturbation and stops on it, a stationary distribution of the field resembling a distorted motionless switching wave forms, though $|E_{\mathrm{hom}}|^2 \neq I_0$. With the fixed parameters of inhomogeneity, stopping of the switching wave occurs in the range of the incident radiation intensities including Maxwell's value $I_0$ and depending on the parameters of the inhomogeneity.

The specific feature of switching of a wide-aperture bistable system with inhomogeneities is the other aspect of the same phenomenon (see Sect. 2.9). When incident radiation intensity is lower than a certain critical value $I_{\mathrm{cr}}$, inhomogeneities cause only local distortions of the field in the nonlinear interferometer, which can be interpreted as switching waves stopped by inhomogeneities. When $I_{\mathrm{in}}$ exceeds $I_{\mathrm{cr}}$, inhomogeneities are unable to stop switching

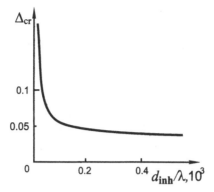

**Fig. 4.24.** Dependence of the critical value of the amplitude of incident radiation inhomogeneity on the inhomogeneity width [321]

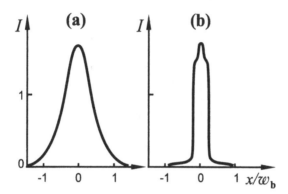

**Fig. 4.25.** Bistability of the beam intensity profile in the interferometer with a saturable absorption: (**a**) the profile forms at a small initial intensity of the field in the interferometer; (**b**) this profile is realized at its high initial level [324]

waves; their fronts move from the inhomogeneity. Propagating in this way, the waves switch consecutively the entire scheme to the upper state, though $I_{in} < I_{max}$. The dependence of the critical intensity $I_{cr}$ on the inhomogeneity parameters is of the type shown in Fig. 2.25. Initiation of switching of a wide-aperture nonlinear interferometer by inhomogeneities in its optical path was studied in [318], in which an estimation of the probability of the switching by inhomogeneities with Gaussian statistics of parameter distribution was also given.

### 4.6.2 Spatial Hysteresis

When the power of the incident radiation beam changes, spatial hysteresis manifests itself again. For the same final power (and the same shape) of the incident beam, a smooth profile of the output radiation intensity forms in the interferometer (Fig. 4.25a) in the case of a low initial level of the radiation

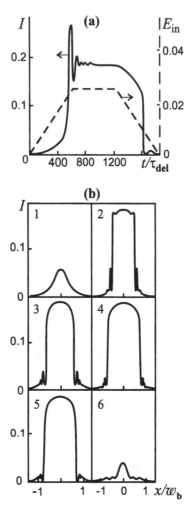

**Fig. 4.26.** (a) Temporal variation of the incident beam amplitude (*dashed line*), output intensity at the beam axis (*solid line*) and (b) the kinetics of the beam profile in an interferometer with Kerr nonlinearity: $t/\tau_{\mathrm{del}} = 540$ (1), 740 (2), 860 (3), 1200 (4), 1260 (5), and 1620 (6) [324]

intensity, whereas in the case of its high level we have a profile with sharp spatial variation of the intensity (Fig. 4.25b). For wide beams, with a known magnitude of Maxwell's intensity value, it is possible to offer a scheme for constructing profiles of the output radiation intensity similar to Fig. 2.18.

The kinetics of spatial hysteresis with slow time variation of the incident radiation amplitude for a wide range of conditions is also analogous to that discussed for bistability with increasing absorption in Chap. 2. It is illustrated in Fig. 4.26. In this case, again, switching up does not occurs simultaneously over the entire section of the beam, but first in its central part and then in a narrow zone propagating as a gradually decelerating switching wave.

### 4.6.3 Oblique Incidence

For a fast nonlinearity and plane-wave radiation, the difference of the direction of the radiation incidence from normal ($\theta \neq 0$) causes only an increase in the longitudinal extension of the interferometer, quadratic in $\theta$ ($\theta^2 \ll 1$). Therefore the conditions of bistability in the framework of the plane-wave approximation practically do not depend (considering small changes of the phase shift $\Delta_{\text{ph}}$ and thickness of the nonlinear element $d$) on the angle of incidence $\theta$ or on the transverse shift of the beams possible with a misalignment of the cavity. However, the stability of the transverse field structure depends on these factors (see Sect. 4.3 and [152]).

In the case of the field structures inhomogeneous in the transverse directions, e.g., for incidence of the radiation beam on the interferometer or for propagation of switching waves, the situation changes. Even in approximations of geometrical optics and a thin nonlinear screen (Sect. 4.1), the relation of the fields of the incident beam and beam formed in the interferometer is not local:

$$E(\boldsymbol{r}_\perp, t) = E_{\text{in}}(\boldsymbol{r}_\perp, t) + R \exp(i\Delta_{\text{ph}}) E(\boldsymbol{r}_\perp - \boldsymbol{r}_1, t - \tau_{\text{del}})$$
$$\times K(|E(\boldsymbol{r}_\perp - \boldsymbol{r}_1, t - \tau_{\text{del}})|^2) . \tag{4.6.2}$$

Here $\boldsymbol{r}_1$ is the transverse shift of a ray during one round-trip in the interferometer. It follows from (4.6.2) that [323]

$$E(\boldsymbol{r}_\perp, t) = \sum_{s=0}^{\infty} E_{\text{in}}(\boldsymbol{r}_\perp - s\boldsymbol{r}_1, t - s\tau_{\text{del}}) R^s \exp(is\Delta_{\text{ph}}) \prod_{n=0}^{s} K_n , \tag{4.6.3}$$

$$K_0 = 1 , \quad K_n = K(|E(\boldsymbol{r}_\perp - n\boldsymbol{r}_1, t - n\tau_{\text{del}})|^2) . \tag{4.6.4}$$

Expression (4.6.3) corresponds to the presentation of the field in the multipass interferometer as a sum of rays, and it gives, for the specified local nonlinear coefficient of transmission $K$ and for a fixed distribution of incident radiation, an unambiguous algorithm for the determination of the field in the interferometer. Therefore bistability does not exist in the framework of (4.6.3). The reason is that, in this case, the field transverse interaction appears to be unidirectional: the field in a point $\boldsymbol{r}_\perp$ depends on the field values at $\boldsymbol{r}_\perp - \boldsymbol{r}_1$, $\boldsymbol{r}_\perp - 2\boldsymbol{r}_1$, etc., but does not depend on the field amplitude at $\boldsymbol{r}_\perp + \boldsymbol{r}_1$, $\boldsymbol{r}_\perp + 2\boldsymbol{r}_1$, etc. The necessary condition of bistability is bidirectional transverse coupling. It can be caused by diffraction or scattering of radiation and by diffusion in the medium (i.e., by a nonlocal property of the coefficient of transmission $K$). Below we will discuss one more possibility for bistability retained at oblique incidence in double-beam schemes.

The evaluation of bistability conditions is as follows: For the diffractive mechanism of the transverse coupling, a typical size of transverse interaction $\sqrt{\lambda L}$ should exceed the transverse shift $r_1 \propto L\theta$. Hence the critical angle of incidence

$$\theta_{cr} \propto \sqrt{\lambda/L} \quad (L \gg \lambda) . \tag{4.6.5}$$

At considerable spatial filtering

$$\theta_{cr} \propto (q_f L)^{-1} . \tag{4.6.6}$$

Bistability remains only at $\theta < \theta_{cr}$, though in the approximation of plane waves it also takes place at $\theta > \theta_{cr}$. In this sense the opinion presented in [179] with regard to hysteresis absence for beams with a finite width has to be refined substantially: bistability is indeed absent if the angle of incidence exceeds the critical value determined by the transverse coupling.

We can draw a similar conclusion from an analysis of velocities of "left" and "right" switching waves (4.4.24), which is valid for any type of nonlinearity at sufficiently small angles of incidence $\theta$. Then splitting of Maxwell's intensity takes place at

$$\theta < \min[v_{max}/(d_g v^{(0)}), \ |v_{min}|/(d_g v^{(0)})] , \tag{4.6.7}$$

where values $v^{(0)}$, $v_{max}$ and $v_{min}$ are taken at normal incidence. In this case "left" and "right" switching waves become motionless at $I_{in} = I_0^l$ and $I_0^r$, correspondingly. At

$$\theta > \theta_{cr} = v_{max}/(d_g v^{(0)}) , \tag{4.6.8}$$

the "left" switching wave does not stop in the entire range of its existence; in this case spatial hysteresis is not possible. Note also that it follows from (4.4.24) that the velocities of "left" and "right" switching waves coincide, both by module and by direction, at $I_{in} = I_0$:

$$v_l(I_0) = v_r(I_0) = d_g v^{(0)}\theta . \tag{4.6.9}$$

Generally, $v_r = v_l$ at $I_{in} = I_e \neq I_0$ [310]. These results are illustrated in Fig. 4.27; they are valid also in the model of a nonlinear screen [343]. For the case of the Fabry–Perot interferometer with a slow thermo-optical mechanism of nonlinearity, consideration of [310] shows that the scheme is not as sensitive to the angle of incidence, as in the case of fast nonlinearity. A similar conclusion is valid if the transverse coupling is determined by diffusion in the medium.

**Spatial Hysteresis at Oblique Incidence.** Let the incident radiation be a beam with the width considerably exceeding the value of the geometric shift $r_1 \propto L\theta$ and the specific width of the switching wave front. Then the incident radiation is locally close to plane waves, and one can use the concepts of the switching waves with slowly varying characteristics (see Chap. 2).

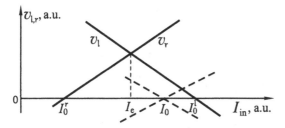

**Fig. 4.27.** The dependence of velocities of "left" and "right" switching waves on the intensity of incident radiation at normal (*dashed lines*) and oblique (*solid lines*) incidence [310]

If the intensities of the branch edges of the transfer function $I_{\min}$ and $I_{\max}$, Maxwell's intensity values $I_0^l$ and $I_0^r$, and the maximum intensity of the incident beam $I_m$ satisfy the inequalities

$$I_{\min} < I_0^r < I_0^l < I_m < I_{\max} , \qquad (4.6.10)$$

then we can again use a construction of the type shown in Fig. 2.18. Spatial bistability corresponds in Fig. 4.28 either to the smooth profile of radiation intensity in interferometer 1 constructed with only the lower branch of the transfer function or to the "combined" profile 2. The central part of profile 2 is determined by the upper branch of the transfer function, and sharp spatial jumps between the branches occur in the neighbourhoods of such $x_r$ and $x_l$ that

$$I_{in}(x_r) = I_0^r , \quad I_{in}(x_l) = I_0^l . \qquad (4.6.11)$$

Even if the intensity distribution of incident radiation is symmetric with respect to $x$, the combined profile 2 of intensity of transmitted or reflected radiation is asymmetric.

Spatial hysteresis under the conditions considered is similar to that for the schemes of bistability at increasing absorption (see Sect. 2.6) and for nonlinear interferometers at normal incidence. With a slow increase in the beam power from small values up to the value $I_m(t) = I_{\max}$, the profile of field intensity in the interferometer will be smooth (profile 1). At $I_m > I_{\max}$, switching to the upper branch of the transfer function takes place in the central part of the beam. The boundaries of the switched region will propagate from the beam axis in the form of right and left switching waves with different velocities. The motion of the wave fronts will slow down and stop upon reaching conditions (4.6.11).

More precisely, spatial hysteresis takes place under less rigid conditions than (4.6.10). Above all, the angle of incidence should be less than the critical value determined by (4.6.8); otherwise bistability and hysteresis are absent for any intensity distributions of incident radiation. But if $\theta < \theta_{cr}$, then spatial hysteresis is possible under conditions

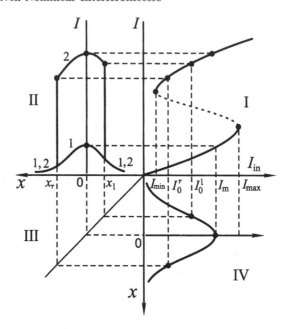

**Fig. 4.28.** Spatial bistability at oblique incidence of radiation

$$I_0^l < I_m < I_{max} \,. \tag{4.6.12}$$

In this case the right switching wave does not stop at any $I_{in}$. Then spatial switching between the branches on the right front $x = x_r$ will take place at $I_{in}(x_r) \approx I_{min}$. In Fig. 4.28 it corresponds to the substitution $I_0^r \rightarrow I_{min}$; a more exact description would require consideration of the nonlocality mentioned above.

An example of spatial bistability at oblique incidence is shown in Fig. 4.29. If the initial field in the interferometer is small, then a smooth profile of

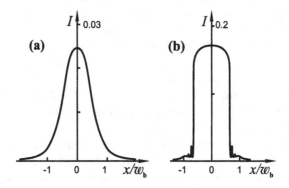

**Fig. 4.29a,b.** Bistability of the intensity profile in an interferometer with Kerr nonlinearity for an oblique incidence of radiation [323]

intensity $I(x)$ forms (Fig. 4.29a), whereas for large initial fields the stationary profile $I(x)$ contains sharp fronts of spatial switching (Fig. 4.29b).

### 4.6.4 "Mechanics" of Dissipative Solitons

In Sect. 4.4 we derived approximate equations of motion of DOSs in interferometers with smooth inhomogeneities and threshold nonlinearity. Here this "mechanical" motion will be studied for interferometers with various types of nonlinearity.

If the intensity $I_{\mathrm{in}}$ and phase $\Phi_{\mathrm{in}}$ of external radiation do not depend on transverse coordinates $r_{\perp} = (x, y)$, a DOS can be localized in any position on the (unbounded) interferometer aperture, i.e., the coordinate of its centre $r_0$ is arbitrary. Let us now assume $I_{\mathrm{in}}$ and $\Phi_{\mathrm{in}}$ to be slowly varying functions of transverse coordinates $r_{\perp}$ [on the scales of the half-width $w_n^{(\pm)}$ of a positive (+) or negative (−) DOS in the $n$th state]. As a single DOS is motionless in a homogeneous external field, the equation of its motion in a smoothly inhomogeneous field should have the form

$$\dot{r}_0 = F(r_0) \,, \tag{4.6.13}$$

where

$$F = -\nabla U \,, \quad U = -(\alpha \Phi_{\mathrm{in}} + \beta I_{\mathrm{in}}) \,. \tag{4.6.14}$$

Coefficients $\alpha$ and $\beta$ were found analytically in Sect. 4.4 for threshold nonlinearity. For the case of fast nonlinearities in systems with the Galilean transformation (4.1.25, 4.1.26), $\alpha = v_{\mathrm{gr}}/k_0$. Determination of these coefficients for an arbitrary nonlinearity needs numerical work [209]. Here we will analyze DOS mechanics on the basis of (4.6.13, 4.6.14), taking the coefficients as given.

In general terms coefficients $\alpha$ and $\beta$ depend on intensity at the location of the DOS: $\alpha = \alpha\big(I_{\mathrm{in}}(r_0)\big)$, $\beta = \beta\big(I_{\mathrm{in}}(r_0)\big)$. We will assume that, in the considered domain of DOS motion, the intensity of external radiation varies within narrow limits. Then, as $\alpha$ and $\beta$ in (4.6.14) are factors at small gradients, one can neglect their variation, assuming them to be constant. The same reasons validate the scalar nature of coefficients $\alpha$ and $\beta$.

One can interpret (4.6.13) mechanically as the equation of two-dimensional motion (in the $x$–$y$ plane) of a lightweight particle in a viscous medium under the action of the force $F$, i.e., as the limit of the Newtonian equation

$$m\ddot{r}_0 = -\dot{r}_0 + F(r_0) \,, \tag{4.6.15}$$

when the particle mass $m \to 0$ (at the unit viscosity coefficient). Note that, according to (4.6.13), DOSs obey not "Newtonian" but "Aristotelian" mechanics, as external actions determine the velocity of their motion (but not acceleration).

If the phase of incident radiation $\Phi_{in}$ is given unambiguously, the "potential" $U$ is also determined unambiguously in terms of (4.6.14). Then a DOS will move along the lines of the quickest descent of the surface $U(x, y)$. With time it will approach either the minimum of the potential $U$ (settling of a stationary regime) or the boundary of existence of DOSs of this type (4.5.3), where (4.6.13) is violated (see below).

We will illustrate such motion first for the case of a transversely one-dimensional (slot) beam of external radiation:

$$I_{in} = I_{in}(x) , \quad \Phi_{in} = \Phi_{in}(x) . \tag{4.6.16}$$

Then $F_y = 0$ and $y_0 =$ const., and it is easy to determine implicitly the DOS trajectory $x_0(t)$ from (4.6.13). Let the incident beam have a bell-like (unimodular) intensity profile $I_{in}(x)$ (Fig. 4.30a) and a plane wavefront (e.g., a Gaussian beam). At normal incidence $d\Phi_{in}/dx = 0$, and the "potential" $U(x)$ (Fig. 4.30b) has the shape of an inverted intensity profile (accurate to insignificant constants; for definiteness we consider the case $\beta > 0$). The DOS will move with time from any of its initial positions $x_0(t_0)$ [within the range of its existence (4.5.3)] to the potential minimum $x = x_s$, coinciding with the maximum of external radiation intensity. One can pictorially present a DOS as a ball rolling down to the bottom of a gently sloping well.

If we now proceed to the case of external radiation with a small angle of incidence $\theta = k_0^{-1}d\Phi_{in}/dx =$ const., then the "potential" $U$ is tilted by the angle proportional to $\theta$. Until the angle $\theta$ is less than some critical value $\theta_{cr}$ (Fig. 4.30c), there is a single stable equilibrium state (a bottom of the potential well, $x = x_s$) somewhat shifted in comparison with the case $\theta = 0$. The

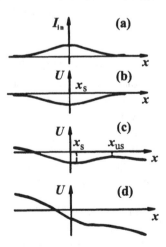

**Fig. 4.30.** Holding radiation intensity profile (a) and "potential curves" (b–d) for DOS movement for normal (b) and oblique [(c): $\theta < \theta_{cr}$; (d): $\theta > \theta_{cr}$] incidence of holding radiation [293, 294]

domain of its attraction is restricted to the position of unstable equilibrium $x_{us}$; so at $x_0(t_0) < x_{us}$ DOS approaches with time stable equilibrium, and for $x_0(t_0) > x_{us}$ DOS moves away from the beam centre to the boundary of its existence. The critical value $\theta_{cr}$ corresponds to the merging of the stable and unstable positions of equilibrium [the inflection point of the dependence $I_{in}(x)$]. For $\theta > \theta_{cr}$ (Fig. 4.30d) the "potential" $U$ has no more minima, and DOS slides to the boundary of its existence from any initial position $x_0(t_0)$.

A potential minimum also arises in the case of a transversely homogeneous distribution of incident radiation intensity, if its wavefront has a bell-like inhomogeneity (e.g., $\Phi_{in} = Cx^2$). Then, depending on the sign of the corresponding effective lens (the sign of the coefficient $C$), DOS either is attracted to the potential minimum (lens axis) or moves away from it.

DOS behaviour in the vicinity of the boundary of its existence depends on the DOS type and on the relation between local intensity $I_{in}$ and Maxwell's intensity value $I_0$. Positive DOS decays to switching waves running apart for $I_{in} > I_0$. At boundaries with $I_{in} < I_0$ it shrinks; the excited DOS turns into a DOS with smaller width, and the DOS in the ground state collapses (disappears). The behaviour of negative DOSs is the same if the signs in the intensity inequalities are changed to the opposite ones.

The new, topological peculiarities of DOS dynamics arise for transversely two-dimensional inhomogeneities when external radiation has wave-front dislocations (vortices, or defects). In the centre of a vortex $I_{in} = 0$, and phase changes by $2m\pi$, where $m = 0, \pm1, \pm2, \ldots$ is the topological index, at path-tracing around the vortex.[3] Under these conditions the potential $U(x,y)$ becomes a multi-valued function of coordinates (whereas the phase gradient $\nabla\Phi_{in}$ and force $\boldsymbol{F}$ are unambiguous).

Let external radiation have the following complex amplitude characteristic of the optical cavity mode:

$$E_{in}(\varrho, \varphi) = A\varrho \exp(-\varrho^2/2w^2) \exp(i\varphi) . \qquad (4.6.17)$$

Here $\varrho$ and $\varphi$ are the polar coordinates, $w$ is the beam width, and the single vortex with the topological index $m = 1$ is situated at the coordinate origin $\varrho = 0$. From projections of (4.6.17) to radial and azimuthal directions (variables are separated for them), we find that DOS motion forms as rotation along a circle of radius $w$ with the angular velocity $\Omega = v_{gr}/(k_0w^2)$ and period $T = 2\pi k_0 w^2/v_{gr}$. DOS trajectories in the phase plane (Fig. 4.31a) are spirals winding round the circle representing the stable limit cycle. Notice that the characteristics of external radiation do not depend on time, and there are no linear elements rotating the field in the scheme. It is important that Pointing's vector for the field (4.6.17) has a nonzero azimuthal component that causes DOS rotation around the vortex.

---

[3] The origin of vortices can be interference, and initially they were studied in linear optical systems [248, 421].

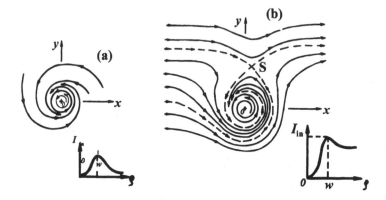

**Fig. 4.31.** DOS trajectories in the case of holding radiation with wavefront disloca-
tion for normal (**a**) and oblique (**b**) incidence; intensity profiles of holding radiation
are also shown [293, 294]

Let the radiation with a vortex fall on the interferometer at the angle
of incidence $\theta$ to its axis in the $x$–$z$ plane. Then the radiation phase $\Phi_{in} =
\varphi + k_0\theta\varrho\cos\varphi$. An example of intensity profile $I_{in}(\varrho)$ providing existence of
the type of DOS chosen on the entire interferometer aperture (excluding
some vicinity of the vortex $O$) is shown in Fig. 4.31b. At a large distance
from the vortex ($\varrho \gg w$) DOS moves parallel to the $x$-axis with velocity
$v_{gr}\theta$. For sufficiently small angles $\theta$ there is a single (unstable) position of
equilibrium (the saddle S with coordinates $y_s = 1/k_0\theta \gg w$ and $x_s \ll w$),
and a stable limit cycle close to the cycle of the radius $w$ (the periodic regime
corresponding to DOS rotation around the vortex). Two separatrices come
into the saddle S from the region $x_0(t_0 = -\infty) = -\infty$ at finite $y_0(t_0 = -\infty)$.
One of two leaving separatrices goes at $t_0 \to +\infty$ to $x_0 = +\infty$ also at finite
$y_0$, and the second one is winding on the stable limit cycle.

As Fig. 4.31b shows, two separatrices coming into the saddle S (dashed
lines with arrows) separate the range of aiming distances, corresponding to
the capture of the DOS, flying upon the vortex, in the vortex's vicinity,
with settling of DOS rotation. For larger values of aiming distance, DOS is
scattered by the vortex. For large angles of incidence, DOS is shifted by the
vortex for any aiming distance.

For a finite interferometer aperture, there are diffractive oscillations of
the field in the vicinity of the mirror edges. These oscillations prevent the
DOS from going away from the interferometer, and they stop the DOS close
to the mirror edge.

"Aristotelian mechanical equations" for a system of distant DOSs weakly
interacting due to overlap of their tails are a natural generalization of (4.6.13)
[293, 294]:

$$\dot{r}_k = F(r_k) + \sum_n F_{int}(r_k - r_n). \qquad (4.6.18)$$

Here $F_{\text{int}}$ represents pairwise centre-force interaction depending on the distance between the centre coordinates $r_k$ of the individual DOSs. The modulus of the interaction force is close to the field asymptotics at the DOS periphery; therefore the corresponding potential typically includes diffraction oscillations in dependence on the distance. This confirms the simple recipes for construction of stationary multi-soliton configurations given above. Note, however, that this approximation does not describe slow motion and rotation of asymmetric solitonic structures connected with non-pairwise type of interaction of the individual DOSs (Sect. 4.5.3).

### 4.6.5 Dissipative Solitons and Spatial Hysteresis

The existence of DOSs drastically changes the kinetics of the local field perturbations in bistable interferometers, as compared with the case of one-component schemes (see Sect. 2.6). This is connected to the fact that although DOSs are, to a certain degree, analogous to the "critical nuclei" they are stable and have an excitation threshold.

A DOS's existence does not influence the kinetics of switching on, when the power of the holding radiation beam (e.g., Gaussian) increases from small values, because DOSs do not form at this stage. If the intensity of the beam is great enough, so that the width of the region switched to the upper state considerably exceeds the width of the switching wave front, then the fronts of spatial switching serving as the boundaries of this region will be far away from one another. For a slow temporary decrease in the maximum beam intensity $I_{\text{m}}$, these fronts will approach and form a "positive" DOS. The DOS will remain during the further decrease in $I_{\text{m}}$ up to the boundary of DOS existence $I_1$. Therefore switching down will occur not at $I_{\text{m}} = I_0$, as was the case in the one-component distributed bistable systems, but at $I_{\text{m}} = I_1 < I_0$.

Let us note one more circumstance in the kinetics of switching down. It can be recalled that the maximum intensity in the centre of a "positive" DOS exceeds the intensity of the transversely homogeneous state corresponding to the upper branch of the transfer function. Therefore in the process of switching down, a slow temporal decrease in the maximum intensity of the incident beam in the region $I_{\text{m}}(t) \approx I_0$ is accompanied by an increase in field axial intensity in the nonlinear interferometer, as shown by curve 1 in Fig. 4.15.

The possibility of artificial excitation of different types of DOSs leads to multivariance of spatial hysteresis in wide-aperture nonlinear interferometers with diffractive transverse coupling. A situation as shown in Fig. 4.28, where the transverse distribution of the characteristics of output radiation are determined only by the upper and lower branches of the transfer function, is justified only in the absence of DOSs. In the general case, in the region of the aperture where the local intensity of incident radiation falls within the range of DOS existence, $I_1 < I_{\text{in}}(r_\perp) < I_2$, the intensity profile of the output radiation can also include (or not include) one or several local perturbations of the DOS type, for both the lower and upper branches of the transfer function.

## 4.7 Switching Waves and Solitons in Conditions of Instability

Under the conditions discussed in Sect. 4.2, the stationary states of the non-linear interferometer can be unstable, irrespective of the field transverse structure, and will be replaced by different types of nonstationary regimes. For the McCall instability [223], the stationary regime of relaxation oscillations is periodic in time. Regimes of the multistable interferometers with competing nonlinearities were studied theoretically (in the framework of point and distributed models) and experimentally in [35, 36, 351, 130, 408, 266, 268].

For considerable longitudinal extension of the interferometer (the time delay $\tau_{del}$ exceeds the relaxation time $\tau_{rel}$) and a slow variation, e.g., of the intensity of incident radiation, Feigenbaum's scenario of transition to chaos at period-doubling bifurcations (Ikeda's instabilities [146]) is typical, as for the hybrid devices discussed in Chap. 3. Under conditions of bistability or multistability, when one of several stationary or nonstationary regimes can form at the fixed intensity of incident radiation, depending on initial conditions, switching waves between not only stationary but also nonstationary states are possible for transversely distributed systems. Similarly, switching waves can exist between states subjected to instabilities of the transverse field structure (see Sect. 4.3). Thus we come to the generalization of the notion of switching waves [332, 320], which, in turn, enlarges the variety of manifestations of the phenomenon of spatial hysteresis.

### 4.7.1 Plane-Wave Excitation

Now we will consider a ring interferometer with a nonlinearity corresponding to the two-level scheme (4.1.35) and a substantial frequency detuning $w_0$. Incident radiation is a plane wave with intensity $I_{in}$ propagating along the interferometer axis (the angle of incidence $\theta = 0$). An example of the transfer function with Ikeda instability of the stationary states on the portion of the upper branch is shown in Fig. 4.32a. With an increase in intensity $I_{in}$, the stationary regime becomes unstable and is replaced, in the form of consecutive bifurcations, by periodic regimes with periods $T = 2\tau_{del}, 4\tau_{del}, \ldots$ and finally by a stochastic regime.

If in the initial moment we form the field close to the stationary state for the lower branch of the transfer function on one half of the interferometer aperture ($x < 0$), and the field close to the nonstationary state for the same intensity $I_{in}$ on the second half ($x > 0$), then a wave of modulation (self-pulsation, Fig. 4.33a) or stochasticity (Fig. 4.34a) will form and will

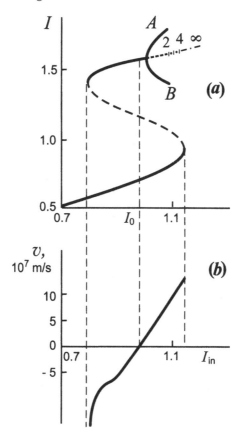

**Fig. 4.32.** Transfer function of a nonlinear interferometer under conditions of Ikeda's instability (**a**) and the dependence of velocity of generalized switching waves on the intensity of incident radiation (**b**) [332, 320]

propagate in the transverse direction. Modulation depth and the nature of the spectrum of the local field vary with position relative to the wave front. At considerable distances from the switching wave front ($x - vt \to \pm\infty$), these characteristics are close to those for transversely homogeneous regimes. Figure 4.33 presents an example of a wave of switching between a stationary regime and a four-periodic regime (the period $T = 4\tau_{\mathrm{del}}$). In the region occupied by the stationary regime, the spectrum is concentrated in the range of low frequencies ($f \ll f_4 = \pi/2\tau_{\mathrm{del}}$), and in the region of the four-periodic regime the contribution of the frequency $f_4$ corresponding to the period $T = 4\tau_{\mathrm{del}}$ is considerable. There is also the frequency peak $f_2 = \pi/\tau_{\mathrm{del}}$ representing the two-periodic regime localized near the front. Note that the characteristics of such "intermediate" regimes can change noticeably as compared with the case for transversely homogeneous regimes. This is connected with the fact that the intermediate regimes occupy only a limited section

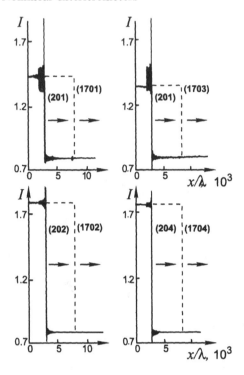

**Fig. 4.33.** Fronts of a modulation wave at different time moments (time $t/\tau_{\rm del}$ is indicated in brackets) [332, 320]

of the wave front. For spatial transition to the stationary state along the switching wave front, the modulation depth decreases, and the level of the noise component in the spectrum decreases for waves of stochasticity, or dynamical chaos (Fig. 4.34b). The velocity of switching waves of different types depends smoothly (without jumps in bifurcation points) on the intensity of the incident radiation (Fig. 4.32b).

A new type of the field structure – a standing pulsing distribution – can exist under the conditions of the existence of a periodic regime, e.g., $T = 2\tau_{\rm del}$ (bistability is not necessary in this case). There are two states for such a regime for the fixed intensity $I_{\rm in}$ ($A$ and $B$ in Fig. 4.32a, abruptly replacing each other in the time interval $\tau_{\rm del}$ ($\tau_{\rm rel} \ll \tau_{\rm del}$). A standing pulsing distribution can be excited by the formation in the initial time moment of a field in a state close to $A$ on some section of the aperture, and in a state close to $B$ on the other part of the aperture. As there is no preference between states $A$ and $B$ exchanging at places in a time interval $\tau_{\rm del}$, neither of these states can displace the other with time, and no noticeable displacement of the boundaries of the aperture sections occupied by the different states occurs.

Such pulsating distributions of the field can form with the collision of two waves of switching from a stationary state to a periodic state. Thereby

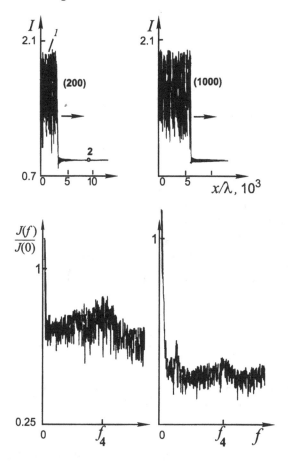

**Fig. 4.34.** Fronts of the wave of dynamical chaos (*top*) and the power spectrum (*bottom*) for front positions 1 and 2 [320]

the colliding waves of modulation (as well as the diffractive switching waves under conditions of DOS existence) are not necessarily annihilated with the field transition to the transversely homogeneous state on the entire aperture. Some section of the aperture can form where the field oscillates periodically (with a period multiple to $\tau_{\mathrm{del}}$) in antiphase with the field in the other part of the aperture. The typical minimum width of such a section is several light wavelengths. If the initial width of the section is less than the minimum width, then its boundaries diffuse, and the field distribution approaches with time a transversely homogeneous distribution. Also possible is the formation of a considerable number of sections with different phases of pulsations separated on the aperture [320].

When we take into consideration relaxation in the nonlinear medium, the following changes occur: For periodic pulsations, the period of the oscillations

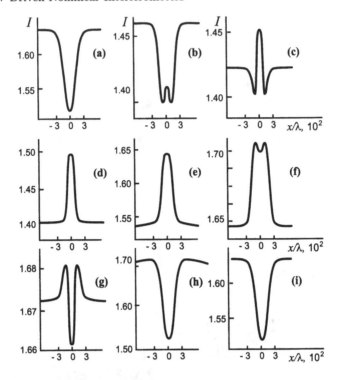

**Fig. 4.35.** Intensity profiles for pulsed field distributions in the time moments $t/\tau_{\mathrm{del}} = 0.125$ (**a**), 0.75 (**b**), 0.875 (**c**), 1 (**d**), 1.5 (**e**), 2 (**f**), 2.125 (**g**), 2.375 (**h**), 2.5 (**i**); $\tau_{\mathrm{rel}}/\tau_{\mathrm{del}} = 0.1$ [332]

increases by about $\tau_{\mathrm{del}}$ (as for hybrid schemes, see Sect. 3.2). Periodic pulsing distributions (with an increased period) retain their existence. An example is presented in Fig. 4.35. The stochastic regimes are especially sensitive to the nonlinearity relaxation time (see Chap. 3), so consecutive analysis of these structures presents a complex problem.

Similar phenomena take place for the different nonstationary (periodic and stochastic) regimes, including conditions of bistability absence. The features of these regimes allow us to observe the spatio-temporal development of self-pulsations and dynamic chaos.

### 4.7.2 Excitation with a Radiation Beam

In the case of incidence of a wide radiation beam on the nonlinear interferometer, separate sections of the interferometer aperture may correspond to different nonstationary regimes. Therefore, with slow temporal variation of the external beam power, transient processes in the form of waves of self-pulsations and stochasticity are possible. Different nonstationary regimes will spatially coexist in the settled state [343, 320].

Figure 4.36 presents settled (with periodic time variation) transverse profiles of radiation intensity in the interferometer; in the absence of the transverse coupling (approximation of independent light tubes), a two-periodic regime with period $T = 2\tau_{\text{del}}$ would form for the central part of the beam, and a stationary regime would be realized for its periphery. In the time moments $t = n\tau_{\text{del}}$, profile 1 is realized at odd $n$, and profile 2 corresponds to even $n$. It follows from calculations that spatial filtering smooths out the temporal oscillations. With a decrease in maximum intensity of the beam $I_{\text{m}}$, stationary regimes form, and with an increase in $I_{\text{m}}$ the period multiplicity increases. Figure 4.37 shows the intensity profiles for a regime with period $T = 4\tau_{\text{del}}$. A further increase in $I_{\text{m}}$ leads to the irregular temporal variation of intensity transverse profiles in the interferometer (Fig. 4.38), which can be considered stochastic because of a decrease in the autocorrelation function.

A consecutive description of the kinetics of the stochastic processes needs to take into account the finite time of relaxation of the nonlinear medium $\tau_{\text{del}}$. This causes fragmentation of radiation in the interferometer into pulses with

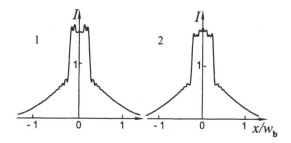

**Fig. 4.36.** Temporal variation of the radiation intensity profile; period $T = 2\tau_{\text{del}}$ [343]

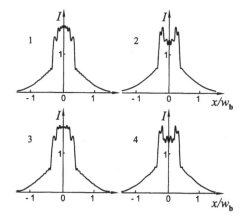

**Fig. 4.37.** Temporal variation of the radiation intensity profile; period $T = 4\tau_{\text{del}}$ [343]

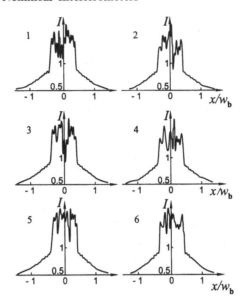

**Fig. 4.38.** Stochastic variation of the radiation intensity profile [343]

a typical duration (correlation time) of the order of $\tau_{\text{rel}}$ [191], as it was in the hybrid bistable systems (Chap. 3). The possibility of observing stochastic regimes and their features is substantially determined by spatial filtering of radiation in the interferometer. The spatio-temporal chaos described was obtained in calculations for both diffraction [343, 234] and diffusion [111] mechanisms of transverse coupling.

With a decrease in the interferometer transverse sizes to magnitudes comparable to the width of the switching wave front, the asymptotic representations developed above lose their applicability; the description by means of mode expansions close to the case of point systems becomes sufficient. For the interferometers with small Fresnel's numbers ($N_F < 100$) we should take into account the diffractive losses which effectively decrease (as compared with the case of unrestricted aperture) the coefficient of reflection of the mirrors, as well as the additional phase detuning $\delta'$, biasing the bistability region [344]. For limited power of the incident radiation and small $N_F$, bistability is absent because of considerable diffractive losses in the interferometer (phase shift $\delta'$ is not so important, as it can be compensated by the choice of the interferometer base). For somewhat greater $N_F$, the transverse modes still differ substantially by their losses. Therefore hysteresis of the total power (over the cross-section) is realized, whereas the field transverse distribution in the lower and upper states corresponds to the same mode (with the lowest losses). With a further increase in $N_F$, the effective number of the modes increases, and spatial hysteresis is properly observed.

### 4.7.3 Effect of Transverse Instability

If no special measures are taken, instabilities of the transverse structure are usually primarily developed in wide-aperture nonlinear interferometers (see Sect. 4.3). This fact hinders observation of Ikeda's instabilities and, correspondingly, of the waves of modulation and chaos. The presence of transverse instabilities modifies features of switching waves and DOSs [322, 89].

Let us assume, for instance, that transversely homogeneous states are stable for the lower branch of the transverse function and are modulationally unstable for the upper branch, i.e., bistability corresponds to homogeneous (lower branch) and inhomogeneous transversely modulated states (upper branch). Then a narrow positive DOS (in the ground state) still forms from a narrow initial local perturbation (Fig. 4.39a). Modulation instability (deduced for infinite plane waves) is inessential for this DOS, as the region in the field distribution occupied by the states adjoining the unstable upper branch is comparatively narrow. However, if the initial perturbation is sufficiently wide, it produces not a wider (excited) DOS, but perturbation decay into several DOSs in the ground state (Fig. 4.39b). The number of DOSs formed is proportional to the width of the initial perturbation. Note that the field in the structures presented in Fig. 4.39 tends in its peripheral part (at considerable distances from the central part of the positive DOS) to the stationary level corresponding to the stable lower branch. Naturally, the same results are valid for local perturbations which form negative DOSs, when the lower branch is modulationally unstable and the upper branch is stable.

If transversely homogeneous states corresponding to both the lower and upper branches are unstable, the development of a local perturbation is somewhat different. Even weak perturbations against the background of transversely homogeneous states increase, causing the replacement of such states by transversely modulated distributions with the period determined by the spatial frequency of perturbations with maximum growth rate. For the case

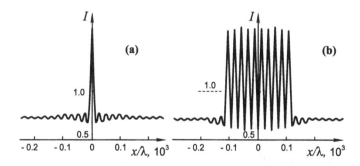

**Fig. 4.39.** Steady-state intensity profiles under conditions of a stable lower branch and a modulationally unstable upper branch of the transfer function; the width of the initial local perturbation $w_{\text{init}} = 24\,\lambda$ (a) and $160\,\lambda$ (b) [322, 89]

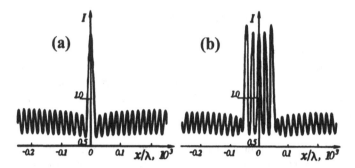

**Fig. 4.40.** Steady-state intensity profiles under conditions of modulation instability of the lower and upper branches of the transfer function for narrow (a) and wide (b) initial local perturbations [322, 89]

of a large local perturbation, modulation growth is accelerated in the vicinity of its fronts. This resembles propagation of the modulation wave from the centre of the perturbation to the periphery; however, the front of such a wave gradually diffuses. Under conditions of instability of the upper and lower branches, the DOS in the ground state forms from a narrow initial local perturbation with sufficiently high amplitude, and a wide perturbation decays into several DOSs (Fig. 4.40). In this case, in contrast to the case of the stable lower branch (Fig. 4.39), a transversely modulated state corresponding to instability saturation forms in the peripheral regions. The structures resemble "static stratification" (see [395]).

Instabilities of the transverse structure in the bistability range can affect not the entire upper branch, but only part of it. Then "waves of spatial modulation" convert smoothly into standard waves of switching between stable homogeneous states. The waves of spatial modulation exhibit themselves in the character of spatial hysteresis for interferometer excitation by a wide radiation beam. In particular, for constant parameters of the system and external radiation, the field profile can either be smooth, determined by the lower branch of the transfer function, or contain considerable spatial modulation in the central part of the beam, depending on initial conditions. The kinetics of hysteresis variations includes the stage of propagating switching waves in this case, too.

## 4.8 Effect of Componency

Taking into account radiation polarization, or the vector nature of the electric field, means an increase in the number of degrees of freedom, or the scheme componency. Similarly, under certain phase-matching conditions, several beams with essentially different carrier frequencies and/or wavevectors can propagate inside nonlinear interferometers. In these cases, the governing equation takes the form of a set of coupled equations for a number of scalar components of the field envelope. The greater the componency (a number of degrees of freedom, or a number of these equations), the richer the variety of spatio-temporal patterns in the corresponding nonlinear interferometer. In the present section we will study these patterns, citing the example of a magneto-optic interferometer with the Kerr nonlinearity [342].

### 4.8.1 Model of a Magneto-optic Interferometer

The system we consider is an interferometer driven by external coherent hold-ing radiation of arbitrary polarization. The interferometer is filled with media with the Kerr nonlinearity and a linear magneto-optic (Faraday) effect.[4] The basic equations governing propagation of polarized monochromatic radiation through a planar magneto-optic waveguide are [48, 49]

$$
i\frac{\partial E_1}{\partial z} + \frac{i}{v_{\rm gr}}\frac{\partial E_1}{\partial t} + \frac{\partial^2 E_1}{\partial x^2} + 2\alpha(|E_1|^2 + \mu|E_2|^2)E_1 - QE_1 = 0 \,,
$$

$$
i\frac{\partial E_2}{\partial z} + \frac{i}{v_{\rm gr}}\frac{\partial E_2}{\partial t} + \frac{\partial^2 E_2}{\partial x^2} + 2\alpha(|E_2|^2 + \mu|E_1|^2)E_2 + QE_2 = 0 \,. \quad (4.8.1)
$$

Here $E_{1,2}$ are two circularly polarized components of the electric field en-velope, while the electric field longitudinal component is negligibly small. The longitudinal coordinate $z$ is normalized to the diffraction length; the transverse coordinate $x$ is normalized to the typical width of the waveguide soliton. In the other transverse direction $y$ a single-mode regime is realized, and (4.8.1) are obtained by averaging the initial paraxial equation over $y$. The value $Q$ is an effective magneto-optic parameter proportional to the ex-ternal static magnetic field. The parameter $\alpha = 1$ or $-1$ for self-focusing or self-defocusing media. The parameter $\mu$ is determined by the nonlinearity mechanism, and we will use in the following calculations the value $\mu = 2$ (the electronic mechanism). A special case of coincident propagation constants for the field $x$- and $y$-components was chosen (in the notation of [48] $\nu = 0$). Frequency dispersion is neglected. By scaling of time ($t \to tv_{\rm gr}$) we set the group velocity $v_{\rm gr} = 1$.

We will consider a ring interferometer with the axis along $z$, filled in with such a waveguide and driven by a coherent external radiation with amplitudes $E_{\rm in1}$ and $E_{\rm in2}$ for the two polarization states. Our next step is to use the mean-field approximation, or to average (4.8.1) over the longitudinal coordinate $z$ (Sect. 4.1). It is justified if all linear and nonlinear field distortions during a single round-trip over the cavity are small. It means that the cavity length is small compared with the diffraction length, losses are small because of

---

[4] Polarized patterns in the case of an isotropic Kerr medium were considered in [144].

high reflectivity of the cavity mirrors ($|1 - R| \ll 1$, where $R$ is a product of the amplitude coefficients of reflection of the mirrors), phase detunings $\Delta_{1,2}$ for the two polarizations are small, and nonlinear phase incursion over the medium length is small compared with unity. Such averaging, with the boundary conditions taken into account, gives the following equations:

$$\frac{\partial E_1}{\partial t} - i\frac{\partial^2 E_1}{\partial x^2} - 2i\alpha(|E_1|^2 + \mu|E_2|^2)E_1 + (i + \zeta_1)E_1 = E_{in1} ,$$

$$\frac{\partial E_2}{\partial t} - i\frac{\partial^2 E_2}{\partial x^2} - 2i\alpha(|E_2|^2 + \mu|E_1|^2)E_2 + (-i + \zeta_2)E_2 = E_{in2} . \quad (4.8.2)$$

Here $\zeta_{1,2} = [1 - R\exp(i\Delta_{1,2})]/L$, where $L$ is the medium length, amplitudes are renormalized $[E_{in1,2} \to E_{in1,2}/(LQ), E_{1,2} \to E_{1,2}/\sqrt{Q}]$, time $Qt \to t$, and coordinate $x\sqrt{Q} \to x$. Next, it is assumed that frequencies of the external radiation for the two polarizations coincide, then $\Delta_1 = \Delta_2$ and $\zeta_1 = \zeta_2 = \zeta = \zeta_r + i\zeta_i$ (a complex value). Note that $\zeta_r$ is proportional to losses, $\zeta_r \propto (1 - R) > 0$. If transverse variation of the field envelope is not essential (e.g., for a single-mode waveguide), we can average (4.8.2) over the $x$-direction. Then we can consider a lumped (point) system:

$$\frac{dE_1}{dt} - 2i\alpha(|E_1|^2 + \mu|E_2|^2)E_1 + (i + \zeta)E_1 = E_{in1} ,$$

$$\frac{dE_2}{dt} - 2i\alpha(|E_2|^2 + \mu|E_1|^2)E_2 + (-i + \zeta)E_2 = E_{in2} . \quad (4.8.3)$$

### 4.8.2 Homogeneous Stationary and Nonstationary Regimes

**Stationary Homogeneous States.** Introducing intensities $I_n = |E_n|^2$ ($n = 1, 2$), we obtain the following from (4.8.3) for such states:

$$\{[2\alpha(I_1 + \mu I_2) - 1 - \zeta_i]^2 + \zeta_r^2\}I_1 = I_{in1} ,$$

$$\{[2\alpha(I_2 + \mu I_1) + 1 - \zeta_i]^2 + \zeta_r^2\}I_2 = I_{in2} . \quad (4.8.4)$$

This is a system of two coupled cubic equations in $I_1$ and $I_2$. Therefore, the number of solutions for fixed scheme parameters and fixed intensities of external radiation may be up to nine. For small intensities ($I_{1,2} \ll 1, |\zeta|$) there is only one solution with intensity linear dependencies:

$$I_1 = \frac{I_{in1}}{(1 + \zeta_i)^2 + \zeta_r^2} , \quad I_2 = \frac{I_{in2}}{(1 - \zeta_i)^2 + \zeta_r^2} . \quad (4.8.5)$$

Further we assume $I_{in1} = I_{in2} = I_{in}$ and $I_{in}$ to be a control parameter. For very large intensities in the case $-1 < \mu < 3$, there is also only one solution of (4.8.4):

$$I_1 \approx I_2 \approx \left(\frac{I_{in}}{4(1 + \mu)^2}\right)^{1/3} . \quad (4.8.6)$$

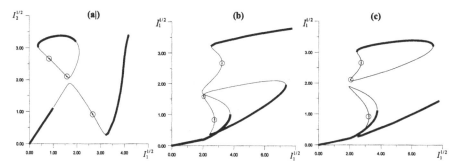

**Fig. 4.41a–c.** Relations between intensities in a magneto-optic interferometer; *thick lines* and interiors of the *circles* correspond to stable regimes, and *thin lines* represent unstable solutions

Instead of directly solving the algebraic equation of the ninth degree, let us combine the two equations (4.8.4) to exclude $I_{\text{in}}$:

$$\{[2\alpha(I_2+\mu I_1)+1-\zeta_i]^2+\zeta_r^2\}I_2-\{[2\alpha(I_1+\mu I_2)-1-\zeta_i]^2+\zeta_r^2\}I_1 = 0 . \quad (4.8.7)$$

It is easy to find the relation between $I_1$ and $I_2$ from this cubic equation. The result is presented in Fig. 4.41a for $\zeta_r = 2$, $\zeta_i = 20$, and $\alpha = 1$. Note that dependence in Fig. 4.41a consists of an "infinite" curve [for large intensities it approaches the straight line $I_1 = I_2$ according to (4.8.6)] and an isolated loop. For fixed $I_1$ and $I_2$ we find only one value of $I_{\text{in}}$ using (4.8.4). The corresponding relations between $I_1$, $I_2$ and $I_{\text{in}}$ are given in Fig. 4.41b,c. One can see that, depending on $I_{\text{in}}$, there are one to nine branches of dependence of $I_{1,2}$ on $I_{\text{in}}$ representing stationary states. Their stability is discussed below.

**Nonstationary Regimes.** Nonstationary regimes of a lumped (point) system are described by (4.8.3). To examine the linear stability of a stationary solution (index 0), let us introduce a small perturbation from it:

$$E_n(t) = E_{n0}[1+\delta E_n(t)] , \quad |\delta E_n|^2 \ll 1 , \quad |E_{n0}|^2 = I_n , \quad n = 1, 2 . \quad (4.8.8)$$

By the standard procedure we obtain for $\delta E_n$ a set of linear differential equations with constant coefficients. Its solution has the form

$$\delta E_n = u_n \exp(\gamma t) + v_n^* \exp(\gamma^* t) . \quad (4.8.9)$$

For $u_n$ and $v_n$ we have a system of four linear algebraic equations, and the requirement that the determinant of the system $D = 0$ gives us an equation for the perturbation growth rate $\gamma$. Then we obtain a biquadratic equation in $G = \gamma + \zeta_r$:

$$G^4 + aG^2 + b = 0 , \quad (4.8.10)$$

$$a = A_1^2 + A_2^2 - J_1^2 - J_2^2 , \quad J_n = 2\alpha I_n ,$$
$$b = A_1^2 A_2^2 - A_1^2 J_2^2 - A_2^2 J_1^2 - 4\mu^2 A_1 A_2 J_1 J_2$$
$$+4\mu^2 A_1 J_1 J_2^2 + 4\mu^2 A_2 J_2 J_1^2 + (1 - 4\mu^2) J_1^2 J_2^2 ,$$
$$A_1 = 2J_1 + \mu J_2 - 1 - \zeta_i , \quad A_2 = 2J_2 + \mu J_1 + 1 - \zeta_i . \qquad (4.8.11)$$

A stationary regime is stable if all four roots $\gamma$ have a non-negative real part, and it is unstable if there is a root with a positive real part, $\mathrm{Re}\,\gamma > 0$. For small intensities, the real parts of all four roots coincide: $\mathrm{Re}\,\gamma_m = -\zeta_r < 0$; thus the regime is stable. Results for the stability of different regimes are represented in Fig. 4.41. It can be seen that there are cases of mono-, bi- and multistability.

Closer examination shows that there are two variants of stability loss. For the first variant, we have a real root $\gamma = 0$ at the boundary of stability, and it changes its sign when we cross the boundary. This is a saddle-node bifurcation corresponding to the edges of branches presented in Fig. 4.41b,c. More interesting is the second variant when there is a pair of complex-conjugated roots with a nonzero imaginary part, and the real part changes its sign crossing the boundary. This is the Andronov–Hopf bifurcation, when creation of a stable periodic regime in the vicinity of an unstable stationary regime is possible. An example is presented in Fig. 4.42. Note that we have here a generalized bistability (coexistence of stationary and periodic regimes).

For multistability, contrary to bistability, the kinetics of intensities $I_{1,2}$ with a slow change of intensity of external radiation $I_{in}$ is not evident. The answer to this question can be found by direct solution of (4.8.3) for a pulse of external radiation with the time-dependent amplitude $E_{in}(t)$. One can see from Fig. 4.43 that, for a slowly varying amplitude of external radiation, the dependence $I_{1,2}(I_{in})$ follows the lower branch up to its end, and then jumps to the branch going to infinity for infinite $I_{in}$. Thus, in such a way we cannot

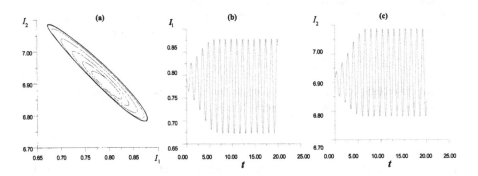

**Fig. 4.42.** Transient to periodic regime for the Andronov–Hopf bifurcation for driven cw radiation with intensity $I_{in} = 56.16$: (a) phase plane $I_1$, $I_2$; (b) temporal dependence of intensity $I_1(t)$; (c) temporal dependence of intensity $I_2(t)$

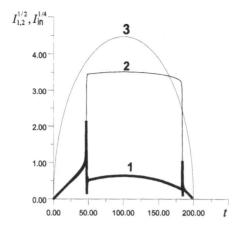

**Fig. 4.43.** Dynamics of intensities $I_1^{1/2}$ (*thick line* 1) and $I_2^{1/2}$ (*thin line* 2) for pulse of driven radiation $I_{in}^{1/4}$ (*thin line* 3)

obtain regimes corresponding to the loop, which needs special conditions of initiation.

### 4.8.3 Modulation Instability

Now we consider a transversely distributed system, which is described by (4.8.2). Let external radiation be transversely independent (a plane wave). Then there are transversely independent solutions, whose characteristics are given by (4.8.3). However, even if these regimes are stable for the point scheme, they can be unstable for the distributed system due to modulation instability (see Sect. 4.3 and [48]). The analysis is similar to those given in the previous section with the following addition: perturbations have a transverse dependence of the form $\exp(\pm i\kappa x)$, where $\kappa$ is the perturbation transverse spatial frequency. The perturbation growth rate $\gamma$ is found from the same equations (4.8.10-4.8.11) with the only replacement being $A_{1,2} \to A_{1,2} - \kappa^2$. Calculations show that for conditions of Fig. 4.41, only the lower branch remains stable for these perturbations within the entire range of its existence ($0 < |E_{in}| < 14.148$), while all other branches, though stable for the point model, suffer modulation instability. In Fig. 4.44 we present a typical dependence of $\gamma$ on the square of the spatial frequency $\kappa$. Perturbations grow only in a certain finite range of these frequencies: $\kappa_{min} < \kappa < \kappa_{max}$, $\kappa_{min} > 0$. Therefore, it is possible to suppress modulation instability by spatial filtering inside the cavity (see Sect. 4.3). Direct simulations show that, as a result of modulation instability, complex (chaotic) spatiotemporal structures arise.

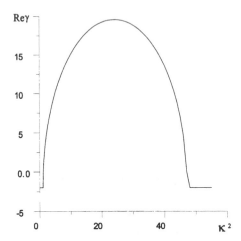

**Fig. 4.44.** Dependence of perturbation growth rate $\mathrm{Re}\,\gamma$ on the square of the spatial frequency $\kappa$ for homogeneous field distribution with amplitudes $|E_1| = 3.29$ and $|E_2| = 0.3$

### 4.8.4 Polarized Dissipative Optical Solitons

As was demonstrated in Sect. 4.7, DOSs exist even in the absence of classical bistability, when there are no two stable homogeneous distributions of the field for the fixed parameters of the nonlinear interferometer. Therefore, narrowness of ranges of the control parameter – the holding radiation intensity $I_{\mathrm{in}}$ – where the homogeneous distributions are stable, does not preclude formation of DOSs, at least in the ground state.

Let us consider the formation of polarized DOSs from initial field distributions corresponding to a stable lower branch at the periphery ($|x| > w$) and one of the other (modulationally unstable) branches with a small perturbation in the central part ($|x| < w$), where $w$ is the typical width of the perturbation. For formation of a single DOS, the width $w$ must not be too large; otherwise a number of coupled DOSs arises (see Sect. 4.7). It follows from calculations that the range of the DOS existence is surprisingly wide: $5.8 < |E_{\mathrm{in}}| < 12.2$. It is close to the entire range of bi- and multistability for the point scheme, with a small total shift to the side of smaller intensities of the holding radiation.

Calculations show the existence of two types of polarized DOSs, shown for $|E_{\mathrm{in}}| = 8$ in Fig. 4.45. Here transverse profiles of intensities $I_{1,2}$ and of real and imaginary parts of the envelopes $E_{1,2}$ for the steady-state DOSs of types I and II are given. Intensity oscillations in the vicinity of the lower branch correspond to local destructive interference connected with complex phase profiles for the field components $E_{1,2}$; there are no such oscillations for $\mathrm{Re}\,E_{1,2}$ and $\mathrm{Im}\,E_{1,2}$. The transverse profile shapes indicate that the DOSs are fundamental, without intensity oscillations in the vicinity of the maximum

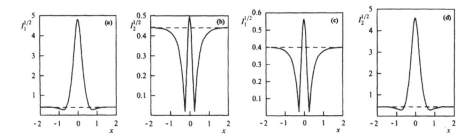

**Fig. 4.45.** Transverse profiles of intensities for steady-state polarized dissipative solitons of type I (**a,b**) and II (**c,d**)

intensity, as in the case of scalar DOSs under the conditions of modulation instability (Sect. 4.7). The presence of the two types of the fundamental DOSs is a manifestation of the polarization degree of freedom, which enriches substantially the variety of single and coupled DOSs. Note also that the abnormally wide range of the polarized DOSs existence is an important factor that is promising for applications in optical information processing (see Appendix C).

### 4.8.5 Related Schemes

**Two-Beam Interferometers.** In these schemes, a nonlinear interferometer is driven by two beams of holding radiation with different incidence angles: $\theta_1 > 0$ and $\theta_2 < 0$, $\theta_{1,2}^2 \ll 1$ [392, 190, 222, 121, 164, 261, 183]. In these schemes, the critical dependence on the angle of incidence for fast nonlinearity can be reduced, because optical transverse coupling becomes bidirectional.

Spatial modulation of the total field does not produce new types of instabilities or fairly high spatial frequencies corresponding to beams with combination angles $m\theta_1 + n\theta_2$ ($m, n$ are integer) if the difference between the incidence angles exceeds the value $\sim \sqrt{|\delta\varepsilon|}$. Then, for a symmetric incidence of the two beams ($\theta_1 = -\theta_2$) with coinciding intensity profiles [$I_{\mathrm{in}1}(x) = I_{\mathrm{in}2}(x)$], we have in fact the same equations as in the case of a single beam. However, now, due to the equivalence of directions $x$ and $-x$, Maxwell's intensity value will exist, for which the switching wave is motionless. Correspondingly, spatial bistability and spatial hysteresis will take place for wide radiation beams.

Close to this scheme and to the scheme of a nonlinear layer [310] are planar nonlinear optical waveguides; the differences relate mainly to the method of input of external radiation. In the literature the eigenmodes of such waveguides with a nonlinear refraction index were studied in an approximation of the paraxial equation (A.21), and a conclusion was made as to the bistability of the scheme. The argument in favour of bistability, e.g., in [53], was the existence in the waveguide of two nonlinear eigenmodes with different constants of propagation and different transverse profiles at the same total power of radiation. Similar considerations were also expressed as applied to the regimes of self-trapping in a continuous medium with a sharp intensity dependence of the nonlinear refraction index ("bistable solitons") [161, 83].

As was already explained in Chap. 1, in the terminology we accepted, absolute bistability of such schemes is possible only in the presence of sufficiently strong feedback. Besides, instability of radiation propagation in the nonlinear medium with respect to field modulation in direction $y$ is essential. Absolute bistability could appear in the presence of feedback mechanisms, e.g., of diffusion type [thermo-optical nonlinearity, $\delta\varepsilon = \delta\varepsilon(T)$]. To prove absolute bistability, it is necessary to illustrate not only the existence, but also the sufficiency of the feedback mechanism for system switching. Convective bistability does not need any mechanism of spatial feedback; however, in this case it is necessary to prove the existence of several stable attractors which serve final states of initial beams with different transverse profiles for large longitudinal distances ($z \to \infty$).

**Two-Frequency Interferometers.** When the medium inside an interferometer is anisotropic and offers optical nonlinearity of the second order in the electric field, then, under certain phase-matching conditions, effective generation of the second harmonics occurs, and radiation presents a superposition of fields with the fundamental optical frequency $\omega$ and its second harmonic with frequency $2\omega$, even if the holding radiation is monochromatic (with the fundamental frequency $\omega$).

In the simplest variant of the mean-field approximation, the governing equations for complex dimensionless amplitudes for the fundamental frequency $E_1$ and for the second harmonic $E_2$ are [383]

$$i\frac{\partial E_1}{\partial t} + \Delta_\perp E_1 + (i\gamma_1 + \delta_1)E_1 + E_1^* E_2 = E_{\text{in}} ,$$

$$i\frac{\partial E_2}{\partial t} + \alpha\Delta_\perp E_1 + (i\gamma_2 + \delta_2)E_2 + E_1^2 = 0 . \tag{4.8.12}$$

Here $\Delta_\perp$ is, as before, the transverse Laplacian, $\gamma_{1,2}$ are radiation decay rates in the interferometer at frequencies $\omega$ and $2\omega$, respectively, $\delta_{1,2}$ are interferometer detunings with respect to the frequency of the holding radiation with the amplitude $E_{\text{in}}$. The parameter $\alpha$ is determined by the ratio of refractive indices at $\omega$ and $2\omega$, and for the most realistic case $\alpha = 1/2$. It is in this case that the set of equations (4.8.12) has the Galilean transform (see Sect. 4.1). Equation (4.8.12) also has translation symmetry for any parameters.

One- and two-dimensional DOSs presenting coupled states of light with two essentially different frequencies were predicted and investigated theoretically for such interferometers in [85, 259, 201, 199, 383, 90]. Similar three-frequency structures in optical parametric oscillators were studied, also theoretically, in [200, 376, 389, 182, 252, 192, 385, 386, 361].

**Nonlinear Layer with a Feedback Mirror.** If the longitudinal extension of nonlinear medium $d$ and of interferometer $L$ is small as compared with typical lengths of diffraction, dispersion, and nonlinear distortions (see Sect. 4.1 and Appendix A), then averaging of the field in the longitudinal direction $z$ and use of the model of purely transverse distributivity (mean-field approximation) is possible. In this model, there are no essential distinctions between the schemes with unidirectional (ring interferometers) and bidirectional (two-mirror, or Fabry–Perot interferometers) propagation of radiation.

Beyond the framework of the model of transverse distributivity, the following new factors influence the features of the stationary regimes. First, a standing wave of radiation induces a distributed mirror in the nonlinear medium which serves as an additional feedback. Under ordinary conditions the effective coefficient of reflection from this distributed mirror is noticeably smaller than the reflection coefficient of the interferometer mirrors $R$: $|\delta\varepsilon|d/\lambda \ll R$, where $\delta\varepsilon$ is the nonlinear part of the medium refractive index and $\lambda$ is the light wavelength. Besides, inhomogeneity of the medium and, correspondingly, the reflection coefficient of the distributed mirror decrease considerably in the presence of the diffusion processes, with diffusion length exceeding the light wavelength. The distributed reflection can be neglected under such conditions.

Second, in the schemes with counterpropagating beams, the redistribution of intensity over the beam section can be substantial. This is the factor responsible for the mechanism of "transverse optical bistability" mentioned in Sect. 1.4. Under certain conditions, this effect can show itself also in the Fabry–Perot interferometers, leading to specific hysteresis phenomena. For sufficiently wide beams and, correspondingly, small intensity gradients, these effects disappear. An additional distinction is the dependence of the kinetics of Fabry–Perot interferometers on the position of the nonlinear layer [63]. At the same time, under actual conditions the differences of hysteresis phenomena in ring and Fabry–Perot interferometers are rather of quantitative than qualitative nature.

Optical schemes consisting of a nonlinear layer and a single feedback mirror (Fig. 4.46) were proposed by Firth [94]. Depending on the type of the medium nonlinearity, they are monostable [94] or bistable [354] and allow development of modulation instability and formation of DOSs. We will demonstrate this following [334, 304].

In the scheme shown in Fig. 4.46a, radiation with a complex amplitude $E_{in}$ is incident on a layer of nonlinear medium with amplitude (complex) transmission coefficient $K$; it passes through a linear spacing and reflects backward from a feedback mirror M, with amplitude reflection coefficient $r$, which is placed at a distance $L/2$ from the layer. The layer is assumed to be thin: its thickness $d$ is much smaller than the diffraction length $l_{dfr} = w^2/\lambda$, where $\lambda$ is the light wavelength and $w$ is the typical scale of the field transverse variation. On the other hand, the width of the layer is much greater than the

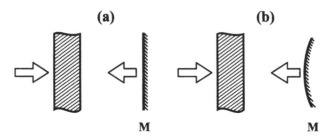

**Fig. 4.46a,b.** Schemes with counterpropagating beams: the layer of nonlinear medium is *hatched*

light wavelength $(d \gg \lambda)$, so that the interference of direct and reflected waves is negligible. Then, for fast nonlinearity, the transmission coefficient of the layer is determined by the sum of instantaneous intensities of the incident wave and the reflected wave with the amplitude $E$: $K(t) = K(|E_{\rm in}|^2 + |E(t)|^2)$.

Let us analyze the field dynamics at discrete moments in time $t_n = nL/c$, $n = 1, 2, 3 \ldots$, separated by the round-trip time $L/c$. At the moment $t_n$, the amplitude of the field passing through the layer in the forward direction (the direction of propagation of the external radiation) is given by $F_n = K(|E_{\rm in}|^2 + |E_n|^2)E_{\rm in}$. The field amplitude of the backward wave at the layer output is $E_n = r\hat{D}F_{n-1}$, where $\hat{D} = \exp\left(i\frac{L}{2k}\Delta_\perp\right)$ is the operator of diffraction field transformation (see Appendix A), $k = 2\pi/\lambda$ is the wavenumber, $\Delta_\perp = \partial^2/\partial x^2 + \partial^2/\partial y^2$ is the transverse Laplacian, and $x$ and $y$ are the transverse coordinates. Then the equation of discrete dynamics of the field transverse structure is

$$\hat{D}^{-1}E_{n+1} = rK(|E_{\rm in}|^2 + |E_n|^2)E_{\rm in} , \qquad (4.8.13)$$

where $\hat{D}^{-1} = \exp\left(-i\frac{L}{2k}\Delta_\perp\right)$ is the inverse diffraction operator.

Now assume, first, that the length of the linear medium $L/2$ is smaller that the diffraction length $l_{\rm dfr}$, so that the approximation $\hat{D}^{-1} \approx 1 - i\frac{L}{2k}\Delta_\perp$ is justified. Second, the field variation during one round-trip is assumed to be small, which makes it possible to approximate the discrete dynamics by continuous dynamics: $E_{n+1} = E(t + L/c) \approx E(t) + (L/c)(\partial E/\partial t)$. As a result, we obtain the governing equation:

$$\frac{L}{c}\frac{\partial E}{\partial t} - i\frac{L}{2k}\Delta_\perp E = -E + rK(|E_{\rm in}|^2 + |E|^2)E_{\rm in} . \qquad (4.8.14)$$

Equation (4.8.14) is close in form to the equation of the model of purely transverse distributivity (mean-field approximation) (4.1.23). Both these equations have the same symmetry properties (for $E_{\rm in} = \text{const.}$), including the translation invariance and the "Galilean transformation" (see Sect. 4.1), i.e., if $E(x, y, t)$ is a solution of (4.8.14) for the external field amplitude $E_{\rm in}$, then for $E_{\rm in} = E_{\rm in}\exp(ik\theta x)\exp(-i\nu t)$ there is a solution $E = E(x - vt, y, t)\exp(ik\theta x)\exp(-i\nu t)$, where $v = c\theta$ and $\nu = ck\theta^2/2$.

Note that the description of stationary homogeneous regimes (monochromatic plane waves with normal incidence) is the same in the discrete (4.8.13) and continuous (4.8.14) versions. In this case, the intensity $I = |E|^2$ is described by

$$I = RK_i(I + I_{in})I_{in} ,\qquad (4.8.15)$$

where $I_{in} = |E_{in}|^2$ is the intensity of external radiation, $R = |r|^2$ is the intensity reflection coefficient of the mirror, and $K_i = |K|^2$ is the intensity transmission coefficient of the layer. Relation (4.8.15) is conveniently rewritten in the form

$$\frac{1}{R}\frac{I}{I_{in}} = K_i\left[I_{in}\left(1 + \frac{I}{I_{in}}\right)\right] ,\qquad (4.8.16)$$

or

$$I_{in} = \frac{1}{1 + \frac{I}{I_{in}}}K_i^{-1}\left(\frac{1}{R}\frac{I}{I_{in}}\right) .\qquad (4.8.17)$$

Here, $K_i^{-1}$ is the inverse function for $K_i(I)$. Varying the ratio $I/I_{in}$ in the range $0–R$, we uniquely find, by means of (4.8.17), $I_{in}$ and, therefore, $I$ as well. The resulting dependence $I(I_{in})$ can be ambiguous, which corresponds to the possibility for bistability or multistability.

The closeness of (4.8.14) to the equation of the mean-field approximation (4.1.23) for driven nonlinear interferometers leads to a similar inference about the presence and features of optical structures in the scheme under consideration. For transparent media $K_i = 1$, and in the plane-wave approximation, the scheme is monostable. However, even in such schemes development of modulation instability and formation of corresponding "infinite" (in contrast to localized) structures are possible [94]. Let us present a simple demonstration of the modulation instability in the case of a thin layer with an instantaneous Kerr nonlinearity following [334].[5] We will find conditions of development of transverse instabilities in the scheme Fig. 4.46a (without beam magnification) and demonstrate that these instabilities are suppressed in the scheme Fig. 4.46b (with magnification).

The linear analysis of stability is close to the cases given in Sect. 4.3, but is more simple because the phase relations are inessential here. For the scheme without beam magnification (Fig. 4.46a), the unperturbed field consists of two plane waves: the incident wave with intensity $I_{in} = |E_{in}|^2$ and the backward wave with intensity $I_b = RI_{in}$. Note that the unperturbed field is unambiguously determined by the intensity of incident radiation.

As before, it is convenient to expand small perturbation into the spectrum of plane waves. Therefore, it is sufficient to consider the modulation of the intensity of the counterpropagating waves within the thin layer of the medium with the only spatial frequency $q$:

---

[5] See [71, 256, 405] and references therein for more detailed consideration, taking into account the nonlinearity finite relaxation time, inhomogeneous broadening, standing wave effects, etc.

$$I_b = RI_{in}[1 + a_n \cos(\mathbf{q}\mathbf{r}_\perp)] , \quad a_n^2 \ll 1 , \quad 0 < z < d . \qquad (4.8.18)$$

The radiation propagating in the forward direction acquires, after passing through the nonlinear layer, phase modulation:

$$E_{z=d}^{(n)} = E_{in} \exp\left(i\frac{kd\varepsilon_2}{2\varepsilon_0}(I_{in} + 2I_b^{(n)})\right)$$

$$= E_{in} \exp(i\varphi_0) \exp\left(i\frac{I_{in}}{2I_{thr}} a_n \cos(\mathbf{q}\mathbf{r}_\perp)\right)$$

$$\approx E_{in} \exp(i\varphi_0)\left(1 + i\frac{I_{in}}{2I_{thr}} a_n \cos(\mathbf{q}\mathbf{r}_\perp)\right) , \qquad (4.8.19)$$

$$\varphi_0 = kd(1 + 2R)\frac{\varepsilon_2 I_{in}}{2\varepsilon_0} , \quad I_{thr} = \frac{\varepsilon_0}{2kdR\varepsilon_2} . \qquad (4.8.20)$$

Phase modulation is partially transformed into amplitude modulation when radiation diffracts in the free interval between the layer and the mirror. Therefore we have for the field of the backward wave in the region of the thin layer

$$E_b^{(n+1)} \approx R^{1/2} E_{in} \exp(i\varphi_1)$$

$$\times \left[1 + i\frac{I_{in}}{2I_{thr}} \exp\left(-i\frac{q^2 L}{k}\right) a_n \cos(\mathbf{q}\mathbf{r}_\perp)\right] ,$$

$$I_b^{(n+1)} = RI_{in}[1 + a_{n+1}\cos(\mathbf{q}\mathbf{r}_\perp)] , \qquad (4.8.21)$$

$$a_{n+1} = a_n \frac{I_{in}}{I_{thr}} \sin\frac{q^2 L}{k} . \qquad (4.8.22)$$

It follows from (4.8.22) that for $I_{in} < I_{thr}$ the modulation depth decreases with time for any spatial frequency $q$. This means stability of the plane-wave regime. But if $I_{in} > I_{thr}$, then there are zones of instability, with an increase in the modulation depth of perturbations, with the spatial frequencies satisfying the relation:

$$\left|\sin\frac{q^2 L}{k}\right| > \frac{I_{in}}{I_{thr}} . \qquad (4.8.23)$$

We next consider the scheme with beam magnification, e.g., with a convex mirror (Fig. 4.46b). The intensity of the unperturbed backward wave decreases in the region of the thin medium layer: $I_b = \eta R/I_{in}$, where $\eta = 1/M$ is the coefficient of intensity attenuation of this wave due to magnification and $M$ is the magnification coefficient. The spatial frequency of perturbation changes with the number of light round-trips $n$:

$$\mathbf{q}_n = \eta^n \mathbf{q}_0 . \qquad (4.8.24)$$

Therefore instead of (4.8.21) and (4.8.22) we obtain

$$I_b^{(n)} = \eta RI_{in}[1 + a_n \cos(\mathbf{q}\mathbf{r}_\perp)] , \qquad (4.8.25)$$

$$a_{n+1} = a_n \frac{I_{in}}{I_{thr}} \sin\frac{q^2 L}{k} , \quad I_{thr} = \frac{\varepsilon_0}{2kd\eta R\varepsilon_2} . \qquad (4.8.26)$$

Now the value $I_{thr}$ does not have the meaning of a threshold value of intensity. Actually, for magnification $(\eta < 1)$ the modulation depth decreases for any intensity at large times $(n \to \infty)$: $a_{n+1}/a_n \propto \eta^{2n+1} \to 0$, although an increase in the modulation depth is possible during the limited number of round-trips. So, instabilities develop only if $\eta \geq 1$ and $I_{in} > I_{thr}$. Nonlinear saturation of perturbation growth results in the formation of "infinite" (periodically modulated, e.g., hexagonal) structures [71, 256, 405].

When the medium absorption coefficient depends on radiation intensity, bistability, switching waves and localized structures become possible. In the framework of (4.8.14), switching waves and DOSs can be demonstrated for the model of threshold nonlinearity (see Sect. 4.4 and [304]). Similarly, one can construct bound states of DOSs and switching waves, as we have demonstrated for the case of driven nonlinear interferometers.

# 4.9 Bibliography; Experiments

The review of the first publications on the effects of spatial distributivity in nonlinear interferometers has already been presented in Sect. 1.4. Now there are so many publications in the field that it is very difficult to present a detailed review of all the original studies. I thus give here references to recent books and reviews [406, 387, 99, 414, 25, 98, 415, 307, 95] and discuss some aspects of the effect of spatial distributivity on optical structures in nonlinear interferometers; special attention will be paid to experimental research.

## 4.9.1 Instabilities Related to Longitudinal Distributivity

To this type can be attributed the self-pulsations predicted by Bonifacio and Lugiato in a Fabry–Perot cavity filled with a two-level medium [54]. The cavity length should be sufficiently large to fit the frequency interval between the adjacent longitudinal modes $c/2L$ with the Rabi frequency $\mu E/\hbar$ ($\mu$ is the dipole moment of transition and $E$ the amplitude of the field). A review of different theoretical approaches to the Bonifacio–Lugiato self-pulsations can be found in [117, 63, 202, 28]. Experimentally such self-pulsations have been observed by Segar and Make [357] in a Fabry–Perot cavity of 182 m length filled with HCN gas.

Ikeda instabilities [146] in the cavities, for which the round-trip time exceeds considerably the medium relaxation time, lead to the replacement of the stationary regimes by periodic and chaotic regimes with a smooth change in the scheme parameters; the conditions for development of these instabilities and the Bonifacio–Lugiato self-pulsations differ mainly by the value of interferometer detuning [63]. More various than in the case of hybrid schemes (see Chap. 3), scenarios of transition to chaos for the nonlinear interferometers have been analyzed in [235, 103]. The calculations carried out in [191] illustrate the essential effect of relaxation time on the dynamics, as well as insufficiency precision of the purely difference equations of the type (4.1.48) for chaotic regimes; such a conclusion follows also from the analysis of hybrid schemes (Sect. 3.2). A consistent consideration of "longitudinal" instabilities and the optimization of their threshold for a ring cavity filled with a Kerr medium was performed by Firth and Sinclair [101].

### 4.9.2 Wide-Aperture Semiconductor Microresonators

The transverse phenomena have been demonstrated most clearly in experiments with driven interferometers filled with semiconductor media. The nonlinearity mechanism can be thermal (slow) or electronic (fast). Here I describe the setup of recent experiments with fast nonlinearity, hereafter referred to for shortness as KW microresonators, for Kuszelewicz's and Weiss' teams [380, 181, 112, 414].

The microresonator includes two Bragg mirrors ($Ga_{0.9}Al_{0.1}As/AlAs$) with 99.5% reflectivity. It is filled with a $3/2$-$\lambda$-thick layer which includes 18 GaAs quantum wells separated by $Ga_{0.7}Al_{0.3}As$ barriers, all of them with 10 nm thickness. The optical resonator length is about 3 μm and varies over the usable sample area ($10 \times 20$ mm), which allows us, by choosing a particular position on the sample, to vary the cavity resonance wavelength in the vicinity of the exciton wavelength. The experiments show a high sensitivity to thickness variations even on the scale of a single atomic layer, due to the great finesse of these resonators.

The source of the coherent radiation is a continuous-wave $Ar^+$-pumped tunable Ti:sapphire ($Ti:Al_2O_3$) laser. For holding radiation, Gaussian beams are used with a width on the sample of about 60 μm. To eliminate thermal effects, all observations are made within times of a few microseconds, repeated at 1 kHz. To introduce local perturbations, smaller areas (with a size about 8 μm within the area illuminated by the holding radiation) can be irradiated by short address pulses ($\leq 0.1$ μs). The address radiation has a polarization orthogonal to the holding radiation polarization, locally and temporarily changing the optical properties of the resonator [231]. All observations are performed in reflection, as the resonator substrate is opaque.

### 4.9.3 Instabilities of the Transverse Field Structure

For ring interferometers with different types of nonlinearity, such instabilities have been analytically researched, following [323], in [343, 228]. Saturation of nonlinearity and effective spatial filtering serve, because of the finiteness of the interferometer aperture or the beam width, as limiting factors. The settled structures of the field, including hexagon patterns, have been analyzed in [204, 100, 230, 369].

In interferometers with abrupt mirror edges, the field has diffractive oscillations even in the absence of nonlinearity; in the nonlinear case these oscillations can be amplified and serve as a source of deep spatial modulation of radiation. Numerical calculations of the field spatial structures in nonlinear interferometers with a finite aperture driven by Gaussian beams have been performed for transversely one-dimensional [228] and two-dimensional [255] variants. As well as in the case of the usual small-scale self-focusing [42], there is a specific scale of the most rapidly growing perturbations; correspondingly, for a finite aperture a finite number of field filaments remains, this number being proportional to the aperture size. Such separate filaments, representing fragments of the settled spatial structure, were called in [228] "solitary waves".

To avoid misunderstanding, let us underline the qualitative dissimilarity between "solitary waves" [228, 5], which are due to the transverse scheme finiteness and the transverse field instabilities found in [323], and localized structures, or DOSs [319, 332, 333, 321], with hard excitation of single and coupled configurations by an infinite plane wave in a transversely infinite interferometer. If, under the conditions in [228], we go from finite to infinite values of interferometer aperture and holding radiation width, then, as the calculations show, an infinite transversely periodic set of filaments, or the "solitary waves", would form. DOSs also differ radically from "quasisolitons" [109, 110] – nonstationary pulses in bistable ring interferometers.

As transversely two-dimensional calculations [416] show, there is a large number of field metastable structures. For the Fabry–Perot interferometers with fast Kerr nonlinearity, the conditions of transverse instabilities development were deduced by Vlasov [401].

In the KW microresonators, hexagonal positive and negative patterns – lattices of bright and dark spots with a period of about 20 μm – were found in a certain range of detunings [380, 181] (see Sect. 4.3). The results were in qualitative accordance with the previous theoretical studies [100, 230, 369]. However, there was no observable threshold intensity for this modulation instability; this was attributed in [380] to the effect of light scattering and spectral filtering inside the resonator.

Experimental observations of different types of instabilities caused by an additional optical element inside the nonlinear ring interferometer, which provides rotation of the field, were first described in [7]. Formation of the spatial structures in the Fabry–Perot interferometer with a liquid-crystal cell without any additional optical elements was observed in [180], where the important role of fluctuations in the medium in this effect was stressed.

In wide-aperture interferometers, the instabilities of the field transverse structure develop usually prior to the "longitudinal" instabilities, hampering the observation of the latter ones. However, with effective spatial filtering, the Ikeda instabilities and bifurcations of period-doubling have been demonstrated in calculations [343, 234, 103] for an interferometer driven with a wide beam (see Sect. 4.7).

Chaos in systems with essential transverse distributivity can be considered to be spatio-temporal; then the question concerning the possibility of purely spatial chaos, with the spatial variables playing the role of time, arises. Such an analogy is sufficiently complete for the "stream" systems representing one-dimensional chains of the elements, each of which is only connected to the previous one [113]. Actually, here, as for "temporal chaos", the elements are ordered along the coordinate, and a small change of conditions at the beginning of the chain can essentially increase for subsequent elements. In other cases the notion of purely spatial chaos needs refinements and stipulations. So, even in the ring scheme of the finite number $N$ of unidirectionally connected elements [253], the presence of boundary conditions essentially distinguishes the spatial coordinate from the time variable. Apparently spatial distributions stationary in time can be considered spatially chaotic only asymptotically at $N \to \infty$. In systems of finite size, the boundary conditions play an important role, and there is usually no spatial direction chosen, which is similar to time. In the absence of temporal instabilities in such systems, one can strictly speak only of multistability, i.e., of a large number of stable stationary spatial structures. It seems that the features of "multiparticle" configurations of DOSs presented in Sect. 4.5 allow us to speak about spatial quasicrystals and spatial chaos. The concept of "spatial chaos" in a system of coupled optical bistable elements has also been analyzed by Firth [93].

### 4.9.4 Switching Waves and Spatial Hysteresis

The simplest transversely distributed model of a bistable interferometer with diffusive transverse coupling, equivalent to the model used in Chap. 2 for the cavityless scheme, was analyzed theoretically in [277, 131, 123, 136, 125, 126, 132, 97, 96, 104]. In [97], a detailed study of the hysteresis behaviour in different parts of the interferometer transverse section was undertaken. In [104], examples are presented for numerical calculations of switching of the Fabry–Perot semiconductor interferometers; diffusion is shown to be the main mechanism of transverse coupling for the InSb semiconductor. Stable switching waves between different couples of the states of multistable interferometers were analyzed in [136, 125, 126, 132, 97].

Early experimental research into the effects of transverse distributivity in nonlinear cavities was performed under the conditions of a thermal mechanism of nonlinearity and for the diffusive (also thermal) type of transverse coupling. This is connected with the comparatively low power level of continuous radiation required for the achievement of the switching threshold in the case of thermal nonlinearity. In thin-film semiconductor interferometers (ZnS, ZnSe), with the index of medium refraction growing with temperature, spatial hysteresis and switching waves were obtained in [23]. The kinetics of spatial hysteresis is in complete agreement with the theoretical predictions [277] presented above. In [131, 123, 125, 126, 132] the switching waves were observed in the interferometer formed by a Ge plate with temperature dependence of the index of refraction (and of absorption), and with mirrors applied on the plate facets; the spatial hysteresis was not studied in these works.

Grigor'yants and Dyuzhikov [129] demonstrated spatial hysteresis for the case of electronic nonlinearity and diffusive transverse coupling in an experiment with the InSb interferometer excited by radiation from a CO-laser. The results are in agreement to the theoretical concepts given above. In experiments [26] where the pulse duration was of the order of the relaxation time of electronic nonlinearity, dynamic spatial hysteresis with transient processes was also observed. For the KW microresonators, switching waves with maximum velocity $v_{max} = 2.5$ km/s were demonstrated in [112].

The first calculations of the diffractive switching waves and diffractive spatial hysteresis [323] were performed for a ring interferometer and for slot (transversely one-dimensional) beams by the fast Fourier transform. For wide-aperture interferometers this method is much more effective than the finite difference approach; it was widely used in further calculations, including simulations with the two transverse coordinates [255]. The hysteresis of the beam profile in a nonlinear interferometer was also studied in [102] by another method – numerical calculations by means of field expansion in the Laguerre–Gauss modes. However, for a correct description of the diffractive phenomena at large Fresnel numbers ($N_F \gg 1$), the number of the Laguerre–Gauss modes $n$ taken into account should be great enough, exceeding $N_F$. Therefore the results [102] obtained for the opposite relation ($n = 6$, $N_F \sim 20$) require additional justifications.

In [131, 123, 136, 125, 126, 132, 97, 96, 104, 23] the angle of radiation incidence was small. The case of large angles of incidence when the interferometer can be treated as a planar waveguide has been experimentally studied for the thermal mechanism of nonlinearity [38]; however, there were no experimental data on velocities of the switching waves that could be compared with the theoretical results (see Sect. 4.6.3).

Transverse coupling of bistable elements was investigated theoretically for the diffractive [233] and diffusive [97, 271, 418] coupling mechanisms. The experiments were performed for the elements from InSb [418] and GaAs [151]. In actual schemes the transverse coupling of the elements caused by the thermal flux through the common substrate is fairly essential [352, 1, 2, 149]. The detailed numerical simulation of bistable etalons from GaAs and AlGaAs with small Fresnel numbers accounting for radiation diffraction, diffusion of non-equilibrium carriers and thermal fluxes was performed in [251, 250].

The possibility of organizing multichannel memory by means of spatial hysteresis in a wide-aperture interferometer with a weak spatial modulation of the external radiation characteristics was shown in [344]. For this approach, the existence of switching waves is sufficient, even if dissipative solitons are absent; this is why the main effect takes place even in the simplest one-component bistable scheme with diffusive transverse coupling (Sect. 2.8).

### 4.9.5 Dissipative Optical Solitons

Stationary and pulsing localized structures (DOSs) in wide-aperture non-linear interferometers with the diffraction mechanism of transverse coupling were predicted and demonstrated theoretically in [319, 332]; the possibility of forming pulsating DOSs was also pointed out in [32], see also [33]. Experimentally DOSs were first found in [265, 267], and the first review of DOS features was given in [291].

In the strict sense, DOSs represent homoclinic trajectories in the phase space, and they are not necessarily connected with the presence of two more simple, "infinite" attractors, such as stable homogeneous (or inhomogeneous) regimes. However, for estimations of the conditions in which DOSs exist, it is desirable to interpret them simply, even though this may be of limited precision. A DOS could thus be considered as an island, or a spot of one of the stationary homogeneous states against a background of another state in a bistable wide-aperture interferometer [319]. More instructive is the interpretation of DOSs as coupled states of switching waves, the coupling being due to diffraction oscillations of the field [321]. This interpretation allows us to predict conditions favourable for the existence of DOSs and explain a discrete spectrum of DOS characteristics. However, as pointed out in [321], this interpretation is only approximate (asymptotical), and DOSs can form even under conditions when switching waves and bistability of stable homogeneous states do not exist.

The simplest, analytical description of DOSs was given for an interferometer with threshold nonlinearity in [290]. As we have seen in Sect. 4.4, this model allows us to find the entire discrete spectrum of single and coupled, stable and unstable localized structures; the model gives the conditions for the formation of a DOS in the "ground" and "excited" states, predicts the motion of asymmetric DOSs, permits us to find DOS distortions in the presence of various inhomogeneities of the scheme and holding radiation, and leads to a "mechanical" description of DOS motion under smooth inhomogeneities.

The latter model conforms with the interpretation of DOSs as coupled states of switching waves, but it is not applicable for DOSs beyond conditions of classical bistability. As shown in [322, 89], DOSs exist even if homogeneous states corresponding to the lower and/or the upper branches of the transfer function are unstable (see Sect. 4.7.3). Similar conclusions were also presented in [384], where results equivalent to those in [322, 89] were obtained by numerical solution of the approximated Swift–Hohenberg equations. Let us recall once again that, strictly speaking, these DOSs also cannot be presented as a spot of a spatially modulated state (e.g., a fragment of a hexagon pattern) against a background of another state, because of the spot's finite size and the infinite size of the periodic system of the hexagons; ranges of existence of DOSs and infinite modulated structures can differ.

Experimentally, the most clear-cut results on DOSs have been obtained with KW microresonators [380] for defocusing nonlinearity. In this case modulation instability is absent; we have classical bistability with two stable transversely homogeneous states and standard switching waves with oscillating "tails". Correspondingly, the DOS interpretation as a coupled state of switching waves is justified under these experimental conditions, and the main experimental conclusions are in close agreement with the simplest variant of our theory presented above. For a quantitative agreement it should of course be necessary to perform more detailed numerical simulations, taking into account the concrete form of semiconductor nonlinearity, the effect of free carrier diffusion, etc. [230, 369].

Related structures can also occur when the diffusive (not diffractive) mechanism of transverse coupling is dominant – but in two-component systems described by a system of the two (or more) coupled diffusion equations of the type of (2.11.4) (a "reaction-diffusion system"). As applied to nonlinear interferometers, it corresponds to competing mechanisms of nonlinearity. The running pulses ("diffusion" autosolitons), the structure self-completion and stratification, i.e., the formation of stationary spatial structures which fill consecutively the entire interferometer aperture, were found theoretically under such conditions [35, 36, 351, 130, 408, 266, 268]. Especially interesting is the process of self-completion in a waiting regime, when the system is initially in a transversely homogeneous state. A hard local excitation leads to the formation of a single stratum. Its appearance excites the adjacent regions so that new strata arise in them, and the process is repeated. Certain other "autowave" phenomena known from the analysis of systems of another physical nature [395, 166] are possible in such interferometers.

New aspects of the spatio-temporal structures are revealed when the inertia of nonlinear optical response is essential. It is equivalent to an increase in the system componency. Self-induced transparency [224, 18] belongs among phenomena in which the resonance response of the medium polarization to pulses of coherent radiation has an oscillatory nature. A peculiar kind of pulses of self-induced transparency in bistable wide-aperture interferometers was considered in [288].

In closing the chapter, the nature of optical patterns and spatial hysteresis for wide-aperture nonlinear interferometers is qualitatively confirmed in experiments, but a number of problems are still unresolved. Among them are the following: the formation of diffractive switching waves and spatial hysteresis under conditions of prevalence of diffractive transverse coupling; the formation of high-order ("excited") dissipative solitons and of their bound symmetric and asymmetric structures; research of dissipative solitons for oblique incidence of external radiation and, more generally, the effect of small-scale and smooth inhomogeneities. Experimental research of the kinetics of switching of the wide-aperture systems from a metastable to a stable state for its initiation by local perturbations created artificially seems also to be important. The results in this sphere would become a helpful reference point for the physics of phase transitions of the first kind. The theory presented admits generalization on a wide field of phenomena for objects of other physical nature.[6] An important impetus for further research in this entire field are the potential prospects for applications to the problem of parallel optical information processing (see Appendix C).

---

[6] For example, as shown in [31], the equation of the mean-field approximation (Sect. 4.1) also describes optical excitation of coherent states of excitons in a semiconductor layer; therefore, in the latter cavity-less scheme, the transverse patterns considered in this chapter are also possible, including modulation instability, dissipative solitons, etc.

# 5. Nonlinear Radiation Reflection

When a plane wave is incident on the interface of two linear transparent media, the regime of transmission or total internal reflection (TIR) is realized depending on the relation between the refractive indices of the media and the angle of incidence. Taking the optical nonlinearity into account requires the statement of the principle of radiation for the field asymptotics far away from the medium interface. Consistent analysis is based on employing "the principle of limiting absorption", i.e., the introduction of absorption into the bulk of the nonlinear medium which can be made arbitrarily small in the final results. It turns out that in addition to the two regimes known in linear optics, two new regimes – a hybrid regime and an oscillation regime – are also possible. This point has an impact on the form of bistability or multistability conditions.

In addition to longitudinal extension, which is essential in the problem of plane-wave reflection, consideration of transverse distributivity is also important in the analysis of nonlinear reflection. It results in local nonlinear breakdown of TIR of beams in media with self-focusing nonlinearity (radiation filamentation). Then radiation transmitted into the nonlinear medium decays into separate intensive filaments, and corresponding intensity gaps appear in the reflected radiation. The number of these filaments and gaps increases with the power of incident radiation. Correspondingly, the observation of bistability is hampered for media with self-focusing nonlinearity. For the media with self-defocusing nonlinearity in which similar instabilities are not observed, numerical solution of the wave equation demonstrates bistability and spatial hysteresis of the nonlinear reflection of a beam, but at relatively high levels of radiation intensity.

## 5.1 Reflection of Plane Wave from Transparent Nonlinear Media

### 5.1.1 Nonlinear Layer

In the present section we mainly follow [272]. The geometry of the problem is shown in Fig. 5.1, where the hatched area is the nonlinear medium with the dielectric permittivity

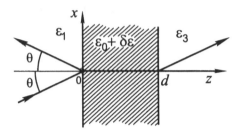

**Fig. 5.1.** Schematic of nonlinear reflection

$$\varepsilon = \varepsilon_0 + \delta\varepsilon(|\boldsymbol{E}|^2) \ . \tag{5.1.1}$$

The time of nonlinearity response is not essential for the analysis of the steady-state reflection of monochromatic radiation. We consider the case of s-polarization of radiation with the only nonzero component of electric field $E_y$, for which we use the following wave (Helmholtz) equation, omitting the time factor $\exp(-i\omega t)$:

$$\frac{\partial^2 E_y}{\partial x^2} + \frac{\partial^2 E_y}{\partial z^2} + \frac{\omega^2}{c^2}\varepsilon E_y = 0 \ . \tag{5.1.2}$$

We are interested in the solutions with the same dependence on the coordinate $x$ as for a plane wave:

$$E_y = E(z)\exp(ik_x x) \ . \tag{5.1.3}$$

Then (5.1.2) is reduced to an ordinary differential equation:

$$\frac{\mathrm{d}^2 E}{\mathrm{d}z^2} - \frac{\omega^2}{c^2}\Delta_0 E + \frac{\omega^2}{c^2}\delta\varepsilon E = 0 \ , \tag{5.1.4}$$

where

$$\Delta_0 = (k_x c/\omega)^2 - \varepsilon_0 \ . \tag{5.1.5}$$

It is convenient to introduce the real field amplitude $A$ and phase $\Phi$ for its solution

$$E(z) = A(z)\exp[i\Phi(z)] \ . \tag{5.1.6}$$

Substituting (5.1.6) into (5.1.4), we obtain the integral of motion which reflects the energy conservation law in the transparent nonlinear medium:

$$C = A^2\frac{\mathrm{d}\Phi}{\mathrm{d}z} = \mathrm{const.} \tag{5.1.7}$$

Value $C$ represents the radiation power flux in the direction $z$. An equation for the field amplitude has the form

$$\frac{\mathrm{d}^2 A}{\mathrm{d}z^2} = F \ . \tag{5.1.8}$$

Here, taking into account (5.1.7),

$$F = F(A) = \frac{C^2}{A^3} + \frac{\omega^2}{c^2}[\Delta_0 - \delta\varepsilon(A^2)]A . \tag{5.1.9}$$

The form of (5.1.8) envisages the use of the mechanical analogy, whose clearness has already been demonstrated in Chap. 2. In the framework of this analogy (5.1.8) is treated as the Newtonian equation of motion of a particle with unit mass under the action of force $F$, amplitude $A$ corresponding to the particle coordinate, and coordinate $z$ to time $t$. The "force" $F$ is related to the "potential energy":

$$U = -\int^A F(A)\,dA . \tag{5.1.10}$$

From the "energy" integral

$$\frac{1}{2}\left(\frac{dA}{dz}\right)^2 + U(A) = W = \text{const.} , \tag{5.1.11}$$

it follows that

$$\frac{dA}{dz} = \pm\sqrt{2[W - U(A)]} . \tag{5.1.12}$$

Dependences $A(z)$ are associated with mechanical trajectories in the allowed regions $U(A) < W$. Between the turning points where $U(A) = W$, this dependence is implicitly determined by

$$\int^A \{2[W - U(A)]\}^{-1/2}\,dA = \pm(z - z_0) , \quad z_0 = \text{const.} \tag{5.1.13}$$

Then we find the field phase according to (5.1.7):

$$\Phi(z) - \Phi(z_0) = C\int_{z_0}^z A^{-2}(z')\,dz' . \tag{5.1.14}$$

Relations (5.1.13) and (5.1.14) represent the exact general form for a plane-wave field in a transparent medium with an arbitrary type of nonlinearity of the refraction index. In the framework of the mechanical analogy, the trajectory $A(z)$ is determined by the shape of potential curves $U(A)$. In the important case of Kerr nonlinearity,

$$\delta\varepsilon = \varepsilon_2 A^2 \tag{5.1.15}$$

(which is considered below), "potential energy" has a simple form:

$$U = \frac{1}{2}\left[\frac{C^2}{A^2} - \frac{\omega^2}{c^2}\left(\Delta_0 A^2 - \frac{1}{2}\varepsilon_2 A^4\right)\right] . \tag{5.1.16}$$

In this case the left-hand side of (5.1.13) is reduced to elliptic functions.

To solve the problem completely, we should use the condition of continuity of the tangential components of the electric and magnetic fields at the interface between the media. In our case it means continuity of $E$ and $dE/dz$ at $z = 0$ and at $z = d$. For simplicity we consider $\varepsilon_1$ to be real, i.e., the linear medium ($z < 0$) is transparent. For this medium (5.1.4) has the form

$$\frac{d^2 E}{dz^2} + \left(\frac{\omega^2}{c^2}\varepsilon_1 - k_x^2\right) E = 0 . \tag{5.1.17}$$

In the case we are most interested in,

$$k_x^2 < (\omega/c)^2 \varepsilon_1 . \tag{5.1.18}$$

Then we can assume

$$\begin{aligned}
&E = E_{\text{in}} \exp(ik_{z1}z) + E_r \exp(-ik_{z1}z) \quad (z < 0) , \\
&k_x = (\omega/c)\sqrt{\varepsilon_1} \sin\theta , \quad k_{z1} = (\omega/c)\sqrt{\varepsilon_1} \cos\theta , \\
&\Delta_0 = \varepsilon_1 \sin^2\theta - \varepsilon_0 .
\end{aligned} \tag{5.1.19}$$

Here $\theta$ has the meaning of an incidence angle for a plane wave with the given amplitude $E_{\text{in}}$, and $E_r$ is an unknown amplitude of the reflected wave. Similarly, in the last linear medium ($z > d$)

$$E = E_{\text{tr}} \exp(ik_{z3}z) , \quad k_{z3} = (\omega/c)\sqrt{\varepsilon_3 - \varepsilon_1 \sin^2\theta} , \quad \text{Re}\, k_{z3} > 0 . \tag{5.1.20}$$

In the case opposite to (5.1.18),

$$k_x^2 > (\omega/c)^2 \varepsilon_1 , \tag{5.1.21}$$

the field in the linear medium decays exponentially with distance from its boundary:

$$E = E_0 \exp(\Gamma z) , \quad \Gamma = \sqrt{k_x^2 - \omega^2 \varepsilon_1/c^2} , \quad z < 0 . \tag{5.1.22}$$

In this case the field in the nonlinear medium exists, though the incident wave (from the region $z < 0$) is absent. Actually, even if nonlinearity of the medium is not essential, the waveguide statement of the problem is possible, in which radiation propagates in the intermediate medium in the direction $x$ and in the coating ($z < 0$ and $z > d$) the field decreases exponentially. Nonlinearity extends over the range of situations of the waveguide type. For example, in the problem with a single interface between the linear and nonlinear media ($d = \infty$), nonlinear surface waves can propagate along $x$ [15, 198]. The general relations (5.1.13-5.1.16) are also valid for them in the nonlinear medium (with additional limitation $C = 0$), but the consequences of the continuity conditions at the interface $z = 0$ differ from the case of (5.1.18). We will discuss this situation at the end of the present section.

When (5.1.18) is satisfied, the conditions at the interfaces $z = 0$ and $z = d$ lead to the following relations:

$$E_{in} + E_r = A_0 \exp(i\Phi_0) , \tag{5.1.23}$$

$$ik_{z1}(E_{in} - E_r) = \left(i(C/A_0) \pm \sqrt{2[W - U(A)]}\right) \exp(i\Phi_0) , \tag{5.1.24}$$

$$ik_{z3} = \left(i(C/A_d) \pm \sqrt{2[W - U(A)]}\right)/A_d . \tag{5.1.25}$$

Here $A_0 = A(0)$, $\Phi_0 = \Phi(0)$, $A_d = A(d)$, and the signs $(\pm)$ are coordinated with the signs in (5.1.12-5.1.13); their selection should lead to a positive value of the layer thickness $d > 0$.

Three complex equations (5.1.23–5.1.25) allow us to determine (to express in terms of $A_d$, to be more exact) the complex amplitude of the reflected wave $E_r$ and four real values of $C$, $W$, $A_0$ and $\Phi_0$ by means of given values of $k_{z1}, k_{z3}$ and the amplitude of an incident wave $E_{in}$. The value $A_d$ itself is found from the value of the layer thickness $d$ using the relation of the type of (5.1.13). If the third medium ($z > d$) is transparent (Im$\varepsilon_3 = 0$) and $\varepsilon_3 - \varepsilon_1 \sin^2 \theta > 0$ then $W = U(A_d)$ and $C = k_{z3}A_d^2$ follow from (5.1.25). In the case $\varepsilon_3 - \varepsilon_1 \sin^2 \theta < 0$, TIR is realized independently from other conditions; the (intensity) reflection coefficient equals unity for TIR: $R = |E_r/E_{in}|^2 = 1$.

The obtained relations serve as the general solution for the problem of reflection (and transmission) of a plane wave from a nonlinear layer or a transparent nonlinear medium [272]. The results are determined by the type of nonlinearity and by the shape of the corresponding potential curves $U_C(A)$. Note that the functions $U_C(A)$ are even by $A$ and by $C$ (therefore, it is enough to consider only the region $A > 0$) and have power divergence at $A \to 0$ if $C \neq 0$. Practically it is convenient to inverse the problem, specifying (and varying) the amplitude of the transmitted wave and then finding the corresponding value of the amplitude of the incident wave $E_{in}$. In view of the absence of fundamental differences between the nonlinear medium layer and the Fabry–Perot interferometer (see Fig. 1.5 and Chap. 4), the possibility of bi- and multistability is obvious enough here. Calculations of hysteresis dependence of the reflection coefficient from a layer of the medium with Kerr nonlinearity for a plane wave (see (5.1.15)) are given in [51]. Analogous is the analysis of multi-layered systems, which include nonlinear media [391].

## 5.1.2 Nonlinear Half-Space

More complex is the problem of reflection of a plane wave from the half-space of a transparent nonlinear medium ($d = \infty$). In this case (5.1.25) is absent, and we must invoke the corresponding conditions at infinity ($z \to \infty$) to solve the problem. Note that the natural requirements for the field to be finite in the nonlinear medium and energy flux to be directed towards the depth of the nonlinear medium (positive sign of the $z$-component of the phase velocity at $z \to \infty$) are not always sufficient to obtain correct conclusions. As we will

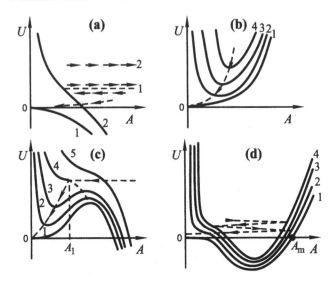

**Fig. 5.2.** "Potential curves" and "trajectories" $A(z)$ (*dashed lines*): (a) $\varepsilon_2 < 0$, $\Delta_0 > 0$, (b) $\varepsilon_2 > 0$, $\Delta_0 < 0$, (c) $\varepsilon_2 < 0$, $\Delta_0 < 0$, (d) $\varepsilon_2 > 0$, $\Delta_0 > 0$

see in Sect. 5.2, a consistent treatment is possible, if we take into account a weak absorption in the nonlinear medium [273]. Therefore, in the present section, devoted to analysis of the transparent Kerr media only, we have to exclude from our analysis the regions of large intensity of incident wave in a number of cases.

We thus assume condition (5.1.18). The meaning of the parameter $\Delta_0$ introduced in (5.1.5) is the following: For linear media ($\varepsilon_2 = 0$) and for $\Delta_0 > 0$, TIR is realized when the (intensity) reflection coefficient $R = 1$, with which field intensity in the medium decreases with distance from the boundary ($A^2 \to 0$ at $z \to \infty$). At $\Delta_0 < 0$ we have the transmission regime ($R < 1$, field intensity in the nonlinear medium does not depend on $z$). The shape of potential curves $U$ determined by (5.1.16) is shown in Fig. 5.2. Longitudinal dependence $A(z)$ is associated with the segments of the horizontal straight lines $W = $ const located in allowed regions $U(A) < W$, and the arrows near the segments show the direction of $z$ increase.

Let $\varepsilon_2 < 0$ and $\Delta_0 > 0$ (Fig. 5.2a). In correspondence with (5.1.16), the potential curves do not have an extremum at $C \neq 0$, and at $C = 0$ they have a single maximum with "energy" level $W = 0$. The requirement of field finiteness in the nonlinear medium at $z \to \infty$ is violated in the case of $C \neq 0$, for example, for all trajectories of types 1 and 2 (Fig. 5.2a), and is satisfied only under the conditions:

$$C = 0, \quad W = 0, \quad \frac{\mathrm{d}A}{\mathrm{d}z} < 0. \tag{5.1.26}$$

Then at $z \to \infty$ the trajectory $A(z)$ comes to the apex (maximum) of the potential curve for $C = 0$, corresponding to $A = 0$. In all other cases the field amplitude $A$ increases unrestrictedly at $z \to \infty$ either monotonously (if at the initial point of the trajectory $dA/dz > 0$) or after an initial decrease ($dA/dz < 0$ in the trajectory segment before its reflection from the potential well edge).

Using (5.1.26), we find the following from (5.1.23-5.1.24):

$$E_{\text{in}} \exp(-i\Phi_0) = \frac{A_0}{2} \left( 1 + i \frac{\sqrt{\Delta_0 - \varepsilon_2 A^2/2}}{\sqrt{\varepsilon_1} \cos \theta} \right) ,$$

$$E_{\text{r}} \exp(-i\Phi_0) = \frac{A_0}{2} \left( 1 - i \frac{\sqrt{\Delta_0 - \varepsilon_2 A^2/2}}{\sqrt{\varepsilon_1} \cos \theta} \right) . \tag{5.1.27}$$

Then the intensity reflection coefficient is

$$R = |E_{\text{r}}/E_{\text{in}}|^2 = 1 , \tag{5.1.28}$$

so the regime corresponds to TIR. Relations (5.1.27) represent the parametric dependence of the amplitude of reflected wave $E_{\text{r}}$ on the amplitude of the incident wave $E_{\text{in}}$ through the value of $A_0$ having the meaning of the field amplitude on the interface of the media $z = 0$. The intensities $I_{\text{in}} = |E|^2$ and $I_{\text{b}} = A_0^2$ are connected by a monotonous dependence:

$$I_{\text{in}} = \frac{I_{\text{b}}}{4} \left( 1 + \frac{\Delta_0 - \varepsilon_2 I_{\text{b}}/2}{\varepsilon_1 \cos^2 \theta} \right) . \tag{5.1.29}$$

The amplitude coefficient of reflection,

$$r = \frac{E_{\text{r}}}{E_{\text{in}}} = \frac{\sqrt{\varepsilon_1} \cos \theta - i\sqrt{\Delta_0 - \varepsilon_2 I_{\text{b}}/2}}{\sqrt{\varepsilon_1} \cos \theta + i\sqrt{\Delta_0 - \varepsilon_2 I_{\text{b}}/2}} , \quad |r|^2 = R = 1 , \tag{5.1.30}$$

is also uniquely determined by the value of intensity $I_{\text{in}}$ (phase of reflection coefficient $\arg r$ varies monotonously from the value corresponding to the linear theory at $I_{\text{in}} = 0$ to $-\pi$ at $I_{\text{in}} \to \infty$). Therefore bistability is absent in this case. Note that we have determined all the characteristics of reflection without the use of the explicit form of the field in the nonlinear medium, although it is not a problem to obtain it in the present case anyway:

$$A(z) = \frac{\sqrt{2\Delta_0/|\varepsilon_2|}}{\sinh[(\omega/c)\sqrt{\Delta_0}(z - z_0)]} . \tag{5.1.31}$$

Constant $z_0$ ($z_0 < 0$) is connected with the value $A_0$ by

$$A_0 = \frac{\sqrt{2\Delta_0/|\varepsilon_2|}}{\sinh[(\omega/c)\sqrt{\Delta_0}|z_0|]} \tag{5.1.32}$$

and, correspondingly, is determined uniquely by the intensity of the incident wave. Thus, although nonlinearity changes the longitudinal distribution of the field and the phase of the reflection coefficient in the present regime of TIR, these changes are only quantitative.

Let us consider now the case in which $\varepsilon_2 > 0$ and $\Delta_0 < 0$ (Fig. 5.2b). Each of the potential curves in the region $A \geq 0$ has the only minimum which represents the field state with constant amplitude, i.e., the transmission regime, the trajectory $A(z)$ in Fig. 5.2b degenerating into a point. All other trajectories correspond to the periodical variation of $A(z)$ and therefore do not satisfy natural requirements at infinity ($z \to \infty$). It follows from (5.1.4) for the transmission regime that

$$E(z) = A_0 \exp\left[i\left(\Phi_0 + (\omega/c)\sqrt{-\Delta_0 + \varepsilon_2 A_0^2}\, z\right)\right], \tag{5.1.33}$$

so that

$$C = \frac{\omega}{c} A_0^2 \sqrt{-\Delta_0 + \varepsilon_2 A_0^2}. \tag{5.1.34}$$

Then we the following from (5.1.23-5.1.24):

$$E_{\text{in}} \exp(-i\Phi_0) = (1+p)A_0/2, \quad E_{\text{r}} \exp(-i\Phi_0) = (1-p)A_0/2, \tag{5.1.35}$$

and correspondingly

$$r = E_{\text{r}}/E_{\text{in}} = (1-p)/(1+p), \tag{5.1.36}$$

where

$$p = \frac{\sqrt{-\Delta_0 + \varepsilon_2 A_0^2}}{\sqrt{\varepsilon_1} \cos\theta}. \tag{5.1.37}$$

Relations (5.1.35–5.1.37) give the parametric dependence of the reflection coefficient $r$ on the intensity of the incident wave $I_{\text{in}} = |A_{\text{in}}|^2$ through the value $I_{\text{b}} = A_0^2$, the intensity of the field at the media interface $z = 0$. The relation of intensities $I_{\text{in}}$ and $I_{\text{b}}$,

$$I_{\text{in}} = \frac{I_{\text{b}}}{4}\left(1 + \frac{\sqrt{-\Delta_0 + \varepsilon_2 I_{\text{b}}}}{\sqrt{\varepsilon_1} \cos\theta}\right)^2, \tag{5.1.38}$$

corresponds to a monotonous dependence in the present case. Therefore at the given value of intensity of the incident wave $I_{\text{in}}$, one can uniquely determine $I_{\text{b}}$ from (5.1.38), then $p$ from (5.1.37) and, finally, the amplitude reflection coefficient $r$ from (5.1.36). With a large radiation intensity ($p \gg 1$) the reflection coefficient $r \to -1$. Under the condition

$$\varepsilon_1 \sin^2\theta < \varepsilon_0 < \varepsilon_1, \tag{5.1.39}$$

the reflection coefficient $r$ changes its sign upon variation of intensity $I_{\text{in}}$ (Fig. 5.3) and turns into zero at $\varepsilon_2 I_{\text{in}} = \varepsilon_1 - \varepsilon_0$ (*nonlinear bleaching* [157]). Bistability is, in this case, absent.

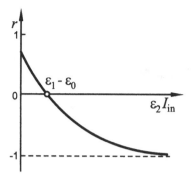

**Fig. 5.3.** Dependence of the amplitude reflection coefficient on the intensity of incident radiation under the condition (5.1.39)

The case in which $\varepsilon_2 < 0$ and $\Delta_0 < 0$ is more complex. The potential curves $U_C(A)$ (Fig. 5.2c) at $C^2 > C_1^2 = (4/27)\omega^2|\Delta_0|^3/c^2\varepsilon_2^2$ have no extremes; therefore only $C^2 \leq C_1^2$ correspond to restricted (at $z \to \infty$) field distributions. The "energy" $W$ is chosen so that the trajectory $A(z)$ goes to the apex of the potential curves at $z \to \infty$:

$$W = U(A_{C,\infty}) , \quad A_{z\to\infty} = A_{C,\infty} , \quad E_{z\to\infty} \approx A_{C,\infty} \exp(iCA_{C,\infty}^{-2}z) .$$
$$(5.1.40)$$

The condition $C > 0$ corresponds to a wave moving from the interface of the media to $z \to +\infty$. Thus, natural conditions at infinity are satisfied at

$$0 \leq C \leq C_1 . \tag{5.1.41}$$

Two types of solutions are possible here. The first corresponds to the regime of transmission and is characterized by a constant value of the field amplitude in the nonlinear medium ($dA/dz = 0$, minima and maxima of the corresponding potential curves in Fig. 5.2c). Expressions (5.1.36) and (5.1.37) for the amplitude coefficient of reflection $r$ are valid again for the regimes of transmission. However, the intensity relation $I_{in}$ and $I_b$ ($0 < I_b < |\Delta_0/\varepsilon_2|$) corresponds now to nonmonotonous dependence with only one maximum. This dependence is represented by curve 1 in Fig. 5.4a, where

$$I_b^{(m)} = \frac{\varepsilon_1 \cos^2\theta}{8|\varepsilon_2|}\left(\sqrt{1+8d_0} - 1 + 4d_0\right) ,$$

$$I_{in}^{(m)} = \frac{1}{64}I_b^{(m)}\left(\sqrt{1+8d_0} + 3\right)^2 , \tag{5.1.42}$$

$$I_{in}^{(1)} = \left|\frac{\Delta_0}{4\varepsilon_2}\right| , \quad d_0 = \frac{|\Delta_0|}{\varepsilon_1 \cos^2\theta} . \tag{5.1.43}$$

At incident wave intensities $I_{in} < I_{in}^{(1)}$ there is one regime of transmission; in the range $I_{in}^{(1)} < I_{in} < I_{in}^{(m)}$ there are two such regimes (the problem of their

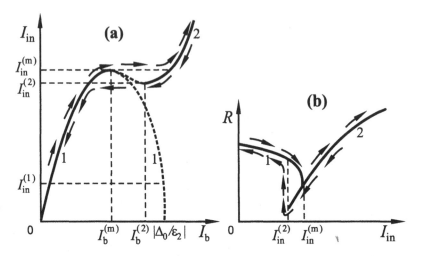

**Fig. 5.4.** Relation between intensities $I_{in}$ and $I_b$ in conditions of existence of hybrid regimes (**a**) and hysteresis variation of the reflection coefficient (**b**)

stability is not discussed here); and the transmission regimes do not exist at $I_{in} > I_{in}^{(m)}$.

The second type of regime, which we will call the hybrid regime, corresponds to the trajectories which come asymptotically (at $z \to \infty$) to the apexes of the potentials curves (Fig. 5.2c). Correspondingly, the value of "energy" $W$ for them is connected uniquely with the value $C$ by parametric dependence through the asymptotic (at $z \to \infty$) value of the amplitude $A_{C,\infty}$:

$$W = U(A_{C,\infty}), \quad C^2 = \left(\frac{\omega}{c}\right)^2 A_{C,\infty}^4 (-\Delta_0 + \varepsilon_2 A_{C,\infty}^2) . \qquad (5.1.44)$$

The term for the regime is justified as follows: The field amplitude in the nonlinear medium near to the interface $z = 0$ decreases (as in the regime of TIR) and then, at large distances from the boundary, approaches the constant value $A_{C,\infty}$ (as for the regime of transmission). It follows from (5.1.23-5.1.24), and (5.1.16) that the intensities of the incident [upper sign on the right-hand side of (5.1.45)] and reflected waves (lower sign) are

$$I_{in\,(r)} = \pm\frac{C}{2k_{z1}} + \frac{W}{2k_{z1}^2} + \frac{1}{4}\left(1 + \frac{\omega^2}{c^2}\frac{\Delta_0}{k_{z1}^2}\right)I_b - \frac{1}{8}\frac{\omega^2\varepsilon_2 I_b^2}{c^2 k_{z1}^2} . \qquad (5.1.45)$$

In accordance with the energy conservation law we obtain the following from (5.1.45):

$$k_{z1}(I_{in} - I_r) = C .$$

It should be recalled that in the absence of condition (5.1.25), natural conditions at infinity lead only to restriction (5.1.41) and to the asymptotic form of the field (5.1.40) in the discussed hybrid regime. And only two complex equations (5.1.23-5.1.24) remain for the determination of the complex amplitude $E_r$ and three real values $A_0$, $\Phi_0$ and $C$ [according to (5.1.44), "energy" $W$ is uniquely connected to the flux $C$]. Hence the so-called continuum problem arises, which has no analogues in linear optics: at a fixed incidence angle and parameters of the media, an infinite number (continuum) of regimes correspond to the same (sufficiently high) intensity of the incident wave [272]. However, in Sect. 5.2 we will show that consideration of locally small absorption in the nonlinear medium leads to the consequence that only one hybrid regime, characterized by the largest energy flux, corresponds to our problem:

$$C = C_1 , \quad \frac{\mathrm{d}A}{\mathrm{d}z} < 0 . \tag{5.1.46}$$

At the same time integrally small absorption (over the entire length of the nonlinear medium) does not lead to the disappearance of the continuum of solutions. Another consistent approach for solving this problem is connected to taking into consideration the transverse finiteness of radiation beams (see Sect. 5.3). It is important that at substantial distances from the boundary ($z \to \infty$) the field should vanish, so that the medium effectively becomes linear. Then the radiation condition at $z \to \infty$ is formulated in the usual way.

Under conditions (5.1.46) we find

$$A^2(z) = A^2_{C_1,\infty} + \frac{2}{|\varepsilon_2|(\omega/c)^2(z - z_0)^2} , \quad A^2_{C_1,\infty} = \frac{2}{3}\left|\frac{\Delta_0}{\varepsilon_2}\right| . \tag{5.1.47}$$

Constant $z_0 < 0$ is uniquely expressed, by means of (5.1.47), through the value of radiation intensity at the media interface, $I_b = A_0^2$, entering (5.1.45). With fixed values $C$ and $W$, the reflection coefficient is expressed by (5.1.45) parametrically through $I_b$. The bistability conditions at nonlinear reflection will be refined in Sect. 5.2 after it is substantiated why (5.1.46) was chosen.

The physical meaning of the longitudinal dependence of the field in hybrid regimes is as follows: At $\Delta_0 < 0$ the transmission regime corresponds to small intensities of the incident wave $I_{\text{in}}$. As $\varepsilon_2 < 0$, then at sufficiently large intensity $I_{\text{in}}$ the nonlinear medium in the boundary layer will have a lower effective dielectric permittivity corresponding to the TIR regime:

$$\varepsilon_0 + \varepsilon_2 A^2 < \varepsilon_1 \sin^2\theta . \tag{5.1.48}$$

Then in correspondence with the nature of the TIR regime, the amplitude $A(z)$ in the nonlinear medium will decrease with increasing $z$. Therefore, at some distance from the medium boundary the sign of inequality (5.1.48) will be reversed. As a result, the hybrid regime has the character of TIR in the vicinity of the medium boundary and the nature of the transmission regime far from it.

Note that the field distribution of the hybrid regime can arise in the problem of wave reflection not only from a nonlinear half-space, but from the layered nonlinear medium as well, and also in the waveguide statement of the problem. In a number of cases, restriction (5.1.46) can be absent. For example, at zero energy flux $C = 0$ we obtain

$$A(z) = |\Delta_0/\varepsilon_2|^{1/2} \, \frac{\tan}{\coth} \left( |\Delta_0/2|^{1/2}(\omega/c)(z - z_0) \right) . \tag{5.1.49}$$

In the last variant left for consideration, $\varepsilon_0 > 0$ and $\Delta_0 > 0$, all potential curves (Fig. 5.2d) have one minimum (at $A > 0$); in addition, at $C = 0$ and $A = 0$ they also have a maximum. As we will see in Sect. 5.2, the minima of the potential curves do not correspond to the regimes satisfying the principle of limiting absorption. Therefore here we can discuss only the TIR regimes, for which

$$C = 0 , \quad W = 0 , \quad A < A_{\mathrm{m}} = \sqrt{2\Delta_0/\varepsilon_2} , \quad R = 1 . \tag{5.1.50}$$

The trajectories $A(z)$ in Fig. 5.2d are located on the axis segment $OA$ ($0 < A < A_{\mathrm{m}}$). At $z \to \infty$ they approach the maximum of the potential curve $A \to 0$ ($C = 0$). There are two types of trajectories among them:

For the first type, $A' = (dA/dz)_{z_0} < 0$ is true at the trajectory origin point. Then the amplitude $A$ decreases monotonously with an increase in $z$, i.e., (5.1.26) is satisfied. Correspondingly, we deal here with the regime of nonlinear TIR discussed above, for which relations (5.1.27–5.1.30) are valid, but with the additional restriction $A_0 < A_{\mathrm{m}}$. Dependence of intensity $I_{\mathrm{in}}$ on $I_{\mathrm{b}}$ (5.1.29) can be nonmonotonous now. Actually, under the conditions

$$0 < \varepsilon_{10} < 2\Delta_0 , \tag{5.1.51}$$

where

$$\varepsilon_{10} = \varepsilon_1 - \varepsilon_0 , \tag{5.1.52}$$

the dependence $I_{\mathrm{in}}(I_{\mathrm{b}})$ (parabola) has one maximum at $I_{\mathrm{b}} = \varepsilon_{10}/\varepsilon_2$ inside the allowed interval $0 < I_{\mathrm{b}} < 2\Delta_0/\varepsilon_2$. Then at $I_{\mathrm{in}} < I_{\mathrm{in}}^{(\mathrm{max})}$ with

$$I_{\mathrm{in}}^{(\mathrm{max})} = \frac{\varepsilon_{10}}{4\varepsilon_2} \left( 1 + \frac{2\Delta_0 - \varepsilon_{10}}{2\varepsilon_1 \cos^2 \theta} \right) , \tag{5.1.53}$$

two different TIR regimes correspond to one value of $I_{\mathrm{in}}$ (stability of the regimes is not discussed). If one particular inequality (5.1.51) is violated, then intensity $I_{\mathrm{in}}$ also increases nonmonotonously with $I_{\mathrm{b}}$ in the allowed interval, reaching at $I_{\mathrm{b}} = 2\Delta_0/\varepsilon_2$ its maximum value of

$$I_{\mathrm{in}}^{(\mathrm{max})} = \Delta_0/(2\varepsilon_2) . \tag{5.1.54}$$

Thus, the discussed regimes of nonlinear TIR are possible only at the restricted [by (5.1.53) or (5.1.54)] values of incident wave intensity $I_{\mathrm{in}}$. At higher intensities $I_{\mathrm{in}}$, more complex regimes are realized, as described in Sect. 5.2.

For the second type of trajectories, $A' > 0$. Therefore, with an increase in $z$, amplitude $A$ first increases to the value $A_m$ (the turning point of the trajectory) and then decreases monotonously, as for the first type of trajectory. Reversing the sign of $A$ leads to substitution of the right-hand sides of (5.1.27) and (5.1.30) by complex-conjugated ones, and (5.1.29) remains true. Therefore intensities $I_b = A_0^2$ and $I_{in}$ do not depend on the sign of $A'$, and the amplitude reflection coefficient for the second type of the trajectories changes only the sign of its phase (as before, the TIR regime with the intensity reflection coefficient $R = 1$ remains). Thus, either four different TIR regimes [under condition (5.1.51)] or two of them [if (5.1.51) is violated] can correspond to the same incident wave intensity $I_{in}$. This circumstance underlines the importance of the analysis of regime stability or the consideration of weak absorption in the nonlinear medium (Sect. 5.2).

One more interpretation of the second type of the trajectories is as follows: In this variant, the propagation of the nonlinear surface wave mentioned above is possible. Namely, the considered trajectory with $A' > 0$ corresponds to a nonmonotonous field distribution of such a wave, with one maximum in the nonlinear medium bulk. The field distribution in the linear medium is expressed by (5.1.22). The boundary condition is

$$E_0' = [A_0' + i(C/A_0)] \exp(i\Phi_0) = \Gamma E_0 . \qquad (5.1.55)$$

Because the value $\Gamma$ is real [see (5.1.22)], the flux $C = 0$, and the amplitude $E(z)$ can be considered real ($\Phi_0 = 0$). This allows us to substitute (5.1.55) with

$$A_0' = \Gamma A_0 . \qquad (5.1.56)$$

As incident radiation is absent, it makes no sense to speak about the angle of incidence, and we have to return to the initial expression (5.1.5) for the value $\Delta_0$. As a result, the nonlinear surface wave exists under the conditions

$$\varepsilon_2 > 0 , \quad \Delta_0 > 0 , \quad \varepsilon_{10} > 0 . \qquad (5.1.57)$$

A more thorough analysis, which is beyond the scope of our discussion here, shows the instability of surface wave propagation with respect to the development of initially slight modulation of the field in direction $y$ [410, 411].

## 5.2 A Weakly Absorbing Nonlinear Medium

Here, following [273], we will consider the same problem as in Sect. 5.1 assuming only that the nonlinear medium has low linear absorption[1]:

$$\varepsilon = \varepsilon_0 + \varepsilon_2 |\boldsymbol{E}|^2 = \varepsilon_0' + i\varepsilon_0'' + \varepsilon_2 |\boldsymbol{E}|^2 , \quad \varepsilon_0'' > 0 . \qquad (5.2.1)$$

---

[1] For high absorption the hysteresis phenomena are depressed. The case of nonlinear absorption is treated in [215] and references therein.

As before, we introduce real field amplitude $A$ and phase $\Phi$ [see (5.1.6)] and the values

$$C = A^2 \frac{d\Phi}{dz}, \quad W = \frac{1}{2} \left( \frac{dA}{dz} \right)^2 + U, \tag{5.2.2}$$

where

$$U = \frac{1}{2} \left( \frac{C^2}{A^2} - \frac{\omega^2}{c^2} \Delta' A^2 + \frac{\omega^2}{2c^2} \varepsilon_2 A^4 \right). \tag{5.2.3}$$

Taking into account absorption, we have

$$\Delta_0 = \Delta' + i\Delta'', \tag{5.2.4}$$

$$\Delta' = \varepsilon_1 \sin^2 \theta - \varepsilon_0', \quad \Delta'' = -\varepsilon_0'' < 0. \tag{5.2.5}$$

In an absorbing medium, values $C$ and $W$ are no longer integrals of motion. Indeed, substituting (5.1.6) into (5.1.4) and separating the real and imaginary parts in it we obtain

$$\frac{dC}{dz} = \frac{\omega^2}{c^2} \Delta'' A^2, \quad \frac{dW}{dz} = \frac{\omega^2}{c^2} \Delta'' C. \tag{5.2.6}$$

Longitudinal variation of the field amplitude is described by

$$\frac{dA}{dz} = \pm \sqrt{2[W - U(A, C)]} \tag{5.2.7}$$

or

$$\frac{d^2 A}{dz^2} - \frac{C^2}{A^3} - \frac{\omega^2}{c^2} (\Delta' A - \varepsilon_2 A^3) = 0. \tag{5.2.8}$$

The values $C$ and $W$ represent respectively the energy flux in the direction of axis $z$ and the "energy", which are not constant in the presence of absorption. In the problem considered the energy flux is in the positive $z$-direction, so $C \geq 0$. As seen from (5.2.6), in the absorbing medium ($\Delta'' < 0$) the flux $C$ and the "energy" $W$ decrease with an increase in $z$. Introduction of the value $\Delta'' < 0$ with the following transition $\Delta'' \to 0$ corresponds to the "principle of limiting absorption" [362]. The field in the medium decays due to absorption at substantial distances from its boundary ($z \to \infty$), and the medium becomes linear. Correspondingly, the field approaches the plane wave:

$$E(z) \propto \exp \left[ -(\omega/c) \sqrt{\Delta_0} z \right]. \tag{5.2.9}$$

The root branch is chosen in the following way:

$$\sqrt{\Delta_0} = q' + iq'', \quad q' > 0, \quad q'' < 0. \tag{5.2.10}$$

Therefore the conditions at infinity lead to asymptotics

$$z \to \infty, \quad A \to 0, \quad \frac{dA}{dz} \to -q'A, \quad C \to -q''A^2. \tag{5.2.11}$$

If nonlinearity and absorption are absent ($\varepsilon_2 = 0$, $\Delta'' = 0$), then at $\Delta' > 0$ we have the TIR regime, and at $\Delta' < 0$ the transmission regime. For fields close to a unidirectional wave (transmission regime), it is convenient to introduce the values $I = A^2$ and $\kappa^2 = -\Delta' + \varepsilon_2 I$. Then, considering the fact that $\kappa^2 > 0$, we obtain in approximation of the slowly varying amplitudes instead of (5.1.4)

$$\frac{d(\kappa I)}{dz} = (\omega/c)\Delta'' I . \tag{5.2.12}$$

Hence at $\Delta' < 0$ we find the relation for a more exact [than (5.2.9)] determination of the field:

$$(\omega/c)\Delta''(z - z_0) = 3\kappa + (-\Delta')^{1/2} \ln \left| \frac{(-\Delta')^{1/2} - \kappa}{(-\Delta')^{1/2} + \kappa} \right| , \quad z_0 = \text{const.} \tag{5.2.13}$$

The boundary conditions at the interface $z = 0$ retain the form (5.1.23-5.1.24) with only the substitution $C \to C_0 = C_{z=0}$. The following relations arise from these conditions for incident wave intensity $I_{in} = |E_{in}|^2$ and for intensity reflection coefficient $R = |r|^2 = |E_r/E_{in}|^2$:

$$I_{in} = Q^{(+)}/4 , \quad R = Q^{(-)}/Q^{(+)} . \tag{5.2.14}$$

Here

$$Q^{(\pm)} = \left( A_0 \pm \frac{C_0}{k_{z1} A_0} \right)^2 + \left( \frac{A_0'}{k_{z1}} \right)^2 , \quad A_0 = A_{z=0} , \quad A_0' = \left( \frac{dA}{dz} \right)_{z=0} ; \tag{5.2.15}$$

the value $k_{z1}$ is defined in (5.1.19). To determine the reflection coefficient we should solve the equation for a field in the nonlinear medium (5.2.6) and (5.2.7) [or (5.2.6) and (5.2.8)] with boundary conditions (5.1.23-5.1.24) and (5.2.11). It is important that conditions (5.2.11) remove the "continuum problem" discussed in Sect. 5.1 while retaining the possibility of bistability (see below).

To analyze reflection regimes, it is convenient as before to use the mechanical analogy (see Sect. 5.1) with the following changes induced by the existence of weak absorption: As was shown in Sect. 5.1 (Fig. 5.2), in the case of transparent media the trajectories $A(z)$ correspond to the segments of horizontal straight lines ($W = \text{const.}$), located in the classically allowed regions [over the corresponding potential curve, where $W \geq U(A, C)$]; the extremes of the potential curves and relations (5.1.36-5.1.37) correspond to transmission regimes in which $A(z) = \text{const.}$ The presence of absorption results in a decrease in $C$ and $W$ with an increase in $z$, so that the trajectories should come to the coordinate origin $A = 0$, $W = 0$ at $z \to +\infty$. It is more convenient to watch the "reverse motion" along the trajectory from $z = +\infty$ to the interface $z = 0$. Thus the trajectory should leave the coordinate origin and then rise, in correspondence with (5.2.6). Simultaneously the potential curve will also rise because of an increase in $C$ from the zero at $z = +\infty$ [see

(5.2.6)]. Interruption of the trajectory at some point corresponding to values $A = A_0$, $C = C_0$, $dA/dz = A_0'$, allows us to calculate the corresponding values of incident wave intensity $I_{in}$ and reflection coefficient $R$ by means of (5.2.14-5.2.15).

The subsequent analysis substantially depends on the relation of the signs of the values $\varepsilon_2$ and $\Delta'$, four variants of which are discussed below. The simplest variants are (a) and (b), in which nonlinearity does not change the regime nature, hybrid regimes and bistability being absent.

(a) $\varepsilon_2 < 0$, $\Delta' > 0$ (Fig. 5.2a). In the absence of absorption ($\Delta'' = 0$) the trajectory is the axis $OA$, $C = 0$, $W = 0$, $A \to 0$ at $z \to +\infty$, and the reflection coefficient $R = 1$, so we have TIR, as was explained in Sect. 5.1. Low absorption (the dashed line in Fig. 5.2) does not lead to considerable changes in the regime.

(b) $\varepsilon_2 > 0$, $\Delta' < 0$ (Fig. 5.2b). Without absorption, there are transmission regimes corresponding to the bottom of the potential curves. The reflection coefficient is given by relations (5.1.36-5.1.37). To consider low absorption it is sufficient to use an approximation of the slowly varying amplitudes [see (5.2.12-5.2.13)] which describes a slow decrease down along the minima of potential curves (the dashed line in Fig. 5.2b). In this case weak absorption induces only small variations in the longitudinal field distribution and the reflection coefficient.

(c) $\varepsilon_2 < 0$, $\Delta' < 0$ (Fig. 5.2c). In a transparent nonlinear medium there are transmission regimes (minima and maxima of the potential curves) and hybrid regimes (horizontal straight lines coming into apexes of the potential curves at $z \to \infty$); it is in this variant that the "continuum problem" originated (see Sect. 5.1). Consideration of low absorption leads to the trajectory shown in Fig. 5.2c by the dashed line. At $A^2 < A_1^2 = (2/3)\Delta'/\varepsilon_2$, a drop to the coordinate origin ($A = 0$, $W = 0$) occurs along the minima of the potential curves with an increase in $z$.[2] The corresponding transmission regimes are described, as in the previous variant (b), by (5.2.12-5.2.13). The reflection coefficient is given by (5.1.36-5.1.37). In the neighbourhood of intensity value $A^2 = A_1^2$, approximation (5.2.12) is not valid anymore. Joining with the horizontal straight line (in the limit of zero absorption) takes place here. On the line $C = C_1$, intensity $A^2(z) > A_1^2$, its longitudinal variation is determined by the (5.1.47), and parametric dependence (through the parameter $I_b = A_0^2$) of reflection coefficient $R = |r|^2$ on incident wave intensity $I_{in}$ has the form following from (5.1.45):

$$I_{in} = u + C_1/(2k_{z1}) \,, \quad R = [u - C_1/(2k_{z1})]/I_{in} \,, \tag{5.2.16}$$

where

$$u = (|\Delta'|A_1^2 + \varepsilon_{10}I_b + |\varepsilon_2|I_b^2/2)/(4\varepsilon_1 \cos^2\theta) \,, \quad \varepsilon_{10} = \varepsilon_1 - \varepsilon_0' \,. \tag{5.2.17}$$

---

[2] Maxima of the potential curves in the region $A^2 < A_1^2$, as well as the dashed part of curve 1 in Fig. 5.4a, do not correspond to the regimes with the correct asymptotics at infinity.

At $I_b = A_1^2$ the values of the incident wave intensity and the reflection coefficient calculated by means of formulae (5.1.36-5.1.37) for transmission regimes and (5.2.16-5.2.17) for hybrid regimes, coincide. However, under the conditions

$$\varepsilon_{10} < 0 , \quad |\varepsilon_{10}| < |\Delta'| < (3/2)|\varepsilon_{10}| \tag{5.2.18}$$

dependence of the intensity $I_{in}$ on the parameter $I_b$ is nonmonotonous both for transmission regimes with the correct asymptotics ($I_b < A_1^2$) and for hybrid regimes ($I_b > A_1^2$). Actually, under conditions (5.2.18) this dependence for the transmission regime has a maximum determined by (5.1.43) at

$$I_b = I_b^{(m)} < A_1^2 , \tag{5.2.19}$$

and for the hybrid regime, there is a minimum at

$$I_b = I_b^{(2)} = -\varepsilon_{10}/|\varepsilon_2| > A_1^2 \tag{5.2.20}$$

with the corresponding value of incident wave intensity

$$I_{in}^{(2)} = \frac{1}{4\varepsilon_1 \cos^2 \theta |\varepsilon_2|} \left( \frac{2\Delta'^2}{3} - \frac{\varepsilon_{10}^2}{2} \right) + \frac{C_1}{2k_{z1}} . \tag{5.2.21}$$

Therefore in the range

$$I_{in}^{(2)} < I_{in} < I_{in}^{(m)} \tag{5.2.22}$$

there are both transmission and hybrid regimes at the same incident wave intensity $I_{in}$. The corresponding scheme of hysteresis change in the parameter $I_b$ (having the meaning of the radiation intensity at the boundary of the nonlinear medium) upon an increase or decrease in intensity $I_{in}$ is indicated in Fig. 5.4a by arrows. Hysteresis variation of the reflection coefficient also takes place (Fig. 5.4b). As indicated in Fig. 5.4, the transmission regime is realized at low intensities $I_{in}$. It is retained up to the value $I_{in}^{(m)}$, at which switching to the hybrid regime occurs, accompanied by a jump in the reflection coefficient phase (while retaining the intensity reflection coefficient $R$). With a consequent decrease in intensity $I_{in}$, the hybrid regime preserves its existence up to the value $I_{in}^{(2)}$, at which point a return to the transmission regime occurs, with a jump in the reflection coefficient.

At a sufficiently high intensity of the incident wave, the field in the boundary layer of the nonlinear medium "quickly" decreases because of the TIR caused by nonlinearity. Then, with the increase in distance from the media interface, the field decrease causes smooth transition to the transmission regime with a further slow (due to low absorption) decay of the field. The presence of absorption $\varepsilon'' > 0$ in this variant is essential, and it is the introduction of this absorption that permits us to consistently solve the "continuum problem" and to determine the correct form of the field distribution on the basis of a correct statement of the conditions at infinity. We can now proceed to the limit $\varepsilon'' \to 0$ in the final expressions, in the spirit of the "principle of limiting absorption" [362].

(d) $\varepsilon_2 > 0$, $\Delta' > 0$ (Fig. 5.2d). Generally speaking, without absorption there are regimes of transmission (the minima of the potential curves) and TIR (a segment of the axis $OA$, $A^2 < A_2^2 = 2\Delta'/\varepsilon_2$). If absorption is taken into account, transmission regimes are rendered impossible, as the corresponding trajectory $A(z)$ lowering on the potential curve minima (representing the transmission regimes) does not reach with $z \to \infty$ the coordinate origin $A = 0$, $C = 0$; thus the conditions at infinity are not obeyed for them.

Field distributions satisfying the conditions at infinity can be found if we perform the "reverse motion" from the coordinate origin $A = 0$, $C = 0$ at $z = \infty$ to the media interface $z = 0$. In this case the image point moves along the almost horizontal line with a slow rise up to the turning point from the potential curve corresponding to the increased value $C = C^{(1)}$. Then the field amplitude decreases up to the next turning point at which $C = C^{(2)}$, etc. As a result we obtain a trajectory shown qualitatively by the dashed line in Fig. 5.2d. With an increase in the incident wave intensity the number of oscillations of the field in the nonlinear medium increases. The oscillatory nature of the field in the nonlinear medium is similar to that found in [360] as applied to normal incidence of an electromagnetic wave on a conductor. It leads to the specific multi-loop dependence of the reflection coefficient $R$ on the incident wave intensity. This dependence is presented in Fig. 5.5, which was obtained by the numerical solution of the first equation of set (5.2.6) and (5.2.8) with boundary conditions (5.2.11). With a low absorption coefficient, the main features of this dependence can also be obtained analytically [273]. The reflection coefficient remains close to unity with a slow increase in intensity $I_{in}$ from small values up to the critical value $I_{in}^{(1)} \approx \Delta'/(2\varepsilon_2)$. With a further increase in $I_{in}$ the series of consequent "vertical drops" from the edges of the loops shown ($I_{in}^{(n)}$, $R^{(n)}$, $n = 1$, 3, 5, ...) and shifts to the next edges ($I_{in}^{(n+2)}$, $R^{(n+2)}$) occurs. If intensity $I_{in}$ further decreases, then after a smooth decrease in the reflection coefficient (the edges with even $n$) the series of "vertical rises" occurs, and at low $I_{in} = I_{in}^{(2)}$ the last abrupt jump from $R = R^{(2)}$ to the original value of the reflection coefficient $R \approx 1$ occurs.

Longitudinal dependence of the field in the nonlinear medium can be interpreted in the following way: With low intensities of the incident wave, the regime is TIR. As $\varepsilon_2 > 0$, sufficiently intensive radiation violates the TIR regime, and the field penetrates into the nonlinear medium. But absorption leads to a decrease in the field intensity. Therefore at some distance from the media interface, the TIR regime will be restored; however, the boundary of internal reflection will be shifted into the nonlinear medium. Because of the reflection of the incident wave from this boundary, the field in the region between the boundary and media interface will be close to a standing wave (the regime of oscillations). Oscillations of the field intensity and nonlinear

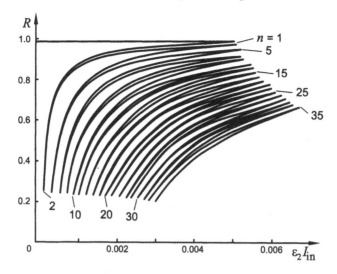

**Fig. 5.5.** Dependence of the reflection coefficient on the intensity of incident radiation [273]

medium permittivity following them result in periodic antireflection (decrease in the reflection coefficient) and to multistability with a variation of the incident wave intensity.

What is the role of absorption in the present variant? In its absence ($\Delta'' = 0$) only the TIR regime is possible, and only at restricted intensity of the incident wave: $I_{in} < I_{in}^{(1)} = \Delta'/(2\varepsilon_2)$. But if intensity $I_{in}$ exceeds this value, no stationary regimes of reflection exist, as is evident from Fig. 5.2**d**. Absorption of course has a finite (nonzero) value in all passive media; therefore it is natural to expect settling of the oscillation regime found when considering absorption. However, the setup time for the oscillation regime will grow with a decrease in absorption (up to the infinite value at $\Delta'' \to 0$). In other words, reflection of the plane wave in this case is essentially nonstationary. The nature of stationary regimes at different combinations of the signs of $\varepsilon_2$ and $\Delta'$ is summarized in Table 5.1. The qualitative form of longitudinal distribution of the field amplitude $A(z)$ in different regimes is illustrated in Fig. 5.6.

Another variant of the realistic conditions corresponds to a finite thickness of the nonlinear medium layer. If integral absorption over the whole thickness of the layer $d$ is not large, then the reflection regime essentially depends on the conditions at the second boundary $z = d$.

To conclude this section, we discuss the feedback mechanism necessary for bistability. For a layer of the nonlinear medium such a mechanism is evident: Fresnel's reflection at the interfaces likens them to mirrors, and the scheme itself is similar to an ordinary nonlinear cavity (see Chap. 4). In the case of a nonlinear half-space ($d = \infty$), only one interface remains, i.e.,

**Table 5.1.** Regimes of nonlinear reflection of a plane wave

|  | $\Delta' > 0$ | $\Delta' < 0$ |
|---|---|---|
| $\varepsilon_2 = 0$ | TIR | Transmission regimes |
| $\varepsilon_2 > 0$ | TIR*, oscillation regimes | Transmission regimes |
| $\varepsilon_2 < 0$ | TIR | Transmission regimes, hybrid regimes** |

* Multistability; ** bistability

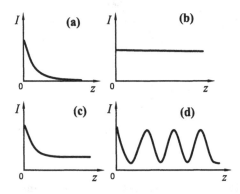

**Fig. 5.6.** Longitudinal distribution of radiation intensity in nonlinear medium for different regimes of reflection: total internal reflection (**a**), transmission regime (**b**), hybrid regime (**c**), oscillation regime (**d**)

one "mirror", which does not provide the required feedback. However, in the TIR regime, as well as in the hybrid and oscillation regimes, there is also the second – distributed – "mirror", induced by field inhomogeneity in the nonlinear medium. Indeed, in these regimes radiation intensity and the effective nonlinear refractive index of the medium change in the longitudinal direction $z$. Longitudinal inhomogeneity of the medium causes reflection of the waves, which is manifested especially sharply in the oscillation regime. To describe feedback, it is necessary to take into account the second-order derivative $\partial^2 E / \partial z^2$, which is neglected when the method of slowly varying amplitudes – commonly used in nonlinear optics – is applied.

## 5.3 Nonlinear Reflection of Beams

### 5.3.1 Boundary Conditions

The necessity to consider nonlinear reflection of beams is due to a number of reasons. First of all, as we have seen in the case of nonlinear cavities (see Chap. 4), bistability in the framework of plane-wave theory does not guarantee its retention for real radiation beams, especially with oblique incidence of the latter. This very problem has been hotly debated in the literature [179, 274, 221, 388, 364]. Second, it is known that the beam transverse structure determines the development of radiation self-focusing and defocusing in a nonlinear medium (see Appendix A). Third, as was demonstrated in Sects. 5.1 and 5.2, medium nonlinearity is especially important if the incidence angle $\theta$ is close to the TIR critical angle, or with grazing incidence ($\theta^2 \ll 1$) and close linear permittivities of the two media ($|\varepsilon_{10}| \ll 1$), when parameter $\Delta'$ is small and comparable to the nonlinear term $\delta\varepsilon$. Under these conditions the beam finite width is most conspicuous even for linear media, causing such effects as transverse beam shift, side modes, etc. [59]. For nonlinear media additional effects can arise, including excitation of the above-mentioned nonlinear surface waves.

At the same time, a consistent analysis of the nonlinear reflection of beams is hampered by the necessity for simultaneous consideration of both transverse and longitudinal variations of the field structure. Therefore material simplifications are necessary here. Thus, in the present section based on [274, 317], we will consider only slit and linearly polarized (s-polarized) beams [$E = E_y(x, z)$], neglecting possible field variations in the $y$-direction. We proceed from the nonlinear Helmholtz equation (5.1.2):

$$\frac{\partial^2 E}{\partial x^2} + \frac{\partial^2 E}{\partial z^2} + k^2 E = 0 , \quad k^2 = \frac{\omega^2}{c^2}\varepsilon . \qquad (5.3.1)$$

In the case of a nonlinear medium layer of thickness $d$, the dielectric permittivity has the form

$$\varepsilon = \begin{cases} \varepsilon_1 , & z < 0 , \\ \varepsilon_0 + \varepsilon_2|E|^2 , & 0 < z < d , \\ \varepsilon_3 , & z > d . \end{cases} \qquad (5.3.2)$$

Generally speaking, (5.3.1) with (5.3.2) should be solved in the entire space $-\infty < z < \infty$. However, it is convenient to formulate the conditions on the layer boundary in such a way that the problem is reduced to the consideration of only the internal region $0 < z < d$. To derive such boundary conditions, we will use the field expansion into the plane wave spectrum and the radiation principle in the linear media ($z > 0$ and $z > d$). So, at $z > d$ the field $E$ is represented in the form

$$E(x, z) = \int_{-\infty}^{\infty} F(k_x) \exp\left(i\sqrt{k_3^2 - k_x^2}\, z\right) \exp(ik_x x)\, dk_x , \qquad (5.3.3)$$

where $k_3 = \omega\sqrt{\varepsilon_3}/c$ (linear media are assumed transparent) and $F(k_x)$ is the density of the field spatial spectrum:

$$F(k_x) \exp\left(i\sqrt{k_3^2 - k_x^2}\, z\right) = \frac{1}{2\pi} \int_{-\infty}^{\infty} E(x, z) \exp(-ik_x x)\, dx . \qquad (5.3.4)$$

At $k_x^2 > k_3^2$ the substitution $i\sqrt{k_3^2 - k_x^2} \rightarrow -\sqrt{k_x^2 - k_3^2}$ is assumed in the corresponding exponents. The form of (5.3.3) provides for the region $z > d$ the existence of only the waves moving away from the layer or decaying waves.

After calculation of the value $\partial E/\partial z$ from (5.3.3), the requirement of continuity of $E$ and $\partial E/\partial z$ at the interface $z = d$ provides the desired boundary condition:

$$\frac{\partial E}{\partial z}(x, d) = \frac{i}{2\pi} \int_{-\infty}^{\infty} dk_x \sqrt{k_3^2 - k_x^2} \exp(ik_x x) \int_{-\infty}^{\infty} dx' \, E(x', d) \exp(-ik_x x') \,. \tag{5.3.5}$$

Similarly, the boundary condition at $z = 0$ is

$$\frac{\partial E}{\partial z}(x, 0) = \frac{\partial E_{\text{in}}}{\partial z}(x, 0) + \frac{i}{2\pi} \int_{-\infty}^{\infty} dk_x \sqrt{k_1^2 - k_x^2} \exp(ik_x x)$$
$$\times \int_{-\infty}^{\infty} dx' \, [E_{\text{in}}(x', 0) - E(x', 0)] \exp(-ik_x x') \tag{5.3.6}$$

or, using the incident field expansion of type (5.3.3),

$$\frac{\partial E}{\partial z}(x, 0) = \frac{i}{2\pi} \int_{-\infty}^{\infty} dk_x \sqrt{k_1^2 - k_x^2} \exp(ik_x x)$$
$$\times \int_{-\infty}^{\infty} dx' \, [2E_{\text{in}}(x', 0) - E(x', 0)] \exp(-ik_x x') \,. \tag{5.3.7}$$

Here $E_{\text{in}}$ is the given field of incident radiation and $k_1 = (\omega/c)\sqrt{\varepsilon_1}$. Note that the field $E$ and its derivatives enter (5.3.5–5.3.7) only by their values at the medium surface ($z = 0, d$). Thus the problem is reduced to an internal problem. Such boundary conditions were proposed in [317]. They are naturally generalized to the case of nonstationary problems and transversely two-dimensional beams. In contrast to Leontovich's approximate boundary conditions [188], (5.3.5–5.3.7) are exact. Simpler approximated relations can be derived from them. Let us assume that the field spatial spectrum is narrow and located close to the carrier wave vector with components

$$k_{x0} = (\omega/c)\sqrt{\varepsilon_1} \sin\theta \,, \quad k_{z1} = \sqrt{k_1^2 - k_{x0}^2} = (\omega/c)\sqrt{\varepsilon_1} \cos\theta \,.$$

Then it is possible to use the expansion in powers of $\kappa = k_x - k_{x0}$ in (5.3.6–5.3.7):

$$\sqrt{k_1^2 - k_x^2} = k_{z1} - \kappa \tan\theta + \ldots \,, \quad \kappa^2 \ll k_{z1}^2 \,. \tag{5.3.8}$$

Retaining only the first term on the right-hand side of (5.3.8), (5.3.7) takes the simplest form

$$\left(\frac{\partial E}{\partial z}\right)_{z=0} = ik_{z1}[2E_{\text{in}}(x, 0) - E(x, 0)] \,. \tag{5.3.9}$$

To improve (5.3.9), let us take into account the term of (5.3.8) linear in $\kappa$ and separate the exponent corresponding to the central wave number $k_{x0}$:

$$E_y(x, z) = E(x, z) \exp(ik_{x0}x) \,. \tag{5.3.10}$$

The boundary condition will now include a derivative with respect to $x$:

$$\left(\frac{\partial E}{\partial z}\right)_{z=0} = \left(ik_{z1} - \tan\theta\frac{\partial}{\partial x}\right)[2E_{\text{in}}(x, 0) - E(x, 0)] \,. \tag{5.3.11}$$

Retention of the terms of higher orders in $\kappa$ ($\kappa^n$) results in the appearance of higher-order derivatives ($\partial^n E/\partial x^n$) in the boundary condition, i.e., expansion of (5.3.7) in series in derivative power. The approximate form (5.3.9) was used in [179], and its more exact variant was proposed in [274].

## 5.3.2 Paraxial Approach

Let the incidence angle of the beam (in the $x$–$z$ plane) determined by the direction of the beam axis be close to the TIR critical angle:

$$\theta = \theta_{\text{cr}} + \delta\theta \,, \quad \delta\theta^2 \ll 1 \,. \tag{5.3.12}$$

Then the beam transmitted into the nonlinear medium propagates mainly in the $x$-direction. Assuming in (5.3.10) that the field amplitude $E$ varies slowly along $x$, one can proceed from wave equation (5.3.1) to a paraxial equation:

$$2ik_{x0}\frac{\partial E}{\partial x} + \frac{\partial^2 E}{\partial z^2} + \frac{\omega^2}{c^2}(-\Delta_0 + \varepsilon_2|E|^2)E = 0 \,, \tag{5.3.13}$$

where the value $\Delta_0$ is introduced from (5.1.19). If the beam width is characterized by value $w_{\text{b}}$ and radiation intensity is negligible, the medium becomes effectively linear at distance $\propto w_{\text{b}}$ from the beam axis:

$$E(x_0, z) = 0 \,, \quad |x_0| \gg w_{\text{b}} \,, \quad x_0 < 0 \,. \tag{5.3.14}$$

Now, in such statement of the problem, the field is determined uniquely for all $x > x_0$. Note that the problem of bi- and multistability predicted by the plane-wave theory (see Sects. 5.1 and 5.2) requires more detailed analysis and is beyond the paraxial approximation. This point will be discussed below.

Paraxial equation (5.3.13) with boundary conditions (5.3.11) and (5.3.14) can be solved numerically. At $x = x_0$ the field was given as zero in accordance with (5.1.14). In [317] calculations were performed for Gaussian incident beams with various width and amplitude $E_{\text{m}} = E_{\text{in}}(x = 0)$ at different incident angles $\theta$, as well as for the "plane-Gaussian" beams, which are Gaussian at $x < 0$ and plane wave at $x > 0$. The code was tested with phenomena well studied in linear optics, including a lateral shift of the reflected beam [59].

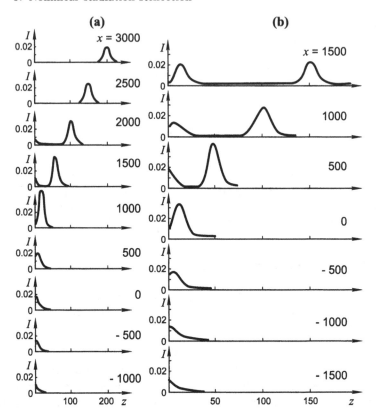

**Fig. 5.7.** Distributions of radiation intensity in a nonlinear medium: $E_\mathrm{m} = 0.062$ (a) and 0.067 (b) [274]

The results given in Figs. 5.7 and 5.8 agree with conclusions of the plane-wave theory (with the exception of the bistability problem). Good agreement was found for transmission regimes between the local reflection coefficient and corresponding values in the plane-wave theory. For media with negative Kerr nonlinearity, the transition from transmission regimes to hybrid regimes was confirmed, which was found in Sect. 5.2; weak absorption was taken into account. However, in general the local reflection coefficient for beams is not determined by the local intensity of the incident beam because of the lateral shift of reflected beam in the TIR and hybrid regimes.

The following phenomenon takes place in media with self-focusing non-linearity ($\varepsilon_2 > 0$) at $\delta\theta > 0$: At low radiation intensities the regime is TIR, the reflected beam being similar to the incident beam shifted in the positive $x$-direction. With an increase in the amplitude $E_\mathrm{m}$ up to the value close to $\sqrt{\Delta_0/2}$, a deep gap appears in the profile of the reflected beam (Fig. 5.8a). It is connected with the TIR nonlinear breakdown by a narrow radiation filament coming from the medium interface in the region of intensity max-

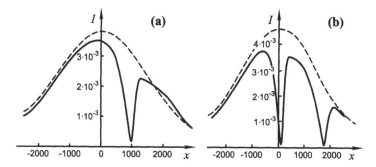

**Fig. 5.8a,b.** Transverse profiles of intensity of incident (*dashed curves*) and reflected (*solid curves*) radiation under the conditions of Fig. 5.7 [274]

imum of the incident beam (Fig. 5.7a). With a further increase in $E_m$, the number of gaps in the reflected beam profile and the number of the filaments in the nonlinear medium bulk increase (Fig. 5.7b and Fig. 5.8b). For the "plane-Gaussian" beams, filaments are generated periodically (in the transverse direction $x$, $x > 0$) and are accompanying by a periodic structure of intensity gaps for the reflected beam. The phenomenon is a threshold one, and the critical value of radiation intensity decreases when the incident angle approaches the TIR critical angle. Bending of the filament in the nonlinear medium could be associated with the asymmetry of its intensity profile [62].

### 5.3.3 Nonparaxial Approach: Bistability of Beam Reflection

As was already shown, paraxial equation (5.3.13) with the boundary conditions presented above has a unique solution at the given parameters of the incident beam. However, this equation is inapplicable under conditions of sharp hysteresis jumps with high transverse gradients of the field. As in Chap. 4, the description of switching waves and boundary layers arising in these regions involves not paraxial, but complete wave equation and/or boundary conditions with the second (or of higher order) field derivative $\partial^2 E/\partial x^2$.

We will demonstrate bistability with nonlinear reflection by numeric solution of (5.3.1) with boundary conditions (5.3.6-5.3.7), following [317]. To avoid beam breakdown into separate filaments, we will consider the medium with self-defocusing (the negative coefficient of nonlinearity, $\varepsilon_2 < 0$). In order to simplify calculations and use the problem's symmetry, we assume radiation incidence to be normal ($\theta = 0$). Fairly high critical values of switching intensity serve as the cost of these simplifications.

Away from the beam axis (at distance $x_0$ noticeably exceeding its width $w_b$), we use a boundary condition of the form

$$\left(\frac{\partial E}{\partial x}\right)_{x=\pm x_0} = 0 \ . \tag{5.3.15}$$

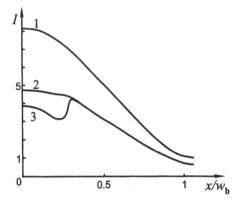

**Fig. 5.9.** Transverse profiles of the intensity of incident (1) and reflected (2,3) radiation beams [317]

For a numerical solution, the nonlinear medium is represented by a layer of finite thickness $d$. Transition to the problem of beam reflection from a half-space is achieved by minimizing reflection from the interface $z = d$. To this end, linear bulk absorption is introduced, $\varepsilon_0'' = \mathrm{Im}\,\varepsilon_0 > 0$, and the value $\varepsilon_3$ is chosen corresponding to antireflection.[3]

We thus consider normal incidence of the beam of the form

$$E_{\mathrm{in}}(x) = \begin{cases} E_0 \exp(-x^2/w_{\mathrm{b}}^2) , & |x| < w_{\mathrm{b}} , \\ E_0/e, & |x| > w_{\mathrm{b}} . \end{cases} \tag{5.3.16}$$

Field nonzero background at $|x| > w_{\mathrm{b}}$ reduces the intensity variation range. Referring the reader for calculation details to [317], we present here the main results.

In Fig. 5.9 the transverse profiles of intensity $I = |E|^2$ of the incident and reflected beams are shown. Depending on initial field distribution, one of the two stationary profiles forms. The first is smooth, and it practically does not change its shape upon incident beam width variation; thus the relation between the local values of intensity of incident and reflected beams corresponds to the plane-wave theory (the transmission regime). The second profile includes an intensity jump, which becomes sharper with an increase in the incident beam width. At sufficiently large $w_{\mathrm{b}}$ this jump represents the boundary layer (a front of the motionless switching wave) between the states of the transmission and hybrid regimes. Therefore, spatial bistability exists in the problem of radiation beam reflection from a nonlinear interface too.

---

[3] Generally speaking, introduction of absorption is not mandatory, as the beam diffraction widening in the transparent medium also results in the required field decay at a great distance from the layer front interface. However, introduction of absorption is more convenient for the calculations and it needs a smaller layer thickness.

## 5.4 Discussion and References

As was already mentioned, the prediction of hysteresis in the case of non-linear reflection and the consistent analysis of normal incidence of the plane electromagnetic wave on the boundary of a medium with a refractive index nonlinearity of the self-focusing type were first presented by Silin [360]. In particular, the regimes referred to as oscillation regimes in Sect. 5.2 were found in [360].

As applied to optics and oblique incidence of light, the possibility of hysteresis was examined again in [52, 157]. In [52] the case of self-focusing nonlinearity only was considered, and two nonlinearity types (self-focusing and self-defocusing) were studied in [157]. However, in both [52] and [157] it was postulated that only regimes known in linear reflection – regimes of transmission and of TIR, see Sect. 5.1 – are possible in the case of nonlinear reflection. Thereby oscillation regimes existing under self-focusing nonlinearity were disregarded in [52, 157]. For defocusing nonlinearity the hybrid regimes omitted in [157] were studied later together with the "continuum problem" in [272]. Note that oscillation regimes were also found (and called "self-reflection regimes") in [348, 215, 216, 150, 106] for a two-level model of the medium and in [163, 207] for semiconductors. Indirect experimental evidence in favour of the existence of oscillation regimes in a system of two-level atoms was presented in [355]. Nonlinear reflection of a plane wave from a photorefractive crystal was considered in [70].

Referring the reader, for further references on this subject, to monograph [51], we note that a consistent theory of plane-wave nonlinear reflection from the boundary of media with Kerr-type self-focusing and self-defocusing non-linearities was given in [273, 178]. For media with defocusing nonlinearity, conclusions consistent with those of [273, 178] were obtained later – but differently – in [158]. A different point of view – that of settling of TIR instead of the hybrid regimes – was suggested in [260] and later entered in monograph [51]; however, this statement is evidently incorrect.

The conclusions of [360, 273, 178] also differ from those in [52, 157] in the question of plane-wave reflection from a medium with self-focusing nonlinearity. In our opinion, this is connected with insufficiently exact interpretation of radiation conditions (at $z \to \infty$) in [52, 157]. However, further discussions of this subject are not so pressing now because of the instability of nonlinear plane waves and radiation filamentation in media with self-focusing nonlinearity found in [179, 274]. This phenomenon hampers the observation of possible bistability for such media and makes inapplicable the plane-wave theory in these cases. Note also that consideration of instabilities in the case of defocusing nonlinearity [303] confirms our solution of the "continuum problem" (see Sect. 5.2).

The question of nonlinear refraction of radiation beams was first discussed in [179, 274]. The unfoundedness of the conclusion of [179], in contradiction to [274], with regard to complete absence of bistability in the case of beams with finite width, has already been explained in Sect. 5.3. As it will be argued at the end of this section (see also [289]), this conclusion should be reformulated as the difficulty to reach bistability in the simplest schematic of nonlinear reflection, Fig. 5.1.

One more question in connection with this discussion refers to the applicability of the paraxial approach to the problem of nonlinear radiation reflection. We have already repeatedly stated that bistability is not described by the paraxial equation (in the absence of other additional mechanisms of bidirectional transverse feedback). But, conceivably, bistability is not the only subject deserving consideration! In particular, the phenomena of filamentation and the lateral shift of radiation beams incident on the nonlinear medium and the excitation of nonlinear surface waves (see Sect. 5.3 and [179, 274, 242, 14, 4, 29, 50, 173, 17]) could be important for surface polariton spectroscopy [6].

Abstracting from possible instabilities, one can imagine the following scenario of nonlinear reflection of radiation beams: The existence of bistability in the plane-wave approximation means that switching waves can propagate in the transversely distributed system. If the velocity of the switching wave changes sign with intensity change in the bistability range,[4] then hysteresis will take place upon the reflection of a wide beam. This is just the case of normal incidence analyzed in Sect. 5.3 and in [317]. In the opposite case, when the switching wave does not stop at any intensity value within the range of its existence, bistability of the reflection of wide radiation beams is excluded. At a sufficiently large intensity of incident radiation, the profile of reflected radiation intensity will include fairly sharp jumps corresponding to spatial switching between the branches of the plane-wave theory.

Neither switching waves nor sharp stationary fronts of spatial switching are described by the paraxial equation. The latter is applied only for regimes which do not include spatial switching. For them, use of the paraxial approximation is justified and reasonable.

As demonstrated in Sect. 5.3, bistability could be realized when we have reflection of a wide radiation beam from a medium with defocusing nonlinearity where modulation instability does not occur. At a small angle of incidence, $\theta$, the critical value of switching intensity is large (nonlinear variation of dielectric permittivity $|\delta\varepsilon| \sim \varepsilon_0$ is necessary). At grazing incidence, when radiation propagates mainly along the interface (in a direction close to the $x$-axis, Fig. 5.1), the transverse coupling becomes one-sided (see Sect. 4.4). Then bistability disappears because of radiation drift along $x$ and the absence of motionless switching waves.

Experimental research of nonlinear reflection has so far only been performed for media with self-focusing nonlinearity. The first of the studies [365, 366] reported some evidence for hysteresis. However, subsequent analy-

---

[4] More exactly, it should be valid for the "left switching wave", see Sect. 4.4.

sis [364] – which was more thorough – demonstrated the absence of bistability and the importance of nonlinearity relaxation.

It would be possible to reach bistability by the introduction of additional two-sided transverse feedback, induced by diffusion or heat transfer in the nonlinear medium (see Sect. 4.4). However, switching would then be fairly slow. Using the approach mentioned in Sect. 4.4, one could modify the nonlinear reflection scheme to reach bistability at small and fast nonlinearities, corresponding to the prediction of the plane-wave theory (see Sect. 5.2). It could be a two-beam scheme of nonlinear reflection (with two beams incident on the nonlinear medium) [289], a scheme with distributed feedback at surface of medium bulk [47], etc. An important requirement for these schemes is the absence of instabilities or the possibility of their suppression. Note that these schemes differ radically from schemes of transverse bistability with counter-propagating beams [159, 117, 197] (for these schemes, bistability is absent for external radiation in the form of plane wave).

In connection with essential longitudinal distributivity, settling of non-stationary regimes of nonlinear reflection is possible, even in the presence of low absorption in the medium bulk. Examples of numerical calculations of nonstationary regimes of plane-wave reflection from the transparent nonlinear medium with self-focusing nonlinearity are presented in [269]. However, it is difficult to apply the conclusions of [269] to any real systems because of the field transverse instability discussed above; and they are not applied to separate filaments forming in the nonlinear medium due to radiation beam breakdown. For the diffraction mechanism of transverse coupling, e.g., in two-beam schemes of nonlinear reflection, conditions of existence of localized structures (*dissipative optical solitons*) and other spatial structures could be provided as well (see Chaps. 4 and 6); their variety increases in regimes with a varying polarization state of radiation. These considerations would hopefully boost further experiments on nonlinear reflection, which could rehabilitate low-threshold and fast bistability with nonlinear reflection (e.g., in the two-beam schemes) and discover new nontrivial spatio-temporal structures.

# 6. Bistable Laser Schemes

In a laser with a saturable absorber, bistability and hysteresis variation of the output power with pump variation exist. If such a laser has a large aperture, and diffraction serves as the dominant mechanism of transverse coupling, formation is possible of laser solitons – stable "bright" islands of lasing on a background of a "black" nonlasing state over the rest of the aperture. Similar dissipative localized structures arise in cavity-less schemes of "convective bistability" – a continuous medium with nonlinear amplification and absorption. The set of laser solitons includes structures with the following characteristics: geometrically one-, two-, and three-dimensional; stationary and oscillating; transversely motionless, moving, and rotating; with a regular wavefront and with screw dislocations of different orders; and single and bound configurations. Interaction of laser solitons and their interaction with the scheme inhomogeneities can lead to changes in soliton type and number.

## 6.1 Model and General Relations

### 6.1.1 Laser Model

A wide-aperture laser (Fig. 6.1) differs from the interferometer considered in Chap. 4 (see Figs. 1.5 and 4.1), first, by the presence of an active medium (with gain) inside the cavity, and, second, by the absence of an external signal. The first distinction leads to certain problems in the choice of model for the wide-aperture laser which sometimes are not properly accounted for in the literature. The point is that in an ideal system with an active medium infinite even in one of the directions (e.g., that perpendicular to the cavity axis), unlimited amplification of spontaneous radiation is inevitable in the absence of angular filtration. Thus, all the regimes found in theory without taking into account this fact can be unstable. The suppression of amplified spontaneous radiation and spurious oscillations in wide-aperture laser schemes is of considerable practical interest; we have not touched on it here.

In this chapter we will restrict our consideration to the analysis of transverse phenomena in a wide-aperture laser with a saturable absorber under conditions of bistability (hard regime of lasing) and in corresponding cavity-less schemes presenting continuous media with saturable gain and absorption.

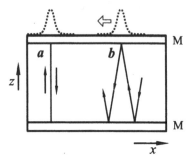

**Fig. 6.1.** Diagram of a laser. The cavity is formed by mirrors M and filled with a medium with amplification and absorption; the light propagates in the directions shown by *arrows*, parallel to the cavity axis $z$ (**a**) or at some angle to it (**b**); *dotted lines* show intensity profiles of transversely motionless (**a**) and moving (**b**) laser solitons

The bistability mechanism was illustrated in Sect. 1.1 in the framework of a point scheme (or in the plane-wave approach) [196]. In the case of the diffusion mechanism of transverse coupling, transverse phenomena such as switching waves and spatial hysteresis have a form close to that presented in Chap. 2 [280]. Therefore we will focus here on the transverse effects under conditions of the diffraction mechanism of transverse coupling. Localized structures – laser solitons – were found theoretically in [312] and were investigated in more detail in [329, 300] and in references therein; see also [11, 414, 415]. Irrespective of lasers, related stable localized solutions of the one-dimensional generalized Ginzburg–Landau equation were noted in [212], see also [68]. Localized structures in nonlinear waveguides with amplification, and in monostable wide-aperture lasers, were considered in [128, 249, 393], but they are unstable (see below).

### 6.1.2 Maxwell–Bloch Equations

We start from (4.1.23) derived for the field envelope $E$ averaged over the longitudinal direction (mean-field approximation) [378]; we neglect frequency dispersion (then the light group and phase speeds $v_0$ coincide) and omit external radiation ($E_{\text{in}} = 0$) and the corresponding phase detuning ($\Delta_{\text{ph}} = 0$). In a dimensionless form it reads

$$\frac{\partial E}{\partial t} - \mathrm{i}\Delta_\perp E = P_{\text{g}} - P_{\text{a}} - E . \tag{6.1.1}$$

Here, as above, $\Delta_\perp$ is the transverse Laplacian, and the dimensionless transverse coordinates $\boldsymbol{r} = (x, y)$ are normalized by the width of the effective Fresnel zone $\sqrt{L/[2k_0(1 - R)]}$, where the cavity length $L$ coincides with the longitudinal extent of the medium, $k_0$ is the light wave number, and $R$ is the product of cavity mirror reflectivities. As in the case of interferometers (Chap. 4), there are no distinctions between ring and two-mirror cavity schemes. Envelopes $P_{a,g}$ of atomic polarizations for the absorber (subscript a) and amplifying medium (subscript g) follow the Bloch equations:

$$\tau_g \frac{\partial g}{\partial t} = g_0 - g - \text{Re}(E P_g^*) , \qquad \tau_a \frac{\partial a}{\partial t} = a_0 - a - b\text{Re}(E P_a^*) ,$$

$$\tau_{\perp g} \frac{\partial P_g}{\partial t} = gE - (1 + i\Delta_g)P_g , \qquad \tau_{\perp a} \frac{\partial P_a}{\partial t} = aE - (1 + i\Delta_a)P_a . \quad (6.1.2)$$

In (6.1.1-6.1.2), the time $t$ is normalized by the cavity transient time, and $g$ and $a$ are population differences in active and passive media, and $g_0$ and $a_0$ are stationary values of the population differences for the nonlasing regime ($E = 0$), which are proportional to small signal coefficients of gain and absorption, respectively. The ratio parameter $b = \tau_a \tau_{\perp a} \mu_a^2 / \tau_g \tau_{\perp g} \mu_g^2$ measures the relative saturability of active and passive media. Here $\mu_{g,a}$ are the atomic dipole momenta, and $\tau_{g,a}, \tau_{\perp g,a}$ are the corresponding relaxation times; dimensionless frequency detunings are $\Delta_g = (\omega_g - \omega_c)\tau_{\perp g}$ and $\Delta_a = (\omega_a - \omega_c)\tau_{\perp a}$, where $\omega_c$ is the frequency of an empty cavity mode, and $\omega_{g,a}$ are the central frequencies of the spectral lines of amplification and absorption. Model (6.1.2) corresponds to the case of homogeneous spectral broadening and the neglecting of particle diffusion.

Let us consider a *class B laser*, for which the media polarization relaxation times are much smaller than the cavity transient time ($\tau_{\perp g, \perp a} \ll 1$), while the population relaxation times $\tau_{g,a}$ are large enough. Following the approach developed in [375], we can take into account the effects of polarization relaxation in the first-order approximation in $\tau_{\perp g, \perp a}$ only. Then we obtain the following system of equations [91]:

$$\frac{\partial \bar{E}}{\partial t} - (i + d)\Delta_\perp \bar{E} = [-1 + (1 - i\Delta_g)\bar{g} - (1 - i\Delta_a)\bar{a}]\bar{E} ,$$

$$\tau_g \frac{\partial \bar{g}}{\partial t} = \bar{g}_0 - (1 + I)\bar{g} , \qquad \tau_a \frac{\partial \bar{a}}{\partial t} = \bar{a}_0 - (1 + \bar{b}I)\bar{a} , \quad (6.1.3)$$

where $I = |\bar{E}|^2$ is the laser field intensity for a normalized amplitude $\bar{E} = E/\sqrt{1 + \Delta_g^2}$, $\bar{g} = g/(1 + \Delta_g^2)$, $\bar{g}_0 = g_0/(1 + \Delta_g^2)$, $\bar{a} = g/(1 + \Delta_a^2)$, $\bar{a}_0 = a_0/(1 + \Delta_a^2)$, $\bar{b} = b(1 + \Delta_g^2)/(1 + \Delta_a^2)$, and the effective intensity-dependent coefficient of diffusion $d = d(I) = 2[\tau_{\perp a} \bar{a}\Delta_a/(1 + \Delta_a^2) - \tau_{\perp g}\bar{g}\Delta_g/(1 + \Delta_g^2)]$.

In the limit of fast nonlinearity ($\tau_{g,a} \to 0$) we find from (6.1.3) $\bar{g}(I) = \bar{g}_0/(1 + I)$, $\bar{a}(I) = \bar{a}_0/(1 + \bar{b}I)$. Then (6.1.3) is reduced to a single equation for the electric field envelope:

$$\frac{\partial \bar{E}}{\partial t} - (i + d)\Delta_\perp \bar{E} = f(|\bar{E}|^2)\bar{E} , \quad (6.1.4)$$

$$f(I) = -1 + \frac{(1 - i\Delta_g)\bar{g}_0}{1 + I} - \frac{(1 - i\Delta_a)\bar{a}_0}{1 + \bar{b}I} ,$$

$$d = d(I) = -2 \left( \frac{\tau_{\perp g} \bar{g}_0 \Delta_{\mathrm{g}}}{(1 + \Delta_{\mathrm{g}}^2)(1 + I)} - \frac{\tau_{\perp a} \bar{a}_0 \Delta_{\mathrm{a}}}{(1 + \Delta_{\mathrm{a}}^2)(1 + \bar{b}\bar{I})} \right) . \tag{6.1.5}$$

Note that unlike the amplitude equations of the Ginzburg–Landau type, where $f(I)$ and $d(I)$ are expanded in power series, (6.1.4) is valid not only in the vicinity of the bistability threshold, and therefore it is more adequate to the experimental situation. The next simplification is to neglect the intensity dependence of the diffusion coefficient $d(I)$; this assumption is justified in Sect. 6.2.2. Note also that if the diffusion coefficient defined in (6.1.5) is not positive, additional terms with fourth-order spatial derivatives must be included in (6.1.4) [375].

In the case when the small diffusion coefficient is neglected ($d = 0$), instead of (6.1.4) we have

$$\frac{\partial \bar{E}}{\partial t} - \mathrm{i} \Delta_{\perp} \bar{E} = f(|\bar{E}|^2) \bar{E} . \tag{6.1.6}$$

A single-mode regime in one of the two transverse directions of the laser can be realized; then geometric dimensionality $D = 1$ ($\boldsymbol{r} = x \boldsymbol{e}_x$, where $\boldsymbol{e}_x$ is the unit vector in the $x$-direction). The same equation (6.1.6) describes a temporal or spatial field structure in a single-mode fiber with amplification and anomalous frequency dispersion (in the reference frame moving with the group velocity, see Appendix A). If both transverse directions are essential then $D = 2$, $\boldsymbol{r} = x, y$. In a continuous medium with frequency dispersion and a fast nonlinearity of amplification and absorption, an equation of the form of (6.1.6) is valid with the geometric dimensionality $D = 3$, $\boldsymbol{r} = x, y, \zeta$ [297]. Let us therefore rewrite this equation in a more general form:

$$\frac{\partial \bar{E}}{\partial \zeta} - \mathrm{i} \Delta_D \bar{E} = f(|\bar{E}|^2) \bar{E} , \quad D = 0, \ 1, \ 2, \ 3 . \tag{6.1.7}$$

Here $\Delta_D$ is the $D$-dimensional Laplacian; $\Delta_0 = 0$ corresponds to a point (nondistributed) scheme. The dimensionless evolution variable $\zeta$ has the meaning of time $t$ or the longitudinal coordinate $z$ (in the moving frame of reference).

### 6.1.3 Energy Balance

When neglecting frequency detunings, the function $f$ is real and determines the difference between gain and total losses; its form is illustrated in Fig. 6.2. In the latter case, the following energetic relation results from (6.1.7) [299]:

$$\frac{\mathrm{d}W}{\mathrm{d}\zeta} = 2 \int f(I) I \ \mathrm{d}\boldsymbol{r} , \tag{6.1.8}$$

where $W(\zeta) = \int I(\boldsymbol{r}, \zeta) \, \mathrm{d}\boldsymbol{r}$ is radiation power or energy (if $D = 3$).[1] It follows from (6.1.8) that for a monostable regime when the function $f(I)$ does not

---

[1] For nonzero frequency detunings, the function $f$ should be replaced by its real part in (6.1.8).

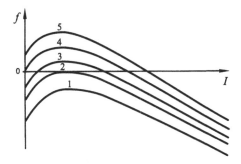

**Fig. 6.2.** Dependence of balance of gain and losses $f$ on radiation intensity $I$ for a number of values of small-signal gain $g_0$. 1: $g_0 < g_{min}$; 2: $g_0 = g_{min}$; 3: $g_{min} < g_0 < g_{max}$; 4: $g_0 = g_{max}$; 5: $g_0 > g_{max}$

change sign with variation of $I$, i.e.,

$$f(I) = 0 \tag{6.1.9}$$

has no positive roots, then the power $W$ changes monotonously with $\zeta$.

The nonlasing regime $E = 0$ is stable [2] with regard to small perturbations if

$$f_0 = f(0) < 0 . \tag{6.1.10}$$

At the same time, inequality (6.1.10) is a necessary condition of stability for "bright" localized structures, because $E \to 0$ at their periphery. For monostable regimes under condition (6.1.10), we have $dW/d\zeta < 0$, and power $W(\zeta) \to 0$ at $\zeta \to \infty$. Thus the nonlasing regime is globally stable, i.e., it will form for any initial conditions. In other words, a necessary condition for the existence of stable localized structures is bistability of the point laser scheme.

### 6.1.4 Symmetry

Equations (6.1.3-6.1.4) and (6.1.6-6.1.7) are invariant under translations and reflections in transverse coordinates, as in the case of interferometers (Chap. 4). New is the symmetry with respect to a phase shift of the field envelope, connected with the absence of external radiation in laser schemes. These symmetries are defined by the transformations, respectively,

$$\boldsymbol{r} \to \boldsymbol{r} + \delta_{\boldsymbol{r}} , \quad r_n \to -r_n , \quad \bar{E} \to \bar{E}\exp(i\delta_0) \tag{6.1.11}$$

---

[2] More precisely, this stability condition is valid for the paraxial approximation (6.1.1). For wide-aperture cavity schemes, this approximation excludes from consideration light rays, e.g., perpendicular to the cavity axis. The nonlasing regime will be stable with respect to amplification of this radiation if, at any frequency, small signal gain is less than small signal absorption.

with arbitrary $\delta_r$, $\delta_0$. Moreover, (6.1.6-6.1.7), describing schemes with fast nonlinearity, exhibit the additional symmetry with respect to the "Galilean transformation" (see Chap. 4) to a reference frame moving in the transverse direction with velocity $v$:

$$\bar{E}(r,\zeta) \rightarrow \bar{E}(r - v\zeta, \zeta) \exp(irv/2 - iv^2\zeta/4) . \qquad (6.1.12)$$

This means that any solution of (6.1.7) generates a family of solutions with additional velocity of transverse motion $v$, spectrum $v$ being continuous (compare Fig. 6.1a and b). Symmetry (6.1.12) follows from relations (4.1.25-4.1.26) describing the oblique incidence of external radiation on an interferometer, for which we take into account the absence of external radiation for a laser. For nonzero values of relaxation times, symmetry (6.1.12) is broken and, hence, optical patterns cannot travel with arbitrary velocities.

### 6.1.5 Monochromatic Plane-Wave Solutions

For such regimes of cavity schemes ($\zeta = t$), after the substitution

$$\bar{E}(r,t) \rightarrow \exp(-i\nu t)\sqrt{\tilde{I}} , \quad \tilde{I} = I(1 + \Delta_g^2) \qquad (6.1.13)$$

and similar substitutions for $P_{a,g}$, we obtain from (6.1.1-6.1.2)

$$-1 + i\nu + \frac{(1 - i\Delta_g + i\nu\tau_{\perp g})g_0}{1 + (\Delta_g - \nu\tau_{\perp g})^2 + \tilde{I}} - \frac{(1 - i\Delta_a + i\nu\tau_{\perp a})a_0}{1 + (\Delta_a - \nu\tau_{\perp a})^2 + b\tilde{I}} = 0 . \qquad (6.1.14)$$

The complex equation (6.1.14) can be separated into two real equations for the wave intensity $I$ and the frequency shift $\nu$; for class B lasers ($\tau_{\perp g,a} \rightarrow 0$) they are

$$\mathrm{Re}\, f(I) = 0 , \quad \nu = -\mathrm{Im}\, f(I) \qquad (6.1.15)$$

with $f(I)$ defined by (6.1.5).[3] The two corresponding solutions of quadratic equation for $I$, if they exist and are positive, determine the upper and intermediate intensity branches (Fig. 6.3). As above, the intermediate branch is related to unstable regimes. Bistability between the nonlasing regime ($E = 0$) and regimes represented by the upper branch, exists in the range

$$g_{\min} < \bar{g} < g_{\max} , \quad g_{\max} = 1 + \bar{a}_0 , \qquad (6.1.16)$$

if saturation intensity for a passive medium is less than that for an active medium, and if the nonlasing regime is stable [condition (6.1.10)].

---

[3] In the latter approximation we neglect off-resonance solutions [218].

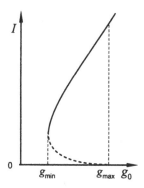

**Fig. 6.3.** Dependence of laser radiation intensity on small-signal gain for the laser point model; the intermediate branch (*thick dashed line*) corresponds to unstable states

### 6.1.6 Modulation Instability

Linearizing (6.1.6) with respect to perturbations of the type

$$\bar{E}(\boldsymbol{r},t) \to \exp(-\mathrm{i}\nu\zeta)\sqrt{I}[1 + a\exp(\gamma\zeta + \mathrm{i}\boldsymbol{\kappa r}) + b^*\exp(\gamma^*\zeta - \mathrm{i}\boldsymbol{\kappa r})]\,, \quad (6.1.17)$$

one can find the dependence of perturbation growth rate $\gamma$ on modulation spatial frequency $\kappa$:

$$\gamma(\kappa^2) = \mathrm{Re}[If'(I)] + \{(\mathrm{Re}[If'(I)])^2 + 2\kappa^2\mathrm{Im}[If'(I)] - \kappa^4\}^{1/2}\,. \quad (6.1.18)$$

It follows from (6.1.18) that the homogeneous regime is unstable in the range $0 < \kappa^2 < 2\kappa_{\mathrm{max}}^2$, when $\kappa_{\mathrm{max}}^2 = \mathrm{Im}[If'(I)] > 0$. The maximum growth rate is given by relation $\gamma(\kappa_{\mathrm{max}}^2) = \mathrm{Re}[If'(I)] + |If'(I)| > 0$. Illustrations of bistability and modulation instability domains in the plane of detunings are given in [91].

## 6.2 Transversely One-Dimensional Structures

### 6.2.1 Fast Nonlinearity

As we have seen, nontrivial stable localized structures of laser radiation are only possible within the bistability range (6.1.16). Settled switching waves and localized structures are characterized by transverse motion of the field as a whole with a constant velocity $v$ (in the transversely one-dimensional laser scheme discussed one can consider that the motion is directed along the $x$-axis and $v$ is a scalar). In the frame of reference moving with the velocity $v$ (substitution $x \to \xi = x - vt$, including the Laplacian $\Delta_D \to \Delta_D'$), for the transformed amplitude

$$\bar{E} = A \exp(-i\nu t + i\nu \xi/2) \,, \tag{6.2.1}$$

it follows from (6.1.7) that

$$\frac{\partial A}{\partial \zeta} = i\alpha A + i\Delta'_D A + f(|A|^2)A \,. \tag{6.2.2}$$

Here $\alpha = \nu + v^2/2$ is the frequency lasing shift. The symmetry relations (6.1.11-6.1.12) are valid for (6.2.2) too. Stationary solutions are obtained from (6.2.2) with a zero left-hand side with asymptotic

$$I_{\xi \to -\infty} = 0 \,, \quad I_{\xi \to +\infty} = I_{\text{st}} \tag{6.2.3}$$

for switching waves and

$$I_{\xi \to \pm\infty} = 0 \tag{6.2.4}$$

for bright localized structures. The latter structures are more important in a laser situation than the switching waves – due to the following reasons: First, the velocity of a single switching wave is not, unlike that for driven interferometers, a single-valued function of the scheme parameters. In a laser with "Galilean invariance" (6.1.12), there is a family of switching waves with a continuously varying velocity of transverse motion. In this case "Maxwell's value" of small-signal gain $g_0$ can be determined for a wide local perturbation consisting of two switching waves with distant (non-interacting) fronts only. Second, for switching waves and for "dark" solitons, there are semi-infinite domains where saturated gain exceeds losses. In these domains the amplification of spontaneous radiation is possible, with the consequences discussed above. This is why we further restrict ourselves to the consideration of "bright" laser solitons with asymptotics (6.2.4).

Let us discuss bifurcations and stability of the *laser solitons* with a negligible diffusion coefficient $(d = 0)$ [399, 91]. With a one-dimensional geometry, (6.2.2) for a stationary soliton takes the form

$$\frac{\mathrm{d}^2 A}{\mathrm{d}\xi^2} + \alpha A - if(|A|^2)A = 0 \tag{6.2.5}$$

and is supplemented by boundary conditions (6.2.4). It is equivalent to a fourth-order set of real ordinary differential equations. However, symmetry with respect to phase shift (6.1.11) allows us to reduce the set to a third-order set by means of introduction of the following real variables [68]:

$$A = a \exp(i\Phi) \,, \quad q = \mathrm{d}\Phi/\mathrm{d}\xi \,, \quad k = a^{-1}\mathrm{d}a/\mathrm{d}\xi \,. \tag{6.2.6}$$

Then it follows from (6.2.5) that

$$\frac{\mathrm{d}a}{\mathrm{d}\xi} = ak \,, \quad \frac{\mathrm{d}q}{\mathrm{d}\xi} = -2qk + \mathrm{Re}\, f(a^2) \,, \quad \frac{\mathrm{d}k}{\mathrm{d}\xi} = -\alpha + q^2 - k^2 - \mathrm{Im}\, f(a^2) \,. \tag{6.2.7}$$

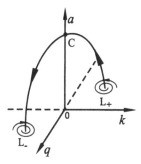

**Fig. 6.4.** Phase space of system (6.2.7): fixed points $L_\pm$ represent nonlasing regimes, and the laser soliton corresponds to the heteroclinic trajectory (*line with arrows*, $L_+CL_-$)

Other symmetry properties lead to the invariance of the vector field given by (6.2.7), with respect to the transformation

$$(\xi, a, q, k) \to (-\xi, a, -q, -k) . \tag{6.2.8}$$

The three-dimensional phase space of (6.2.7) (Fig. 6.4) has two fixed points $L_\pm$ representing the nonlasing regime; their coordinates are: $a = 0$, $q_\pm = \pm\{[(\alpha + f_{02})^2 + f_{01}^2]^{1/2}/2 + (\alpha + f_{02})/2\}^{1/2}$, and $k_\pm = f_{01}/2q_\pm$, where $f_{01} = \mathrm{Re}\, f(0) = -1 + \bar{g}_0 - \bar{a}_0$, $f_{02} = \mathrm{Im}\, f(0) = -(\bar{g}_0\Delta_\mathrm{g} - \bar{a}_0\Delta_\mathrm{a})$. Linearization of (6.2.7) in the vicinity of the fixed points shows that each of the solutions has a single real eigenvalue $\lambda_1^\pm = k_\pm$ and a pair of complex-conjugated eigenvalues $\lambda_2^\pm = -2(k_\pm + iq_\pm)$ and $\lambda_3^\pm = \lambda_2^{\pm*}$. Since in the bistability domain we have $f_{01} < 0$, the fixed points $L_\pm$ are saddle-focuses with a one-dimensional unstable for $L_-$ and stable for $L_+$ manifold and a two-dimensional stable for $L_-$ and unstable for $L_+$ manifold; here stability has the meaning of the features of the trajectories of (6.2.7) and is not connected with the real, temporal stability of any laser regime.

A stationary laser soliton corresponds to a heteroclinic trajectory of (6.2.7) connecting the fixed points $L_+$ and $L_-$ (Fig. 6.4). To find a soliton, we have to find bifurcation points in the parameter space for which (6.2.7) has a heteroclinic trajectory of the type described. In Fig. 6.5 the curve shown demonstrates the relation between parameters $\alpha$ and $g_0$ (all other parameters are fixed) for which "single" localized structures exist. The curve is situated wholly in the region of bistability (6.1.16). It is a spiral curling to the point P which correlates with "Maxwell's conditions" discussed above. Therefore, for fixed parameters a discrete set of "single" structures with different frequency shifts $\alpha$ and different intensity profiles exists (Fig. 6.6).

The existence of a "single" localized structure (soliton) implies the existence of an infinite number of multi-humped, or combined soliton structures, which can be considered as bound states of two or more single solitons [399]. Due to the symmetry properties, after an appropriate shift along the $x$-axis,

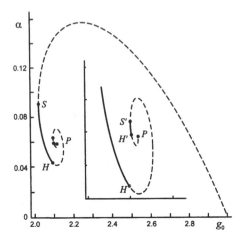

**Fig. 6.5.** Relation between the lasing frequency shift $\alpha$ and small-signal gain $g_0$ for single localized structures. The inset shows the vicinity of the point P. *Solid* and *dashed parts* of the curve correspond to stable and unstable localized structures. S and S' are the points of a saddle-node bifurcation, H and H' are the Andronov–Hopf bifurcation points for localized solutions [399]

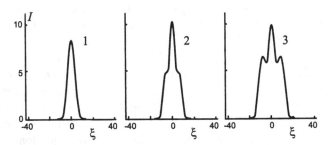

**Fig. 6.6.** Transverse intensity profiles for a number of localized structures [399]

the envelope $A(x)$ of the motionless soliton ($v = 0$, $\xi = x$) can be taken as either an even or an odd function of $x$. The intensity of any stationary soliton structure cannot have more than one zero at a finite value of coordinate $x$ [399]. Hence for an odd function $A(x)$ there is only one zero at $x = 0$, and no zeros for even functions $A(x)$. Note also that for solitons with odd functions $A(x)$ the variable $k$ has a singularity at $a = 0$. In these cases, it is more convenient to use other variables

$$x_1 = |A|^2, \quad x_2 = \left|\frac{\mathrm{d}A}{\mathrm{d}\xi}\right|^2, \quad x_3 = \frac{\mathrm{d}|A|^2}{\mathrm{d}\xi}, \quad x_4 = \frac{1}{i}\left(A\frac{\mathrm{d}A^*}{\mathrm{d}\xi} - A^*\frac{\mathrm{d}A}{\mathrm{d}\xi}\right), \quad (6.2.9)$$

with the following relationships [399]

$$4x_1x_2 - x_3^2 - x_4^2 = 0, \quad x_1 \geq 0, \quad x_2 \geq 0. \quad (6.2.10)$$

Now let us consider the stability of localized structures to small perturbations. The perturbed field has the form

$$A(x,t) = A_0(x) + \delta A(x,t) = A_0(x) + a(x)\exp(\gamma t) + b^*(x)\exp(\gamma^* t) . \quad (6.2.11)$$

Then, linearizing (6.2.2) with respect to $\delta A$, we obtain the linear equation (T denotes transposition, $L_{11} = i\frac{d^2}{dx^2} + i\alpha + f_0 + |A_0|^2 f_0'$, $L_{12} = A_0^2 f_0'$)

$$L\boldsymbol{\Psi} = \gamma\boldsymbol{\Psi} , \quad \boldsymbol{\Psi} = (a, b)^{\mathrm{T}} , \quad (6.2.12)$$

$$L = \begin{pmatrix} L_{11} & L_{12} \\ L_{12}^* & L_{11}^* \end{pmatrix} . \quad (6.2.13)$$

Under conditions of bistability, when the nonlasing regime is stable, the continuous spectrum of operator $L$ is situated in the left half-plane of the complex plane $\gamma = \mathrm{Re}\,\gamma + i\mathrm{Im}\,\gamma$ and has no connections with instability. Therefore, it is the discrete spectrum that determines soliton stability.

Due to the symmetries (6.1.11-6.1.12), the discrete spectrum of operator $L$ includes a triply degenerate zero eigenvalue. Two corresponding eigenvectors, or "neutral modes", are $\boldsymbol{\Psi}_{1,2}(x) = (\mathrm{Re}\,\psi_{1,2}, \mathrm{Im}\,\psi_{1,2})^{\mathrm{T}}$, with $\psi_1 = iA_0$ and $\psi_2 = dA_0/dx$. These eigenvalues obey the equations $L\boldsymbol{\Psi}_{1,2}(x) = 0$. The third zero eigenvalue is associated with the "Galilean symmetry" (6.1.12). It corresponds to the adjoint vector $\boldsymbol{\Psi}_3(x) = (\mathrm{Re}\,\psi_3, \mathrm{Im}\,\psi_3)^{\mathrm{T}}$, with $\psi_3 = -ixA_0/2$, and obeys the equation $L\boldsymbol{\Psi}_3(x) = \boldsymbol{\Psi}_2(x)$. As indicated above, the fundamental soliton envelope is an even or odd function of the coordinate $x$. For a "one-soliton" structure $A(x) = A(-x)$, we have $\boldsymbol{\Psi}_1(x) = \boldsymbol{\Psi}_1(-x)$ and $\boldsymbol{\Psi}_2(x) = -\boldsymbol{\Psi}_2(-x)$. Since $L(x) = L(-x)$, it is possible to study the stability with respect to even (symmetric) and odd (antisymmetric) perturbations separately.

The discrete spectrum of the linear operator $L$ was calculated numerically [399, 397, 91]. For the case of zero detunings (see Fig. 6.5), stable laser solitons correspond to parts of the spiral turns between points S and H, S' and H', etc. Points H, H', ... correspond to the Andronov–Hopf bifurcations (there the real part of the eigenvalue $\gamma$ changes the sign, while the imaginary part of $\gamma$ is nonzero). Calculations show that this bifurcation is supercritical; with an increase in the control parameter – small-signal gain $g_0$ – the stationary localized structure loses stability, and a periodically oscillating laser soliton appears (Fig. 6.7).

With a further increase in $g_0$, periodically oscillating laser solitons lose stability and new types of structures arise with more complex behaviour, up to chaotic behaviour. In Fig. 6.8, the regime of a "leading centre" is illustrated; the central soliton splits periodically into three parts, the peripheral parts moving to the mirror edges, where they perish due to a decrease in the mirror reflection coefficient.

The effect of nonzero frequency detunings is shown in Fig. 6.9. The stationary laser solitons are stable in the central area, including the coordinate

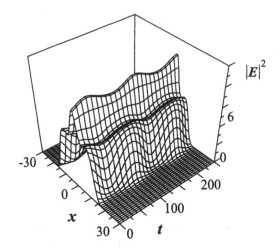

**Fig. 6.7.** Symmetric oscillatory laser soliton occurring above the Andronov–Hopf bifurcation threshold [337, 399, 397]

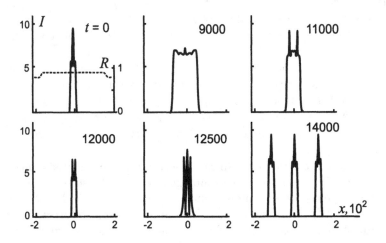

**Fig. 6.8.** Intensity profiles for the regime of a "leading centre"; the decrease in the reflection coefficient in the vicinity of the mirrors' edges is shown by the *dotted line*, which corresponds to the right-hand scale for $t = 0$ [339]

origin $\Delta_g = \Delta_a = 0$. The stability boundaries are given by curves 1 (the Andronov–Hopf bifurcation) and 2 (the saddle-node bifurcation); there is also an additional bifurcation not shown in Fig. 6.9, as corresponding curves are very close to the upper parts of curves 2. In this case, we also have periodically oscillating laser solitons, regimes of a "leading centre", etc. [91].

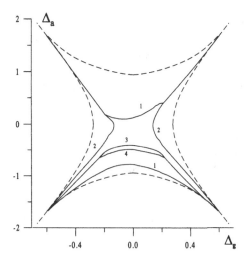

**Fig. 6.9.** Stability boundaries of laser solitons on the plane of frequency detunings. *Dashed lines* indicate the boundaries of the bistability domain of spatially homogeneous distributions. Curves 1 represent the Andronov–Hopf supercritical bifurcation; curves 2 show the saddle-node bifurcation; and curves 3 and 4 are the upper and lower boundaries of the stability domains of an oscillating laser soliton. Below curve 4 an oscillating soliton is transformed into a "leading centre". Bistability of stationary and oscillating solitons takes place between curve 3 and the lower part of curve 1; a narrow bistability domain near the upper curve 1 is not shown [91]

### 6.2.2 Effect of Relaxation Rates

As was noted above, for fast nonlinearity (zero relaxation times of the medium), governing equation (6.2.2) is applicable both to lasers with a saturable absorber and to such a cavity-less scheme as a single-mode fiber with saturable amplification and absorption and quadratic frequency dispersion. The presence of media relaxation makes these two variants different; here we will analyze in more detail the laser variant with a zero value of the diffusion coefficient $d = 0$; the fiber scheme is discussed in the final part of the section.

In the laser scheme, relaxation does not influence the characteristics of a motionless laser soliton. However, it breaks the "Galilean symmetry" (6.1.12). Therefore, transversely motionless and moving laser solitons are not equivalent anymore; even for a motionless soliton, stability depends on relaxation rates and should be examined again. Let us begin by analyzing this point.

**Bifurcations of a Transversely Motionless Laser Soliton.** The analysis is based on (6.1.3) with $d = 0$.[4] Introducing, in addition to the electric field perturbations (6.2.11), perturbations of gain $\delta g$ and absorption $\delta a$, and linearizing (6.1.3), we obtain an equation of type (6.2.12), where now $\boldsymbol{\Psi}$

---

[4] The effect of a nonzero diffusion coefficient on the laser soliton features is considered in [214].

includes additionally perturbations $\delta g$ and $\delta a$. The spectrum of the linear operator $L$ has two zero eigenvalues $\gamma_1 = \gamma_2 = 0$, which correspond with symmetries to the phase shift and to the translation invariance (6.1.11). The third eigenvalue $\gamma_3$, which was equal to zero in the limit $\tau_{g,a} = 0$ due to the "Galilean symmetry" (6.1.12), is now shifted from the origin of the complex plane $\gamma$. The soliton's stability is determined by this shift direction. Therefore, even if the soliton is stable for the fast nonlinearity $\tau_{g,a} = 0$, it may be unstable for arbitrary small nonzero relaxation times.

In Fig. 6.10, bifurcation loci for the motionless laser soliton found by numerical calculation of the discrete spectrum of the operator $L$ are shown. The straight line S indicates the steady-state bifurcation defined by the condition $\gamma_3 = 0$ (an odd eigenvector) (an analytical derivation is given in [91]). For sufficiently small $\tau_{g,a}$ the line S defines the stability boundary of the motionless soliton. When crossing this line from the left, the motionless soliton becomes unstable, giving rise to a soliton slowly moving with a certain definite constant velocity $v$. The velocity value depends on the distance from the stability boundary S ($v = 0$ at the boundary). Since opposite directions of propagation are equivalent, there exist at least two solitons traveling with opposite velocities $\pm v$. If the passive medium relaxes much faster than the active medium, the motionless soliton is always unstable. Indeed, for $\tau_a = 0$ and $\tau_g > 0$ the passive medium is equally saturated by motionless and traveling solitons, while the active medium is less saturated by a traveling soliton. This situation is more favourable for the existence of a stable traveling soliton. With the increase in the relaxation time $\tau_a$, the absorption saturation decreases for a traveling soliton, and for a fixed $\tau_g$ a certain threshold value of $\tau_a$ exists, above which the motionless soliton becomes stable. Thus the absorption relaxation exerts a stabilizing effect on the motionless soliton.

When $\tau_{g,a}$ are large enough, there are different bifurcation scenarios leading to the motionless soliton's instability. Then the stability boundary is associated with the Andronov–Hopf bifurcation (labels H, $H_1$ and $H_2$ in Fig. 6.10). The curve H represents the Andronov–Hopf bifurcation with a pair of pure imaginary eigenvalues $\gamma$ with even eigenvectors $[\boldsymbol{\Psi}(x) = \boldsymbol{\Psi}(-x)]$. This bifurcation leads to an oscillating soliton similar to that shown in Fig. 6.7. Curves $H_1$ and $H_2$ correspond to a pair of pure imaginary eigenvalues with odd eigenvectors $[\boldsymbol{\Psi}(x) = -\boldsymbol{\Psi}(-x)]$. The bifurcation curve $H_2$ terminates in a codimension-2 point (which implies two conditions on the scheme parameters) of the Bogdanov–Takens type [134, 27], marked B in Fig. 6.10. Label Q corresponds to codimension-2 points associated with the interaction between two Andronov–Hopf bifurcations. Under certain conditions quasiperiodic and irregular regimes arise for parameters in the vicinity of these points [134]. Points T denote the intersection of the Andronov–Hopf bifurcations with the pitchfork bifurcation line S. These points correspond to codimension-2 bifurcations with a single zero and two purely imaginary eigenvalues.

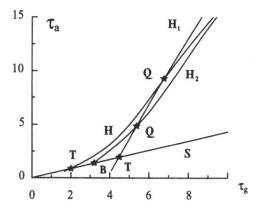

**Fig. 6.10.** Bifurcation diagram for the motionless laser soliton (it is stable above curves S, H, and $H_1$). The steady-state pitchfork bifurcation is shown by the *straight line* S, curves H, $H_1$ and $H_2$ correspond to the Andronov–Hopf bifurcations. *Asterisks* indicate the positions of codimension-2 points B, Q, and T [91]

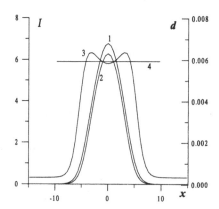

**Fig. 6.11.** "Exact" (curves 1 and 3) and approximate (curve 2 and line 4) stationary laser solitons: transverse profiles of intensity (curves 1 and 2) and diffusion coefficient $d$ (curve 3, line 4, and the right-hand scale) [91]

**Numerical Simulations.** More detailed information follows from a numerical solution of (6.1.3). In Fig. 6.11 the transverse profile of the intensity-dependent diffusion coefficient $d(I)$ is shown, which can be approximated by a constant value in the vicinity of the soliton maximum (line 4). It follows from Fig. 6.11 that this approximation does not distort the soliton shape essentially. Numerical simulations confirm the results of the bifurcation analysis given above. Additionally, they allow us to study dynamic regimes far from the bifurcation threshold, and to find new types of laser solitons.

Stability domains of different soliton types are shown in Fig. 6.12 in the $\tau_g$–$\tau_a$ plane (compare with Fig. 6.10). Steady-state motionless solitons are

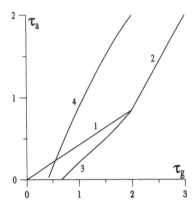

**Fig. 6.12.** Stability domains of laser solitons (see text) [91]

stable to the left of straight line 1 and curve 2. Line 1 corresponds to bifurcation from a motionless soliton to a slowly moving soliton. The latter has an asymmetric intensity profile, since its leading and trailing fronts interact, in a medium with relaxation, with unsaturated and saturated amplification and absorption, respectively. However, when the soliton velocity is small, this asymmetry is practically indistinguishable. Upon crossing the Andronov–Hopf bifurcation curve 2, a transition from a motionless steady-state soliton into a pulsating one occurs. With a further variation of the parameters, a series of successive doubling of the oscillation period is realized (Fig. 6.13), which can be associated with the Feigenbaum scenario for a route to chaos (see Chap. 3).

Numerical simulations reveal the existence of two different types of laser solitons. First, to the right of curve 4, stable solitons moving with a large transverse velocity ("fast solitons") can be formed. These structures are characterized by narrower intensity profiles and much greater peak intensities, as compared with slowly moving solitons (Fig. 6.14). The "fast solitons" are similar to transversely moving domains of Q-switching, typical of lasers with a saturable absorber. Second, solitons with a periodically varying transverse velocity and an oscillating transverse profile can be excited (Fig. 6.15).

Different types of laser solitons can coexist in certain parameter domains. A hysteresis behaviour that takes place with the variation of the population relaxation time in an amplifying medium is illustrated in Fig. 6.16. For the fixed value $\tau_a = 0.3$, a motionless soliton is stable up to $\tau_g = 0.9$. With an increase in $\tau_g$, this soliton loses stability, and a slowly moving soliton occurs. The velocity of a slow soliton increases with $\tau_g$ up to the next bifurcation at $\tau_g = 1.2$, where it is transformed into a soliton with oscillating velocity. As the relaxation time $\tau_g$ is further increased, a hysteresis jump to a branch corresponding to a fast soliton occurs. With a decrease in $\tau_g$, a jump from the branch of the fast solitons to the branch of the motionless ones takes place at $\tau_g = 0.9$.

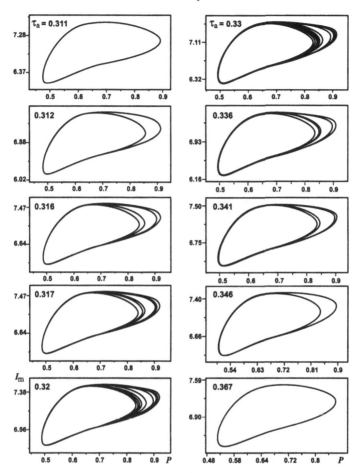

**Fig. 6.13.** Phase portraits of oscillating laser solitons for different relaxation times $\tau_a$; $P$ is the soliton power, and $I_m$ is its maximum intensity [341]

**Fiber Scheme.** We now briefly consider the effect of relaxation rate on dissipative solitons in a scheme of a single-mode optical fiber with saturable amplification and absorption. Instead of (6.1.3), we have the following Maxwell–Bloch equations in the frame of reference moving along the fiber axis $z$ with group velocity $v_{gr}$:

$$\frac{\partial E}{\partial z} - (\mathrm{i} + d)\frac{\partial^2 E}{\partial \tau^2} = (N_g - N_a - 1)E \,,$$
$$\tau_{g,a}\frac{\partial N_{g,a}}{\partial \tau} = N_{0g,a} - N_{g,a}\left(1 + \frac{|E|^2}{I_{g,a}}\right) \,. \tag{6.2.14}$$

Here $\tau = t - z/v_{gr}$; the time $t$ is expressed in the characteristic units of frequency dispersion; the $z$-coordinate is taken in units of the inverse coefficient

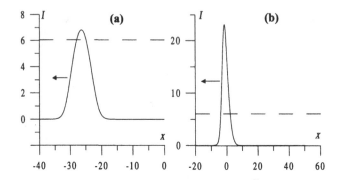

**Fig. 6.14.** Transverse profiles of radiation intensity $I$ for (**a**) "slow" (velocity $v = -0.1$) and (**b**) "fast" ($v = -9.4$) solitons; *arrows* indicate the direction of the soliton motion; *dashed lines* show the intensity of homogeneous lasing [91]

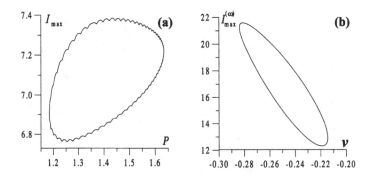

**Fig. 6.15.** Phase portrait of a laser soliton with periodically oscillating velocity and shape. $I_{\max}$ is the maximum radiation near-field intensity and $P$ is the total power in (**a**), $I_{\max}^{(\omega)}$ is the maximum far-field intensity and $v$ is the soliton velocity in (**b**) [91]

of nonresonant absorption; the effective diffusion coefficient $d$ describes the finite width of the spectral line of the absorption and gain; $N_{0g,a}$ are the equilibrium values of the population difference in the absence of radiation ($N_{0g} > 0$ and $N_{0a} < 0$).

Numerical simulations show the existence of "ordinary" dissipative solitons (similar to "slow" laser solitons) and "Q-solitons" characterized by a considerably larger peak intensity and a shorter duration (analogs of "fast" laser solitons) [314, 338]. These solitons exist in different nonoverlapping regions of the fiber parameters (Fig. 6.17). Note that even for sufficiently long pulses, of principal importance are both the finite rate of nonlinear response of the medium and the finiteness of the spectral lines of the gain and absorption; in the absence of these factors, the possibility of the appearance of nonphysical self-accelerating pulses arises.

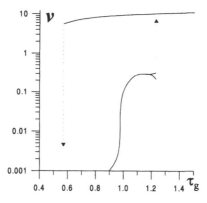

**Fig. 6.16.** Hysteresis change of laser soliton velocity $v$ with the variation of the relaxation time of amplification $\tau_g$ [91]

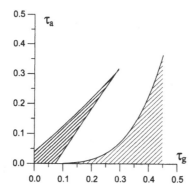

**Fig. 6.17.** Regions of stability of "ordinary" dissipative solitons (*left hatched region*) and "Q-solitons" (*right hatched region*) [338]

## 6.3 Transversely Two-Dimensional Structures

### 6.3.1 Stripes

In Sects. 6.1 and 6.2 we considered transversely one-dimensional structures with a single-mode regime for the second transverse coordinate $y$. For transversely two-dimensional lasers, there exist $y$-independent stripe patterns with the same field dependence on $x$ as for the one-dimensional laser. However, these patterns are modulationally unstable for typical scheme parameters. To find the small perturbation growth rate, we generalize (6.2.11):

$$A(x, y, t) = A_0(x) + a(x)\exp(i\kappa y)\exp(\gamma t) + b^*(x)\exp(-i\kappa y)\exp(\gamma^* t) .$$
(6.3.1)

The calculation results are presented in Fig. 6.18. The modulation instability takes place within a finite range of perturbation spatial frequency $\kappa$.

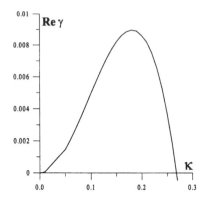

**Fig. 6.18.** Perturbation growth rate $\mathrm{Re}\,\gamma$ versus the perturbation spatial frequency $\kappa$ for a laser stripe structure [91]

### 6.3.2 Cylindrically Symmetric Intensity Distribution

Let us consider, on the basis of (6.1.4), the simplest stationary structures localized in both transverse directions $x, y$ and characterized by a cylindrically symmetric intensity distribution. Then in polar coordinates $\rho, \varphi$ the field reads

$$E = A(\rho)\exp(\mathrm{i}M\varphi)\exp(-\mathrm{i}\alpha t)\,,\quad \rho = \sqrt{x^2 + y^2}\,. \tag{6.3.2}$$

The amplitude $A$ is determined by the ordinary differential equation

$$\frac{\mathrm{d}^2 A}{\mathrm{d}\rho^2} + \frac{1}{\rho}\frac{\mathrm{d}A}{\mathrm{d}\rho} - \frac{M^2}{\rho^2}A + \frac{1}{\mathrm{i}+d}[\mathrm{i}\alpha + f(|A|^2)]A = 0 \tag{6.3.3}$$

with natural requirements of finiteness at $\rho = 0$ and vanishing for $\rho \to \infty$. The frequency shift $\alpha$ serves as a spectral parameter, as in the case of one-dimensional solitons (see Sect. 6.2); the main difference from (6.2.5) is in the presence of the second and third (radial) terms on the left-hand side of (6.3.3). Correspondingly, a bifurcation approach and linear stability analysis for one- and two-dimensional solitons of the type considered are similar. A small perturbation in the stability study is introduced as follows:

$$E = [A(\rho)+a(\rho)\exp(\mathrm{i}m\varphi+\gamma t)+b^*(\rho)\exp(-\mathrm{i}m\varphi+\gamma^*t)]\exp(\mathrm{i}M\varphi)\exp(-\mathrm{i}\alpha t) \tag{6.3.4}$$

$(m = 0, 1, 2, \ldots)$. Then, after linearization of (6.1.4), we obtain the same linear equation (6.2.12) with the following form of the linear operator $L$:

$$L = \begin{pmatrix} (d+\mathrm{i})\Delta_+ + \mathrm{i}\alpha + F_0, & f_0' A^2 \\ (f_0' A^2)^*, & (d-\mathrm{i})\Delta_- - \mathrm{i}\alpha + F_0^* \end{pmatrix}. \tag{6.3.5}$$

Here $\Delta_\pm = \mathrm{d}^2/\mathrm{d}\rho^2 + (1/\rho)\mathrm{d}/\mathrm{d}\rho - (M\pm m)^2/\rho^2$, and $F_0 = f_0 + f_0'|A|^2$. "Neutral modes" – eigensolutions of (6.2.12) with zero eigenvalue $\gamma = 0$ – correspond to symmetry with respect to the phase shift $(m = 0)$ and to shifts of coordinates $(m = 1)$.

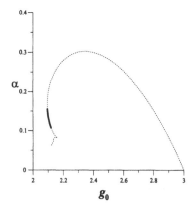

**Fig. 6.19.** Dependence of the frequency shift of a fundamental laser soliton on the small-signal gain; stable solitons are indicated with a *solid line* [92]

**Fundamental Soliton ($M = 0$).** Dependence of the spectral parameter $\alpha$ on the control parameter – a small-signal gain $g_0$ – is given in Fig. 6.19. Corresponding intensity and phase profiles are presented in Fig. 6.20. Solitons are stable within a certain range of gain $g_0$. The left boundary of this range corresponds to a saddle-node bifurcation (merging of stable and unstable localized solutions); only transversely homogeneous regimes are possible for a lesser gain. For typical parameters, a fundamental soliton loses stability with an increase in $g_0$ due to the growth of perturbations with angular harmonic $m = 2$ (Fig. 6.21). In this case $\mathrm{Im}\,\gamma \neq 0$; therefore it is an Andronov–Hopf bifurcation. The answer to the question of which regimes will be realized after this bifurcation will be given below on the basis of numerical simulations.

**Localized Vortices.** Numerical simulations confirm the existence and stability of laser solitons with a cylindrically symmetric intensity distribution and wavefront dislocations ($M = 1, 2, \ldots$). Corresponding intensity and phase distributions are given in Fig. 6.20. Note that solitons with higher topological indices (including $M = 2$) are also stable in some domains of parameters. The reason of such stability – which is impossible in linear optics – is connected to the laser bistability. In the vicinity of the vortex centre, where radiation intensity is small, unsaturated losses are greater than gain. Therefore, a small perturbation arising there will dissolve with time, which results in the stability of localized vortices. There are also narrow domains of parameters where "excited" vortices (with several oscillations in the intensity radial profile) are stable.

### 6.3.3 Asymmetric Rotating Laser Solitons

Even more unusual are the two-dimensional laser solitons shown in Fig. 6.22. Their intensity distribution is not cylindrically symmetric and has, typically, one (Fig. 6.22g) or two (Fig. 6.22a–f) equal maxima. The field distribution rotates around the symmetry axis with a constant angular velocity $\Omega$:

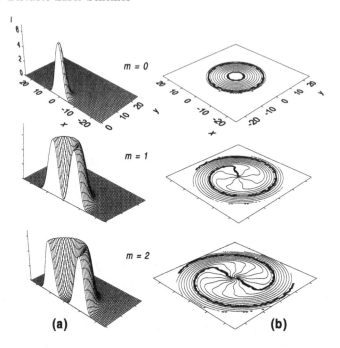

**Fig. 6.20a,b.** Cylindrically symmetric intensity distributions (*left*) and lines of equal phase (*right*) for laser solitons with topological indices $M = 0, 1$ and 2 [330]

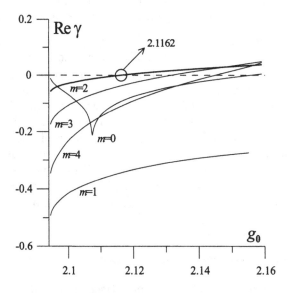

**Fig. 6.21.** Growth rate $\mathrm{Re}\,\gamma$ of perturbations with angular harmonic $m$ versus the small-signal gain for a fundamental laser soliton; the *circle* indicates the instability threshold [92]

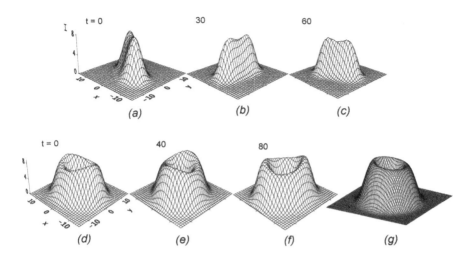

**Fig. 6.22.** Intensity distributions for "two-humped" (**a–f**) and "single-humped" (**g**) rotating laser solitons with topological charge $M = 0$ (*top*) and $M = 1$ (*bottom*) [330]

$$E = F(\rho, \varphi - \Omega t) \exp(-i\nu t) .\qquad(6.3.6)$$

The radiation wavefront can either be regular or include screw dislocations ($M \neq 0$, see Fig. 6.22). The asymmetric rotating laser solitons arise naturally in the case of interactions of symmetric solitons with asymmetric inhomogeneities (see Sect. 6.5) and as a result of the bifurcation of symmetric solitons. Let us discuss the latter variant.

As follows from the analysis given above (see Fig. 6.21), a fundamental soliton loses its stability with an increase in the small-signal gain in the form of the Andronov–Hopf bifurcation. Numerical simulations show that this bifurcation is subcritical. After the bifurcation the fundamental soliton is transformed first into a metastable oscillating structure and then, after a prolonged transient, into an asymmetric rotating laser soliton with zero topological charge.

It follows from calculations that domains of existence of fundamental and rotating laser solitons overlap (Fig. 6.23). With a further increase in the small-signal gain, a rotating soliton exhibits a new bifurcation, transforming into an oscillating (and rotating) asymmetric laser soliton [92] (see also [67]). With a decrease in the gain, a rotating soliton persists in a narrower, as compared with the fundamental soliton, range, turning into the regime of a fundamental soliton with a further decrease in the gain (Fig. 6.23).

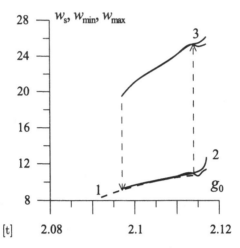

**Fig. 6.23.** Domains of existence of fundamental and asymmetric laser solitons: curve 1: width $w_s$ of a stable fundamental (symmetric) soliton; curves 2 and 3: maximum ($w_{max}$) and minimum ($w_{min}$) width of an asymmetric rotating soliton; the splitting of curve 3 corresponds to the bifurcation to an oscillating rotating soliton [92]

### 6.3.4 Nonparaxial Laser Solitons

The problem of optical soliton width decrease is of great importance now for the applications of optical solitons in information technologies (see Appendix C) and is interesting from a scientific standpoint. A natural scale for the width is the light wavelength. However, when a soliton width approaches this scale, the standard paraxial approximation fails, and we have to include into consideration nonparaxial phenomena. For weakly nonparaxial (wide) beams, the corresponding correction terms are given in Appendix A. Here we will discuss the polarization features of weakly nonparaxial laser solitons following [308].

We will restrict our consideration to the specific case of "convective bistability" and laser radiation propagation through a continuous medium with saturating amplification and absorption. The governing equation is (A.53) for the transverse components of the electric field envelope $\boldsymbol{E}_\perp = E_x, E_y$. The small nonparaxial correction [term $\boldsymbol{Q}_\perp$ on the right-hand side of (A.53)] can be treated using a perturbation approach. In a zero-order (paraxial) approximation we have a fundamental soliton with a cylindrically symmetric field and linear polarization: $E_x = E_0(\varrho)$, $E_y = 0$. In the next order, an orthogonal field component arises: $\delta E_y = \delta E_y(\varrho) \sin(2\varphi)$, where $\varrho$ and $\varphi$ are polar coordinates. Note a simple angular dependence of the nonparaxial correction; normalized radial function $\delta E_y(\varrho)$, found by numerical solution of an inhomogeneous ordinary differential equation, is given in Fig. 6.24. Because $\delta E_y(0) = 0$, radiation polarization remains linear in the beam centre. Off axis it is elliptical and changes over the soliton transverse section. Note

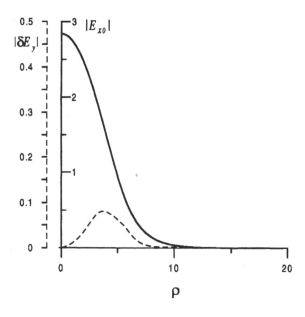

**Fig. 6.24.** Radial profile of the modulus of the amplitude of the paraxial fundamental laser soliton with a linear polarization (*solid line and axis*) and the modulus of the normalized radial function of the nonparaxial component of orthogonal polarization (*dashed line and axis*) [308]

that a similar field structure is typical for "optical needles" – strongly nonparaxial conservative solitons [359, 345]. Because of stability of the paraxial fundamental soliton (see Sect. 6.3.2), small nonparaxial corrections cannot destabilize the soliton considered.

## 6.4 Interaction of Laser Solitons

### 6.4.1 Conditions of "Galilean Symmetry"

The character of soliton interaction depends on many factors; the simplest and basic case corresponds to media with fast nonlinearity and the absence of spatial filtering, when the laser dynamics is governed by (6.1.6). In this case of the "Galilean symmetry", there is a family of laser solitons traveling in transverse directions with arbitrary velocity. The result of soliton interaction depends on the approach speed $v$. For large $v$ we have a regime of weak interaction as the interaction time is small. The condition for the approach speed coincides with (4.5.4); in dimension units it reads ($w$ is soliton width)

$$v \gg c|\delta\varepsilon|w/\lambda . \qquad (6.4.1)$$

This case can be treated analytically using a perturbation approach [409], but only small changes in the initial soliton parameters are possible here. The character of interaction is still retained for smaller approach speeds; a new factor is the appearance of an additional intensity peak during soliton overlapping (Fig. 6.25a, $t = 340$ and $380$). The central peak is associated with the interference of the fields of colliding solitons. The maximum intensity in the peak after collision is small if the interaction duration is small; then the peak dissolves with time (Fig. 6.25a, $t = 600$). For lower speeds the central peak succeeds in growing up to the critical value during collision; thus it is stabilized and transformed into an additional laser soliton (Fig. 6.25b, $t = 420$). For symmetric initial conditions the central soliton is motionless, and the final speeds of two other solitons running away are fairly small in comparison with their initial values. With a further decrease in the approach speed, the central peak grows up to a greater value and begins to dominate. As a result, two colliding solitons merge into a single soliton (Fig. 6.25c). And for very low speeds solitons approach a certain minimum distance and then move apart with decreased values of velocities. Note that the regime of interaction of slowly moving solitons is more sensitive to their phases; it will be considered in more detail below. The regimes of interaction of two-dimensional (Fig. 6.26) and three-dimensional (Sect. 6.6) laser solitons are similar. Note also that, depending on the parameters, additional types of interaction regimes are possible (see Scct. 6.6).

### 6.4.2 Regimes of Weak Overlapping

Here we study the interaction of weakly overlapping solitons in a one-dimensional wide aperture laser with a saturable absorber which is characterized by the purely dissipative type of fast nonlinearity. The electric field envelope $E$ is described by (6.1.4) with a real function $f$ (no frequency detunings). We consider the situation typical of laser systems when the diffusion coefficient $d$ is small and constant. The perturbation approach used was proposed by Gorshkov and Ostrovsky [124, 24]; our presentation is based on [398]; see also review [11] and references therein.

**Equations of Motion.** The perturbation approach uses results of the linear stability analysis of a single stationary soliton with amplitude $A(x)$ obeying (6.2.5) (Sect. 6.2.1). We are reminded that this analysis leads to an eigenvalue problem $\boldsymbol{L\Psi} = \gamma\boldsymbol{\Psi}$ with linear operator (6.2.13)

$$\boldsymbol{L} = \begin{pmatrix} L_{11} & L_{12} \\ L_{12}^* & L_{11}^* \end{pmatrix}. \tag{6.4.2}$$

Further we study the soliton interaction only within the domain of soliton stability.

The symmetry properties of (6.1.4) with $d = 0$ imply that the discrete spectrum of the operator $\boldsymbol{L}$ contains a triply degenerate zero eigenvalue. This

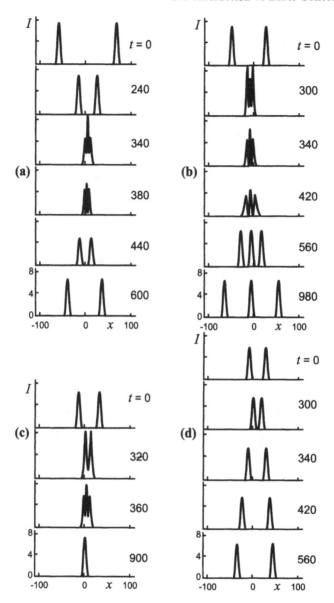

**Fig. 6.25.** Intensity profiles for the collision of one-dimensional laser solitons; approach speed decreases from (**a**) to (**d**) [329]

eigenvalue corresponds to a pair of eigenvectors (neutral modes) $\boldsymbol{\Psi}_{1,2}$ and a root vector $\boldsymbol{\Psi}_3$, which obey the relations

$$L\boldsymbol{\Psi}_{1,2} = 0 , \quad L\boldsymbol{\Psi}_3 = \boldsymbol{\Psi}_1 . \tag{6.4.3}$$

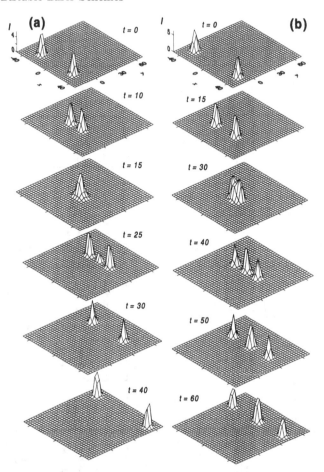

**Fig. 6.26a,b.** Intensity profiles for the collision of two two-dimensional laser solitons [331]

Here $\boldsymbol{\Psi}_s = \left(\psi_s(x), \psi_s^*(x)\right)^{\mathrm{T}}$ with $s = 1, 2, 3$ and

$$\psi_1(x) = \mathrm{d}A/\mathrm{d}x , \quad \psi_2(x) = \mathrm{i}A(x) , \quad \psi_3(x) = -(\mathrm{i}x/2)A(x) . \qquad (6.4.4)$$

Since for a single soliton $A(-x) = A(x)$, the functions $\psi_s$ defined by (6.4.4) have the properties $\psi_{1,3}(x) = -\psi_{1,3}(-x)$ and $\psi_2(x) = \psi_2(-x)$. The neutral modes $\boldsymbol{\Psi}_1$ and $\boldsymbol{\Psi}_2$ are associated with the invariance of (6.1.4) under translations in space and phase shifts respectively, while the root vector $\boldsymbol{\Psi}_3$ is related to the invariance with respect to the *Galilean transformation* to a moving coordinate frame.

In what follows we will use, along with the vectors $\boldsymbol{\Psi}_s$, the vectors $\boldsymbol{\Psi}_s^{\dagger} = \left(\psi_s^{\dagger}(x), \psi_s^{\dagger *}(x)\right)^{\mathrm{T}}$ with $s = 1, 2, 3$, which obey the relations $\boldsymbol{L}^{\dagger}\boldsymbol{\Psi}_{1,2}^{\dagger} = 0$ and

$L^\dagger \Psi_3^\dagger = \Psi_1^\dagger$. Here the adjoint operator $L^\dagger$ defined by the relation $\langle \Phi, L\Psi \rangle = \langle L^\dagger \Phi, \Psi \rangle$ is obtained from (6.4.2) by transposition and complex conjugation, the scalar product being $\langle \Phi, \Psi \rangle = 2 \int_{-\infty}^{\infty} \mathrm{Re}[\varphi^*(x)\psi(x)]\,dx$. In particular, using the definition of $L^\dagger$ and (6.4.3), we obtain $\langle \Psi_1^\dagger, \Psi_1 \rangle = \langle L^\dagger \Psi_3^\dagger, \Psi_1 \rangle = \langle \Psi_3^\dagger, L\Psi_1 \rangle = 0$.

From (6.1.4) with $d = 0$, the asymptotic relations follow

$$A(x) \sim a \exp[\mp(\gamma - i\omega)x]\,, \quad \psi_s^\dagger(x) \sim (\pm 1)^s b_s \exp[\mp(\gamma + i\omega)x]\,, \quad x \to \pm\infty\,. \tag{6.4.5}$$

The coefficients $a$ and $b_s$ can be found numerically. Using inequality (6.1.10) we obtain $\gamma = \mathrm{Im}\sqrt{\alpha - if(0)} > 0$ and $\omega = \mathrm{Re}\sqrt{\alpha - if(0)} > 0$. It follows from (6.4.5) that if the soliton separation is large the overlap can be characterized by the quantity $\exp(-\gamma\zeta_-)$, where $\zeta_- = \zeta_2 - \zeta_1 > 0$ is the distance between soliton intensity maxima. Let the overlap be weak enough, $\exp(-\gamma\zeta_-) = O(\varepsilon^2)$, and the diffusion coefficient be small, $d = O(\varepsilon)$, where the sign $O(\varepsilon^n)$ denotes zero of $n$th order ($\varepsilon \ll 1$). The latter condition assumes that the Galilean symmetry of (6.1.4) is only slightly broken. We write the solution of (6.1.4) describing the interacting pair of solitons in the form

$$E(x,t) = \exp(-i\alpha t)\left[\left(\sum_{j=1}^{2} u_j^{(0)}(x,t)\right) + u^{(1)}(x,t) + u^{(2)}(x,t) + O(\varepsilon^3)\right]\,. \tag{6.4.6}$$

Here the functions $u_j^{(0)}$ are the soliton solutions

$$u_j^{(0)}(x,t) = A_j \exp[-i\varphi_j(t)]\,, \quad A_j = A[x - \zeta_j(t)] \tag{6.4.7}$$

with the coordinates $\zeta_{1,2}$, phases $\varphi_{1,2}$, and velocities $d\zeta_{1,2}/dt$ being slowly varying functions of time, $d\zeta_j/dt = O(\varepsilon)$, $d\varphi_j/dt = O(\varepsilon^2)$, $d^2\zeta_j/dt^2 = O(\varepsilon^2)$. The functions $u^{(m)}(x,t)$, with $m = 1,2$, in (6.4.6) describe the first- and second-order corrections to (6.4.7), $u^{(m)} = O(\varepsilon^m)$. Since the soliton velocities are assumed to be of the order $\varepsilon$, we have $\partial u^{(m)}/\partial t = O(\varepsilon^{m+1})$.

Let us introduce $u^{(m)} = \left(u^{(m)}(x,t), u^{(m)*}(x,t)\right)^T$, $\Psi_{sj} = \left(\exp(-i\varphi_j)\psi_{sj},\right.$ $\left.\exp(i\varphi_j)\psi_{sj}^*\right)^T$ with $\psi_{sj} = \psi_s(x - \zeta_j)$ defined by (6.4.4). Then substituting (6.4.6) into (6.1.4) and collecting the first-order terms in $\varepsilon$ we obtain

$$L\left(\sum_{j=1}^{2} u_j^{(0)}\right)u^{(1)} = -\sum_{j=1}^{2} \Psi_{1j}\frac{d\zeta_j}{dt}\,, \tag{6.4.8}$$

which has a solution

$$u^{(1)} = -\sum_{j=1}^{2} \Psi_{3j}d\zeta_j/dt + O(\varepsilon^3)\,. \tag{6.4.9}$$

Similarly, equating the second-order terms in $\varepsilon$, we obtain

$$L\left(\sum_{j=1}^{2} u_j^{(0)}\right) u^{(2)} = -\sum_{j=1}^{2}\left[-\frac{d\Psi_{3j}}{dx}\left(\frac{d\zeta_j}{dt}\right)^2\right.$$

$$\left. + \Psi_{2j}\frac{d\varphi_j}{dt} + \Psi_{3j}\frac{d^2\zeta_j}{dt^2} + d\Psi_{4j}\frac{d\zeta_j}{dt}\right] - H , \qquad (6.4.10)$$

where $\Psi_{4j} = \left(\exp(-i\varphi_j)\psi_{4j}, \exp(i\varphi_j)\psi_{4j}^*\right)^{\mathrm{T}}$, with $\psi_{4j} = -d^2\psi_{3j}/dx^2$, and $H = \left(h(x,t), h^*(x,t)\right)^{\mathrm{T}}$, in which

$$h(x,t) = \left(\sum_{j=1}^{2} A_j \exp(-i\varphi_j)\right) f\left(\left|\sum_{j=1}^{2} A_j \exp(-i\varphi_j)\right|^2\right)$$

$$-\sum_{j=1}^{2} A_j \exp(-i\varphi_j) f(|A_j|^2) + O(\varepsilon^3) = \sum_{j=1}^{2}\{A_j \exp(-i\varphi_j)$$

$$\times(f_\Sigma + |A_{3-j}|^2 f_\Sigma' - f_j) + A_j^2 A_{3-j} \exp[-i(2\varphi_j - \varphi_{3-j})]f_\Sigma'\}$$

$$+O(\varepsilon^3) , \qquad (6.4.11)$$

where $f_\Sigma, f_\Sigma' = f, f'\left(|A_1|^2 + |A_2|^2\right)$ and $f_j, f_j' = f, f'(|A_j|^2)$. Since the overlap of solitons is weak, the vector $H$ is small $[h(x,t) = O(\varepsilon^2]$.

The equations governing the slow-time evolution of the individual soliton parameters can be obtained from the solvability conditions for (6.4.10). The right-hand side of (6.4.10) must be orthogonal to the solutions of the linear homogeneous equation $L^\dagger\left(\sum_{j=1}^{2} u_j^{(0)}\right)\Psi^\dagger = 0$. These solutions can be estimated as $\sum_{j,s=1}^{2} C_{sj}\Psi_{sj}^\dagger + O(\varepsilon^2)$ with arbitrary constants $C_{sj}$ and

$$\Psi_{sj}^\dagger = \left(\exp(-i\varphi_j)\psi_{sj}^\dagger, \exp(i\varphi_j)\psi_{sj}^{\dagger *}\right)^{\mathrm{T}} , \quad \psi_{sj}^\dagger = \psi_s^\dagger(x - \zeta_j) \quad (s, j = 1, 2) .$$
$$(6.4.12)$$

Using the relations $\langle\Psi_{1,2}^\dagger, \Psi_1\rangle = 0$ and other scalar products found in [398], we finally obtain the following governing equations:

$$\frac{d^2\zeta_-}{dt^2} + D\frac{d\zeta_-}{dt} = r_1 \exp(-\gamma\zeta_-) \cos\varphi_- \sin(\omega\zeta_- + \theta_1) , \qquad (6.4.13)$$

$$\frac{d\varphi_-}{dt} = r_2 \exp(-\gamma\zeta_-) \sin\varphi_- \cos(\omega\zeta_- + \theta_2) + Pv_+v_- , \qquad (6.4.14)$$

$$\frac{dv_+}{dt} + Dv_+ = r_1 \exp(-\gamma\zeta_-) \sin\varphi_- \cos(\omega\zeta_- + \theta_1) , \qquad (6.4.15)$$

$$\frac{d\varphi_+}{dt} = r_2 \exp(-\gamma\zeta_-) \cos\varphi_- \sin(\omega\zeta_- + \theta_2) + P(v_+^2 + v_-^2)/2 , \qquad (6.4.16)$$

with $\zeta_\pm = \zeta_2 \pm \zeta_1$, $\varphi_\pm = \varphi_2 \pm \varphi_1$, $v_\pm = d\zeta_\pm/dt$ $r_s = (-1)^s 4N_s^{-1}|ab_s^*|\sqrt{\gamma^2 + \omega^2}$, $\theta_s = \arg(ab_s^*) + \arg(\gamma - i\omega)$, $D = qd/N_1$, $P = p/N_2$, $N_1 = \langle\Psi_3^\dagger, \Psi_1\rangle =$

$\langle \Psi_1^\dagger, \Psi_3 \rangle$, $N_2 = \langle \Psi_2^\dagger, \Psi_2 \rangle$, $p = \langle \Psi_2^\dagger, d\Psi_3/dx \rangle$, and $q = \langle \Psi_1^\dagger, \Psi_4 \rangle$. The coefficients in (6.4.13–6.4.16) can be evaluated numerically. Equations (6.4.13-6.4.14) govern the time evolution of the distance between the solitons' intensity maxima $\zeta_-$ and the phase difference $\varphi_-$, whereas (6.4.15-6.4.16) determine the centre-of-mass velocity $v_+/2$ and the mean frequency shift $(1/2)d\varphi_+/dt$ of the soliton pair; the latter value does not enter in (6.4.13–6.4.15), which can be solved separately from (6.4.16). Note that the damping terms in (6.4.13) and (6.4.15) are proportional to the diffusion coefficient $d$. We will show below that these terms can stabilize certain in-phase and antiphase bound states of solitons.

**Bound Soliton States.** The steady-state solutions of (6.4.13–6.4.15) correspond to bound soliton states. They are given by

$$BS_0 : \varphi_- = 0, \quad \zeta_- = \frac{\pi n - \theta_1}{\omega}, \quad v_+ = 0,$$

$$\frac{d\varphi_+}{dt} = (-1)^n r_2 \exp(-\gamma\zeta_-) \sin\theta_-, \tag{6.4.17}$$

$$BS_\pi : \varphi_- = \pi, \quad \zeta_- = \frac{\pi n - \theta_1}{\omega}, \quad v_+ = 0,$$

$$\frac{d\varphi_+}{dt} = (-1)^{n+1} r_2 \exp(-\gamma\zeta_-) \sin\theta_-, \tag{6.4.18}$$

$$BS_{\pm\pi/2} : \varphi_- = \pm\frac{\pi}{2}, \quad \zeta_- = \frac{\pi(n+1/2) - \theta_2}{\omega},$$

$$v_+ = \pm(-1)^n \frac{r_1}{D} \exp(-\gamma\zeta_-) \sin\theta_-, \quad d\varphi_+/dt = Pv_+^2/2, \tag{6.4.19}$$

with $\theta_- = \theta_2 - \theta_1$. According to (6.4.17–6.4.19), the bound states with the phase difference $\varphi_- = 0, \pi$ between the solitons are motionless, while the states with $\varphi_- = \pm\pi/2$ travel at the velocity $v_+/2$.

The stability of the bound soliton states can be analyzed in the framework of (6.4.13-6.4.14). For $D, \omega > 0$ the stability conditions for the steady states $BS_0$ and $BS_\pi$ are

$$r_1 r_2 \cos\theta_- > 0, \quad \mp(-1)^n r_1 > 0, \tag{6.4.20}$$

where the $-$ (+) sign corresponds to the state $BS_0$ ($BS_\pi$). It follows from (6.4.20) that all the states $BS_{0,\pi}$ are unstable when $r_1 r_2 \cos\theta_- < 0$. If, on the contrary, $r_1 r_2 \cos\theta_- > 0$, the states $BS_0$ ($BS_\pi$) with $\text{sign}[(-1)^n r_1] < 0$ ($\text{sign}[(-1)^n r_1] > 0$) appear to be stable. According to (6.4.13-6.4.14) for $D = 0$ and $r_1 r_2 \cos\theta_- > 0$ these states, each having a pair of pure imaginary eigenvalues, are neutrally stable. For the parameters typical of class A lasers these bound states are weakly unstable except for the stable antiphase state $BS_\pi$ corresponding to the steady-state solution (6.4.18) with $n = 3$. The latter state corresponds to the minimum possible distance between the solitons.

In the limit $d = 0$, in-phase states $BS_0$ are neutrally stable in the framework of (6.4.13–6.4.15) and weakly unstable according to the numerical solution of the initial equation (6.1.4). The instability could be related to the contribution of the higher-order terms in $\varepsilon$, which were neglected in the derivation of (6.4.13–6.4.16). However, it follows from (6.4.20) that, for $D > 0$, the in-phase bound states become stable when the contribution of the diffusion coefficient to the eigenvalues describing its stability is greater than that of the higher-order terms. Hence, one could expect that there exists a threshold value of the diffusion coefficient $d$ above which the state $BS_0$ with $n = 4$ is stable. This conclusion is illustrated by Fig. 6.27, in which the numerically calculated dependence of the frequency shift of the $BS_0$ state is shown as a function of the diffusion coefficient $d$. Solid and dotted lines indicate stable and unstable bound states, respectively. It is seen from Fig. 6.27 that for $d > d_0$ the state $BS_0$ with $n = 4$ is stable. A similar threshold should exist for the in-phase bound states corresponding to odd $n > 4$ and the antiphase bound states corresponding to even $n > 3$. Since the contribution of the neglected higher-order terms is smaller for greater $n$, the threshold value $d_0$ is expected to decrease with the increase in the number $n$.

Now let us turn to the states $BS_{\pm\pi/2}$ characterized by the phase shift $\pm\pi/2$ between the solitons. For $r_1 r_2 \cos\theta_- > 0$ these solutions are always unstable in the framework of (6.4.13–6.4.15). Moreover, it follows from (6.4.15) that the bound states $BS_{\pm\pi/2}$, traveling with constant velocity, do not exist for $D = 0$ $(d = 0)$. This conclusion is in agreement with the rigorous result presented in Sect. 6.2.1, where it was shown that the only motionless and uniformly moving bound states possible in the diffusionless limit $d = 0$ are in-phase and antiphase states.

When the diffusion coefficient $d$ increases, certain additional terms should be included on the left-hand side of (6.4.13–6.4.16). These terms can stabilize soliton bound states with phase difference $\pm\pi/2$ [213, 12]. Note also that

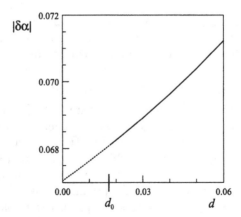

**Fig. 6.27.** Dimensionless frequency shift of a stable (*solid line*) and unstable (*dotted line*) laser soliton in-phase bound state $BS_0$ with $n = 4$ versus the normalized diffusion coefficient $d$; $d_0$ indicates the threshold value of the diffusion coefficient, above which the in-phase state is stable [398]

equations of motion of laser solitons (6.4.13–6.4.16) cannot be derived from a single interaction potential because of the essentially dissipative nature of the medium nonlinearity.

### 6.4.3 Multisoliton Structures

It is natural that multisoliton structures also exist in laser schemes. We are reminded that, for fast nonlinearity, the symmetry properties given in Sect. 6.2.1 impose certain restrictions on the possible types of stationary multisoliton structures. For a proper choice of the transverse coordinate origin, the amplitude $A(x)$ of the motionless soliton can be either an even or an odd function of $x$, and even functions $A(x)$ cannot have zeros for finite $x$. In particular, for the symmetric three-soliton structure shown in Fig. 6.28

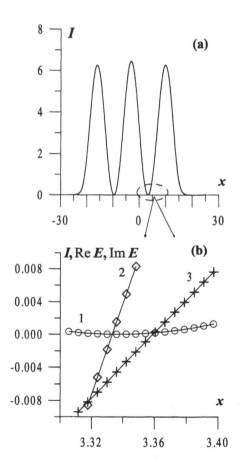

**Fig. 6.28.** Stable three-soliton structure: (a) and curve 1 in (b) are intensity profiles; curves 2 and 3 in (b) show the real and imaginary parts of the field envelope in the vicinity of one of the two intensity minima [91]

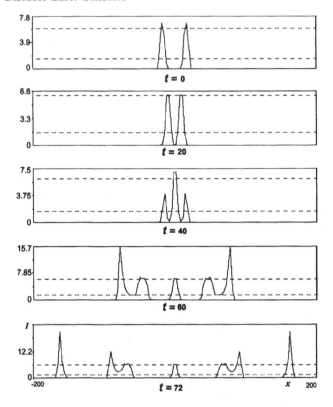

**Fig. 6.29.** The approach of two slow solitons and "triplet" formation [340]

[function $A(x)$ is even], there are no exact intensity zeros; the zeros of the real and imaginary parts of the complex envelope $A(x)$ are slightly split.

As well as in the case of DOSs in driven nonlinear interferometers (Chap. 4), two- and multisoliton complexes in lasers exist even under conditions when single DOSs do not exist; however, usually the ranges of the parameters of their existence differ only slightly.

### 6.4.4 Effect of Relaxation

As we have seen in Sect. 6.2.2, there are various types of laser solitons in the case of media with finite relaxation rates. Most interesting are the regimes of interaction of slow solitons, as the effective interaction time is maximum in this case. Depending on the parameters, the following regimes can be observed:

(1) Two slow solitons approach each other up to a certain minimum distance and then move apart. Although the soliton velocities vary during the interaction, they restore the initial values for large distances between the solitons.

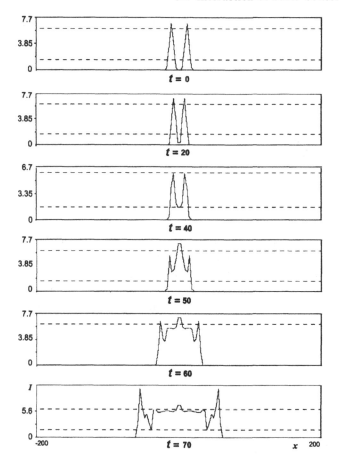

**Fig. 6.30.** Collision of slow solitons with the formation of a widening structure [340]

(2) After collision and merging of two solitons, a "triplet" arises, with a slow soliton in the centre and two metastable structures – the "guns" – moving apart in opposite directions (Fig. 6.29). The "guns" emit a certain number of "fast" solitons (up to tens) and then disappear.

(3) In a narrow range of parameters, there is a similar regime, but in the centre a progressively widening structure arises whose edges are switching waves (Fig. 6.30).

(4) Two colliding slow solitons merge, forming a motionless soliton. These and other types of soliton interaction are considered in [340].

## 6.5 Effect of Smooth and Sharp Inhomogeneities

For a laser with inhomogeneities of the optical path length (in the case of curved mirrors), losses and amplification, we can rewrite (6.1.7) in the form

$$\frac{\partial E}{\partial t} - i\Delta_\perp E = f(|E|^2)E + \tilde{f}(|E|^2, \boldsymbol{r}, t)E . \tag{6.5.1}$$

Here $f$ describes an ideal scheme without inhomogeneities, represented by the additional function $\tilde{f}$. Further we consider smooth and sharp – on the scale of the laser soliton width – inhomogeneities.

### 6.5.1 Smooth Inhomogeneities

The "mechanical equations of motion" can be derived by an approach similar to that presented in Chap. 4 for driven interferometers; historically, they were first obtained by the approximated method of momenta [301]. The soliton parameters, such as the intensity profile $q_{\mathrm{I}}$ and wave front $q_{\mathrm{ph}}$ curvatures, follow the local characteristics of smooth inhomogeneities adiabatically. Then the resulting approximate equation for the soliton centre-of-mass coordinates $\boldsymbol{r}_0$ has the form [302]

$$\ddot{\boldsymbol{r}}_0 = \boldsymbol{B}(\boldsymbol{r}_0, t) + (\dot{\boldsymbol{r}}_0, \nabla)\boldsymbol{D}(\boldsymbol{r}_0, t) , \tag{6.5.2}$$

$$\boldsymbol{B} = 2\nabla f'' - 2\frac{q_{\mathrm{ph}}}{q_{\mathrm{I}}}\nabla f' , \quad \boldsymbol{D} = \frac{1}{2q_{\mathrm{I}}}\nabla f' . \tag{6.5.3}$$

Here $f' = \operatorname{Re} \tilde{f}$ describes inhomogeneities of losses, absorption and amplification, and $f'' = \operatorname{Im} \tilde{f}$ describes inhomogeneities of the cavity optical path length. If there are inhomogeneities only of the optical path length, then $\boldsymbol{D} = 0$, "friction" is absent, and (6.5.2) has the form of the Newtonian equation of motion with the potential "force" $\boldsymbol{B}$. In this case solutions of (6.5.2) include description of oscillatory motion of a laser soliton.

The presence of "friction" ($\boldsymbol{D} \neq 0$) makes possible stable trapping by a local inhomogeneity of a soliton moving toward the inhomogeneity. In Fig. 6.31, a phase plane is given for one-dimensional soliton motion in a laser with a local increase and decrease in losses. In the first case, the trapping regime is unstable. The fixed point (white circle) is a saddle, and its separatrices separate the phase plane into four cells with similar behaviour (Fig. 6.31a). A soliton with an initial velocity exceeding the critical value goes through the inhomogeneity with certain decrease in its velocity. A soliton with a lower velocity is reflected; this is also the case with a decrease in speed.

For the second case (a local decrease in losses), a soliton localization in the minimum of losses is stable. The corresponding fixed point on the phase plane is a node or a focus, depending on the parameters. The phase plane (Fig. 6.31b) is separated into three cells by two specific trajectories at which

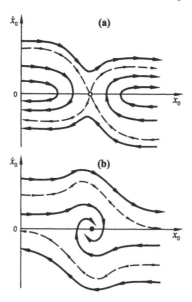

**Fig. 6.31.** Trajectories of soliton motion in a transversely one-dimensional laser with a local increase (**a**) and decrease (**b**) in losses. In the first case, the fixed point (*white circle*) corresponds to the maximum of losses and unstable regime of trapping. In the second case, the fixed point (*black circle*) represents the minimum of losses and stable trapping. The domain of attraction of the latter regime is bounded by two separatrices (*dashed lines with arrows*) [302]

the zero velocity is asymptotically reached (for $x \to \pm\infty$). If the initial velocity exceeds the critical value, the soliton goes through the inhomogeneity with a loss in velocity. The loss is total (the soliton stops) when the initial velocity tends to the critical value. The intermediate cell of the phase plane, situated between the specific trajectories (there the initial speed is smaller than the critical value), serves as the domain of attraction of the stable trapping regime.

## 6.5.2 Sharp Inhomogeneities (Mirror Edges)

The edges of the cavity mirrors present an important example of a sharp inhomogeneity. The finiteness of a real laser aperture is inessential until the soliton approaches the mirrors edges. In the latter case, soliton interaction with mirror edges determines the soliton lifetime and its characteristics. Here we consider a laser model with Galilean invariance (6.1.12).

As is known from the classical Fresnel's problem of a plane-wave reflection from an interface between two media, a sharp inhomogeneity partly reflects and partly transmits the incident radiation. Therefore these regimes of interaction are more various than in the case of smooth inhomogeneities; changes in soliton type and number are now possible. In the ideal scheme (infinite

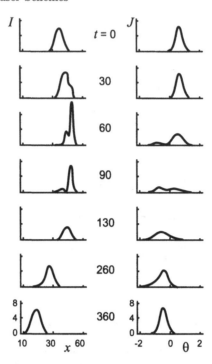

**Fig. 6.32.** Intensity profiles in near- (*left*) and far-field (*right*) zones in the case of laser soliton reflection from the mirror edges with an undercritical incidence angle [329]

aperture), the laser soliton moving with the transverse velocity $v$ is characterized by the angle $\theta = v/c$ between the propagation direction and the cavity axis $z$ (see Fig. 6.1) and by the width of angular (diffraction) divergence, $\theta_d = \lambda/w$ ($\lambda$ is the light wavelength, and $w$ is the soliton width). The relation between these two values determines the result of soliton interaction with the edges. If the angle of incidence is large ($\theta \gg \theta_d$), the geometric drift of the soliton prevails, as compared with diffraction effects. Then the soliton approaching the mirror edges will be removed from the cavity, which means suppression of lasing. In the opposite case ($\theta \ll \theta_d$), the coefficient of diffraction reflection from the edges is close to unity. Therefore the laser soliton will reflect from the edges and move towards the cavity centre. The critical angle for which the effects of soliton geometrical drift and diffraction widening are equalized is $\theta_{cr} = \theta_d$. Note also that solitons with sufficiently small angles $\theta$ will be trapped near the edges due to the field diffraction oscillations.

In Fig. 6.32, the dynamics of soliton reflection from the mirror edges is presented in the case of an undercritical angle of incidence (one-dimensional scheme). With the soliton's approach to the edges, its intensity profile $I(x)$ is deformed up to the appearance of a second maximum ($t = 60$). In this case its far-field intensity distribution $J(\theta)$ acquires the second peak in the region

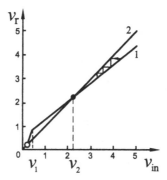

**Fig. 6.33.** The relation between normal components of the velocities of a laser soliton reflected from the mirror edge $v_r$ and incident on the edge $v_{in}$ (curve 1). *Dots* at the intersections of curve 1 with straight line 2 $v_r = v_{in}$ correspond to the unstable ($v = v_1$) and stable ($v = v_2$) equilibrium values of the velocity. The "ladder" *line with arrows* shows the process of temporal variation of the velocity [328]

of negative angles (reflected radiation). The power is gradually transported into the second peak. Since the incidence angle is close to the critical value under the conditions shown in Fig. 6.32, a considerable part of the power is lost after reflection ($t = 130$), but its remainder is sufficient for further soliton restoration ($t = 360$).

The more detailed dependence of the velocity of the reflected soliton on its initial velocity for undercritical angles of incidence is presented in Fig. 6.33 (fundamental two-dimensional solitons). There are two equilibrium values of velocity when it does not change after reflection: $v = v_1$ (unstable regime) and $v = v_2$ (stable regime). A soliton with $v < v_1$ will finally be trapped by the mirror edges. For greater (but undercritical) initial velocities, a soliton acquires the steady-state velocity $v = v_2$ after a number of successive reflections.

Now let us consider the more complicated case of reflection of a laser soliton with a cylindrically symmetric intensity profile and topological charge $M = 1$. For comparatively small velocities of incidence (but nevertheless $v > v_1$), the reflected soliton is of the same type as the incident soliton. But for larger velocities it transforms into an asymmetric rotating soliton with the same topological charge $M = 1$ (Fig. 6.34). As the velocity of incidence increases further, the soliton loses its topological charge upon reflection; the reflected structure is an asymmetric rotating soliton with zero topological charge. For a still higher velocity of incidence, the reflected structure is decomposed into two fragments with zero charge. One of them is destroyed as it goes from the mirror edge, the surviving structure relaxing to a fundamental soliton (Fig. 6.35). Survival of both fragments with their transformation into fundamental solitons is also possible under certain conditions.

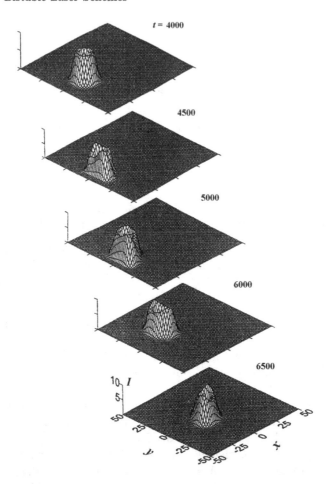

**Fig. 6.34.** Transformation of a vortex laser soliton ($M = 1$) into an asymmetric rotating soliton due to reflection from the mirror edges [328]

## 6.6 Laser Bullets – Three-Dimensional Laser Solitons

An interesting feature of the simplest equation (6.1.7) is the possibility to find its localized solutions with any geometric dimensionality $D = 1$, 2, 3. One- and two-dimensional laser solitons have already been presented above. Here we consider what we call *laser bullets* – three-dimensional laser solitons, or dissipative localized structures in a continuous medium with saturable amplification and absorption and quadratic frequency dispersion. These structures were predicted in [296, 301] and demonstrated numerically in [154, 336, 396, 153]. Note that the scheme considered is cavity-less, characterized by "convective bistability". Strictly speaking, single three-dimensional solitons with a constant shape cannot exist in cavity schemes, where radiation

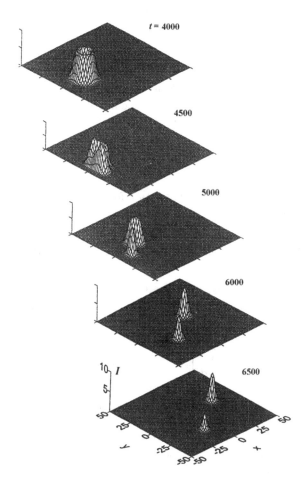

**Fig. 6.35.** Decay of a vortex laser soliton $(M = 1)$ into two fragments due to reflection from the mirror edges [328]

parameters vary periodically due to, e.g., reflections from mirrors. Instead, periodic trains of interacting – due to their tails overlapping – solitons are possible in nonlinear cavities. A decrease in soliton interaction by means of an increase in cavity length contradicts the standard mean-field approximation, which does not allow us to describe single or weakly interacting localized structures with essential variation of the envelope within the cavity.

## 6.6.1 Bifurcation Analysis

In the main variant we adopt the model of fast optical nonlinearity with a simplified account of the relaxation and finite width of spectral lines by means of the introduction of effective diffusion coefficient $d$, which moreover can be neglected in many cases. The corresponding dimensionless governing equation for the electric field envelope $E$ has the form

$$\frac{\partial E}{\partial z} - (i + d)\Delta_3 E = f(|E|^2)E , \quad \Delta_3 = \frac{\partial^2}{\partial x^2} + \frac{\partial^2}{\partial y^2} + \frac{\partial^2}{\partial \tau^2} . \quad (6.6.1)$$

Here $\Delta_3$ stands for the three-dimensional Laplacian which describes diffraction and quadratic frequency dispersion, $\tau = t - z/v_{gr}$ is the time in the frame of coordinates moving along axis $z$ with group velocity $v_{gr}$, and the nonlinear function $f$ (6.1.5) is complex in the case of nonzero frequency detunings.

If we restrict consideration to only stationary spherically symmetric (in the scaling adopted) "bullets" with the envelope of the form

$$E = A(r) \exp(-i\alpha z) , \quad r^2 = x^2 + y^2 + \tau^2 , \quad (6.6.2)$$

then (6.6.1) is reduced to an ordinary differential equation for the radial function $A(r)$ and the eigenvalue $\alpha$ – a shift of the soliton propagation constant

$$\frac{d^2 A}{dr^2} + \frac{2}{r}\frac{dA}{dr} + \frac{1}{i+d}[i\alpha + f(|A|^2)]A = 0 . \quad (6.6.3)$$

The boundary conditions for (6.6.3) consist of the requirement for the finiteness of the amplitude $A$ at $r = 0$ and the decay of $A(r)$ at $r \to \infty$, where $f \to f_0$ and

$$A = \frac{C}{r} \exp\left(-r\sqrt{-\alpha - \frac{f_0}{i+d}}\right) , \quad C = \text{const.} \quad (6.6.4)$$

As in the one- and two-dimensional cases, it is convenient to introduce variables (6.2.6), which allows us to reduce (6.6.3) to the third-order set of real ordinary differential equations. Thus, the procedure for determining fundamental (symmetric) localized structures is practically the same for any dimensionality $D = 1$, 2 and 3. The results are illustrated in Fig. 6.36, where the dependence of the eigenvalue $\alpha$ on the small-signal gain $g_0$ is given for all these dimensionalities for the case of zero detunings.[5] All the curves start from the same point on plane $\alpha, g_0$ ($\alpha = 0$, $g_0 = 1 + a_0$, which corresponds to the stability boundary of the nonlasing regime $A = 0$). Judging by the calculations, all the branches have a tendency to end at the same point (corresponding to the limiting point of a similar spiral for one-dimensional laser localized structures). At this limiting point the width of localized structures tends to infinity; therefore the geometrical dimensionality is not essential.

The next step is to check the stability of the fundamental localized structures found. For three-dimensional laser bullets, as distinct from one- and two-dimensional solitons, it is difficult to apply the procedure of semi-analytical linear stability analysis. More practical is the stability check by means of numerical solution of (6.1.7), with the initial (at $z = 0$) field in the form of

---

[5] More precisely, there is a large number of dependence branches for $D = 2$ and 3, but only the branches shown in Fig. 6.36 include segments corresponding to stable laser solitons.

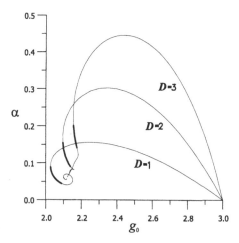

**Fig. 6.36.** Dependence of the spectral parameter of stable (*thick lines*) and unstable (*thin lines*) localized laser structures with dimensionality $D = 1$, 2 and 3 on the small-signal gain coefficient $g_0$ [396]

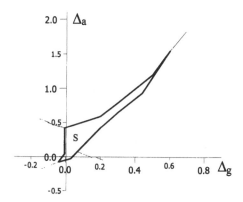

**Fig. 6.37.** Domain of stability S of the fundamental laser bullets on the plane of detunings between the radiation carrier frequency and central frequencies of spectral lines of amplification and absorption [396]

superposition of the radially symmetric soliton and small asymmetric perturbations. The results are shown in Fig. 6.36 for zero frequency detunings and in Fig. 6.37 for arbitrary detunings. We can conclude that laser bullets do exist and are stable in a certain domain of parameters.

## 6.6.2 Formation of Laser Bullets

The dynamics of formation of laser bullets found by numerical solution of (6.1.7) is presented in Figs. 6.38 and 6.39. In the case of Fig. 6.38, we used an initially bell-shaped intensity distribution close to symmetric distribution, the

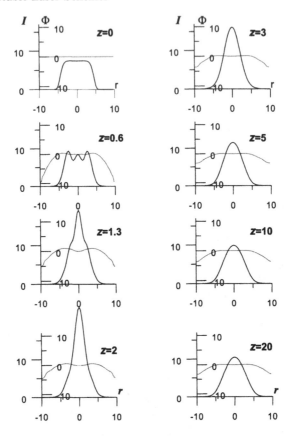

**Fig. 6.38.** Formation of a fundamental (spherically symmetric) laser bullet: radial profiles of intensity $I$ (*solid lines, outer scales*) and phase $\Phi$ (*dashed lines, inner scales*) for different longitudinal coordinates $z$ [154]

phase being taken as constant. The transient includes the stage of formation of switching waves moving in the radial direction and a sharp increase in the peak intensity after these waves have collapsed. The bullet's excitation is hard (threshold-like); the initial low-power radiation structures are gradually dispersed, and, for a sufficiently large initial power, a widening spherical switching wave can form with continuous growth of the size of the switched region. The deformation of the wavefront is related to nonlinear focusing.

From Fig. 6.39 one can see that the transient includes oscillations damped exponentially with the distance $z$. It is one more confirmation of the stability of the laser bullet. Numerical simulations also show the existence of long-living asymmetric pulsing localized radiation structures. Within a certain range of the parameters, such structures are metastable in a homogeneous medium; however, they can be stabilized by the introduction of a weak spatial inhomogeneity [262].

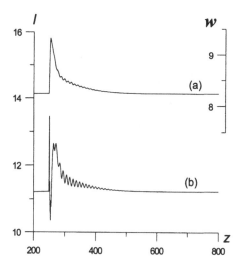

**Fig. 6.39.** Damping of oscillations of the fundamental laser bullets: peak intensity $I$ (**a**) and averaged width $w$ (**b**) versus longitudinal coordinate $z$; an asymmetric initial perturbation was introduced at $z = 250$ [336]

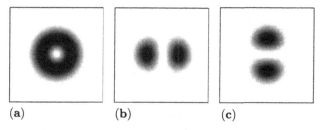

(**a**)          (**b**)          (**c**)

**Fig. 6.40.** Intensity distributions for a metastable topological (toroidal) soliton: (**a**) $xy$ section ($\tau = 0$), (**b**) $x\tau$ section ($y = 0$), (**c**) $y\tau$ section ($x = 0$) [153]

Stable topological (asymmetric) three-dimensional laser solitons have not yet been found. An example of a metastable toroidal laser soliton with topological charge $M = 1$ is given in Fig. 6.40. It is stable when azimuth perturbations are excluded; if they are taken into account, the structure decays with distance $z$, as illustrated in Fig. 6.41.

### 6.6.3 Interaction of Laser Bullets

If the distance between the centres of two bullets significantly exceeds their characteristic width, then their interaction is negligibly small. As for one- and two-dimensional solitons, the result of a collision most of all depends on the velocity of their approach; it depends also on aiming distance and the solitons' phase difference for fairly small speeds. Numerical simulations for the case of head-on collision reveals the following typical scenarios:

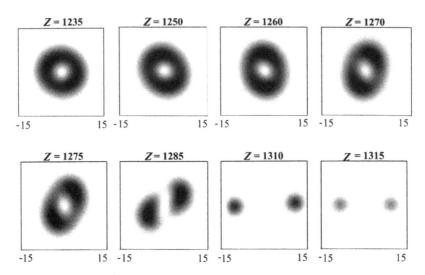

**Fig. 6.41.** Decay of a toroidal soliton: intensity distributions in the $xy$ section $(\tau = 0)$ [153]

1. passing of bullets through one another;
2. repulsion;
3. generation of a new (third) bullet;
4. after collision the switching wave is generated and all the space is filled by radiation.

Scenarios 1–3 are not unexpected: similar regimes were obtained earlier for one- and two-dimensional solitons. The most nontrivial results of collisions – scenarios 3 and 4 – are represented in Fig. 6.42. Scenario 4 (formation and propagation of switching wave after collision) was obtained for the first time for three-dimensional solitons. The existence of such a regime may be explained by the fact that the parameter $g_0$, when the switching wave is generated, exceeds significantly the corresponding "Maxwell's value", for which the velocity of switching wave is equal to zero. For such a value of $g_0$, with the other parameters remaining the same, one- and two-dimensional localized structures do not exist.

### 6.6.4 Nonparaxial Effects

The analysis of weak nonparaxiality for fundamental laser bullets is similar to that presented in Sect. 6.3.4 for two-dimensional laser solitons. The main results are the following [306]: If the unperturbed (paraxial) soliton corresponds to a linear polarization with spherically symmetric distribution $E_x = E_x(r)$, then the next nonparaxial corrections give the field longitudinal and transverse orthogonal components:

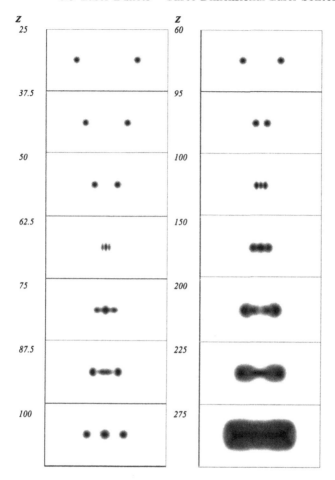

**Fig. 6.42.** Intensity distributions in the $x\tau$ section ($y = 0$) for various values of longitudinal coordinate $z$ for two scenarios of interaction of two laser bullets: generation of a third soliton (*left*), generation of a switching wave (*right*) ([153]

$$E_z = \frac{i}{k_0}\frac{dE_x}{dr}\sin\theta\cos\varphi\,, \quad E_y = R(r)\sin^2\theta\sin(2\varphi)\,. \qquad (6.6.5)$$

Here $r, \theta, \varphi$ are spherical coordinates, and radial function $R(r)$ is determined from an ordinary differential equation and has a shape similar to those presented in Fig. 6.24. Again, the radiation polarization is elliptical and changes over the bullets' volume.

# 7. Conclusion: Comparing Different Types of Optical Patterns

As a conclusion to the book's main text, let us compare the different types of patterns in wide-aperture nonlinear optical systems described by the generalized Ginzburg–Landau equation [300, 68, 297]:

$$\frac{\partial E}{\partial \zeta} = (a + ib)\Delta_D E + f(|E|^2)E + E_{\text{in}} . \tag{7.1}$$

This equation was used, with real constants $a$ and $b$, for transverse pattern analysis in Chap. 2 [heat conductivity equation (2.1.10); geometric dimensionality $D = 1, 2$, $b = 0$, value $E$ and function $f$ are real, $E$ means temperature and $\zeta$ is the time] and Chap. 4 [mean-field equation (4.1.23); $D = 1$, 2, electric field envelope $E$ is complex and $\zeta$ is the time $t$], and it provided a basis for consideration in Chap. 6 ($E_{\text{in}} = 0$, $D = 1, 2, 3$, $\zeta$ is the time $t$ or the longitudinal coordinate $z$). Let us assume that the system does not include inhomogeneities and aperture confinement.

In driven systems (those with "external signal", $E_{\text{in}} = \text{const.} \neq 0$), the intensities of plane-wave regimes $I = |E|^2$ and $I_{\text{in}} = |E_{\text{in}}|^2$ are connected by

$$I|f(I)|^2 = I_{\text{in}} . \tag{7.2}$$

In the case $E_{\text{in}} = 0$ intensity $I$ either vanishes (the nonlasing regime), or is given by the roots of the real part of the function $f$:[1]

$$I\text{Re}f(I) = 0 . \tag{7.3}$$

Both (7.2) and (7.3) can have one or a number of nonnegative roots $I$, which correspond to mono-, bi- or multistability (absolute for $\zeta = t$ and convective for $\zeta = z$). In conservative systems (without dissipation) we have $\text{Re } f(I) = 0$ for any $I$, (7.3) being obeyed identically.

For the diffusion mechanism of the transverse coupling ($b = 0$, $\text{Im } f = 0$), either a plane-wave regime or, under conditions of bistability,[2] a switching wave will be established with time. Any other regimes are not possible there. Therefore we will focus further on more diverse patterns in systems with diffraction coupling, assuming that $a = 0$ or $a$ is a small value.

---

[1] The imaginary part of this function enters into the dispersion relation and determines the nonlinear shift of frequency ($\zeta = t$) or propagation constant ($\zeta = z$).

[2] We do not distinguish bi- and multistability further.

An important class of spatial patterns arises when plane-wave regimes become unstable with regard to small perturbations. In general, even in a point model, when (7.1) transforms to an ordinary differential equation ($\Delta_D = 0$), a stationary regime can lose stability and be replaced by a nonstationary regime (a periodic one; and only for driven systems, when $E_{\text{in}} \neq 0$, see Sect. 4.2). For wide-aperture optical systems more typical is the variant in which the growth of perturbations modulated in transverse directions is the fastest. A spatial pattern established as a result of this modulation instability corresponds to radiation filamentation. The presence of *modulation instability* is sufficient for pattern formation in the case of a monostable system, whereas in a bistable system the instability of one of the plane-wave regimes can induce only switching to another (stable) plane-wave regime.

Convective instabilities of this type were first found in nonlinear optics by Bespalov and Talanov [42] by means of considering plane-wave propagation in a transparent medium with a nonlinear refractive index [(7.1) with $a = 0$, $\text{Re}\, f(I) = 0$, and $E_{\text{in}} = 0$]. First, radiation filamentation is fairly common and is typical of the case with high-power laser radiation; second, it serves as one of the main obstacles in designing laser systems with ultimately high radiation brightness.

For similar reasons, in [323], where *absolute* modulation instability was found for the first time in a driven wide-aperture nonlinear interferometer, also indicated was the possibility of its suppression by means of spatial filtration (see Sect. 4.3). However, in subsequent years many researchers focused their efforts on revealing schemes and conditions where these instabilities are developed most effectively. These studies were not unsuccessful.

It seems that the motivation of these investigations was not always justified. From a scientific standpoint, it is improbable that we would find anything unexpected there. Scenarios of modulation instability development are general and well known; thus, there are no surprises in the similarity of the patterns formed as a result of these instabilities in systems of different nature (see, e.g., [68]). From the point of view of applied physics, the range of problems where radiation filamentation would be of use is very narrow. If we consider applications to optical information processing, an "infinite" pattern is equivalent to only one information bit, which means a sheer waste of space.

In our opinion, the other types of patterns which do not need plane-wave instabilities for their formation deserve more attention. As was found for the first time in [319] (see also Chaps. 4 and 6), a sufficiently large localized (in space and time) perturbation can induce a hard excitation of stable localized structures of optical radiation even without modulation instability. Note that for dissipative systems the nomenclature varies. We think that terms such as "pulses", "defects" or "cavity solitons" are not appropriate in many cases; preferable are the terms "autosolitons" or "dissipative solitons".

Dissipative optical solitons fit into the general scope of self-organization phenomena (synergetics) [247, 138, 140], introducing a specific "optical" streak. The important point is replacement of diffusion typical of traditional self-organization problems in chemistry and biology, by diffraction characteristic in optics, with intrinsic field oscillations. It seems in the optics was perceived a particular richness of localized dissipative structures under conditions of bistability and spatial hysteresis, when instabilities are not necessary. All this stimulated intensive studies of localized optical structures of different types; of special significance are the experiments in [380].

In conservative systems (those without dissipation) there are also stable localized structures – solitons [10, 210, 243, 419]. An important class of solitons is described by (7.1) with $E_{in} = 0$ and $\text{Re} f(I) = 0$ (the "nonlinear Schrödinger equation"). As indicated above, in this case (7.1) has a continuous spectrum of solutions. Similarly, if we fix the system parameters, the conservative solitons are not isolated localized structures, but entire families with continuously varying characteristics, including their peak intensity and width. Small variations in initial conditions induce variations in the parameters of settled solitons.[3] Therefore, drift of the characteristics of a conservative soliton is inevitable under the influence of noise (spontaneous emission). Note also the high sensitivity of conservative solitons to the system's geometric dimensionality. Conservative soliton solutions can be found from more general dissipative soliton solutions if dissipative terms in (7.1) tend to zero. During this limiting process, the transient time for dissipative soliton settling will increase, and metastable structures imitating conservative solitons with intermediate levels of maximum intensity will have a longer and longer lifetime.

For open (dissipative) optical systems which include driven nonlinear interferometers and laser schemes, the spectrum of the soliton's main characteristics is discrete; in this case the maximum intensity is determined by the energetic balance and is close to the intensity of the plane-wave regime. Moderate variations in initial conditions do not change the main characteristics of the settled solitons, including their maximum intensity. Therefore, anomalously high stability and robustness are inherent in dissipative optical solitons. At the same time, the solitons' motion is determined by the scheme's inhomogeneities, which allows us to control and manipulate their positions.

Let us now enlarge on dissipative optical solitons – especially stable light bunches.[4] Note that their hard excitation serves as a manifestation of the spatial hysteresis, because field soliton-free state is also stable. The classification

---

[3] For conservative solitons, the presence of perturbation "internal modes" and anomalously weak (non-exponential) damping is typical, while dissipative solitons are robust, and their perturbations decay exponentially [336].

[4] There is also another type of related localized dissipative structures (also with the hard excitation by an overcritical local perturbation) that is of interest, for example, for the modeling of primary biological objects such as amoebae. They correspond not to "immortal" (stable stationary) solitons, but rather to "living solitons", or "biosolitons". Their living cycle includes growth in size (in the form of widening switching waves), and bisection (e.g., due to modulation instability) into two new objects moving apart from each other; such repetitive processes are possible in lasers with a saturable absorber.

of one-dimensional solitons in interferometers is similar to the one in laser schemes. These solitons correspond to homoclinic or heteroclinic trajectories in a three- or four-dimensional phase space that include a different number of turns in the vicinity of fixed points representing plane-wave regimes. In other words, dissipative soliton "quantum numbers" are the numbers of diffraction oscillations close to maximum and minimum intensity values.

Distinctions between solitons in driven interferometers and those in laser schemes (no external signal) are connected with an additional symmetry of (7.1) for $E_{in} = 0$. Stationary solitons in driven interferometers have the frequency of the external signal and are phase-matched with it, whereas for a laser soliton the frequency is the unknown eigenvalue of the problem, the phase being arbitrary. A single soliton in an ideal driven interferometer is motionless, and it moves with smooth inhomogeneities according to "Aristotelian mechanics" [see (4.6.13)]. A laser soliton in an ideal scheme [with the "Galilean invariance", $a = 0$ in (7.1)] moves with an arbitrary constant velocity and obeys "Newtonian mechanics" in the presence of inhomogeneities [see (6.5.2)]. Degeneracy in soliton velocity is removed in schemes without the "Galilean invariance", e.g., for $a \neq 0$ in (7.1).

Dissipative optical solitons are not only interesting from a scientific standpoint, but also promising in their application, first of all in optical information processing. This is connected with the hard nature of their excitation that excludes the effect of noise, and with the possibilities for manipulating their location, velocities, and structure type (see Appendix C).

Of special interest are three-dimensional dissipative optical solitons – "laser bullets". As follows from a comparison of one- and two-dimensional laser solitons, the set of localized structures abruptly widens with the system geometric dimensionality. Therefore the problem of determination of the full set of three-dimensional dissipative solitons is extremely complicated and presents a challenge to theorists. Neither real or computer (it would be better to say here "machine") experiments have many chances of success since[5]

Its beauty Nature would not unveil;
Try as you might – to no avail.
And no machine shall disentwine
That which your soul would not divine.

(Vladimir Solovjoff, 1853–1900; translated by Stan Voronin)

---

[5] Solovjoff's "machine" is more appropriate for the subject than "Hebeln" (levers) and "Schrauben" (screws) of Goethe's Faust.

# A. Paraxial and Nonparaxial Radiation Propagation

In the entire book, with the exception of Appendix E, radiation is treated *classically*, within the framework of Maxwell's equations [188], disregarding its quantum nature. Going beyond the semiclassical approach (when the electromagnetic field is treated classically and matter is considered in terms of quantum theory) and taking into account the quantum nature of radiation are necessary for weak fields and in the problem of suppression of radiation *quantum fluctuations* (see Appendix E). The quantum nature is important also in the opposite case of high-power radiation, when even a vacuum turns out to be optically nonlinear because of the quantum-electrodynamic effect of its polarization and electron–positron pair creation by photons [41, 16, 295].

Optical radiation in a nonmagnetic medium (with magnetic permittivity $\mu = 1$) is described by the wave equation for the electric field $\tilde{\boldsymbol{E}}$ and Maxwell's equation for electric induction $\tilde{\boldsymbol{D}}$:

$$\nabla \times (\nabla \times \tilde{\boldsymbol{E}}) + \frac{1}{c^2}\frac{\partial^2 \tilde{\boldsymbol{D}}}{\partial t^2} = 0 \,, \tag{A.1}$$

$$\nabla \cdot \tilde{\boldsymbol{D}} = 0 \,, \tag{A.2}$$

where $c$ is the speed of light in vacuum and $t$ is time. The electric induction is connected with the medium dielectric polarization $\tilde{\boldsymbol{P}}$ by

$$\tilde{\boldsymbol{D}} = \tilde{\boldsymbol{E}} + 4\pi \tilde{\boldsymbol{P}} \,. \tag{A.3}$$

**Paraxial Approximation.** Without using the form of the constitutive equation for $\tilde{\boldsymbol{D}}$ or $\tilde{\boldsymbol{P}}$ for a while, we will separate their unperturbed parts and perturbations:

$$\tilde{\boldsymbol{D}} = \tilde{\boldsymbol{D}}_0 + \delta\tilde{\boldsymbol{D}} \,, \quad \tilde{\boldsymbol{P}} = \tilde{\boldsymbol{P}}_0 + \delta\tilde{\boldsymbol{P}} \,,$$

$$\tilde{\boldsymbol{D}}_0 = \hat{\varepsilon}_0 \tilde{\boldsymbol{E}} \,, \qquad \tilde{\boldsymbol{P}}_0 = \frac{1}{4\pi}(\hat{\varepsilon}_0 - 1)\tilde{\boldsymbol{E}} \,,$$

$$\delta\tilde{\boldsymbol{D}} = \delta\hat{\varepsilon}_0 \tilde{\boldsymbol{E}} \,, \qquad \delta\tilde{\boldsymbol{P}} = \frac{1}{4\pi}\hat{\varepsilon}\tilde{\boldsymbol{E}} \,. \tag{A.4}$$

In the situation we are interested in, the operator $\hat{\varepsilon}_0$ corresponds to the dielectric permittivity of the medium in the limit of weak fields; in addition, it is convenient not to include the medium absorption or gain into $\hat{\varepsilon}_0$, relating them to the perturbation $\delta\hat{\varepsilon}$. The operator nature of the linear dielectric permittivity $\hat{\varepsilon}_0$ is caused by *frequency dispersion* of the medium[1] and is determined by the following Fourier expansions:

$$\tilde{\boldsymbol{E}} = \int \tilde{\boldsymbol{E}}_\omega \exp(-\mathrm{i}\omega t)\,\mathrm{d}\omega\ , \tag{A.5}$$

$$\hat{\varepsilon}_0 \tilde{\boldsymbol{E}} = \int \varepsilon_0(\omega)\tilde{\boldsymbol{E}}_\omega \exp(-\mathrm{i}\omega t)\,\mathrm{d}\omega\ . \tag{A.6}$$

In an isotropic medium, $\varepsilon_0(\omega)$ is a scalar and real function of frequency $\omega$ (in the absence of optical absorption and amplification).

Laser radiation has a high degree of monochromaticity and small angular divergence. Therefore, the electric field can be presented in the following form:

$$\tilde{\boldsymbol{E}} = \mathrm{Re}\{\boldsymbol{E}(\boldsymbol{r},t)\exp[\mathrm{i}(k_0 z - \omega_0 t)]\}\ . \tag{A.7}$$

Here the $z$-axis coincides with the radiation beam axis, $\omega_0$ is the carrier frequency of the field in spectral expansion (A.5), $k_0 = k(\omega_0)$, $k(\omega) = \sqrt{\varepsilon(\omega)}\omega/c$ is the wave number in the linear medium, and the *envelope* $\boldsymbol{E}(\boldsymbol{r},t)$ is a *slowly varying* function of time and coordinates on the scales of a period of light oscillations $2\pi/\omega_0$ and light wavelength $\lambda = 2\pi/k_0$, respectively. Thereby we assume

$$\omega_0 \tau_{\mathrm{fr}} \gg 1\ ,\quad r_\| \gg \lambda_0\ ,\quad r_\perp \gg \lambda_0\ , \tag{A.8}$$

where $\tau_{\mathrm{fr}}$ is the duration of the pulse front and $r_\|$ and $r_\perp$ are the characteristic lengths of longitudinal and transverse variations of field amplitude $\boldsymbol{E}$.

Among different nonlinear optical phenomena we will restrict ourselves to the effects of self-action, at which there is no substantial change of radiation frequency. Then for the perturbed component of medium polarization we can introduce *a slowly varying amplitude* similar to (A.7),

$$\delta\tilde{\boldsymbol{P}} = \mathrm{Re}\{\delta\boldsymbol{P}\exp[\mathrm{i}(k_0 z - \omega_0 t)]\}\ , \tag{A.9}$$

and neglect its second-order time derivative

$$\frac{\partial^2 \delta\tilde{\boldsymbol{P}}}{\partial t^2} \approx -\mathrm{Re}\left(2\mathrm{i}\omega_0 \frac{\partial \delta\boldsymbol{P}}{\partial t} + \omega_0^2 \delta\boldsymbol{P}\right)\exp[\mathrm{i}(k_0 z - \omega_0 t)]\ . \tag{A.10}$$

Now we proceed to slowly varying amplitudes in the following term, linear in the field:

$$\frac{1}{c^2}\frac{\partial^2}{\partial t^2}(\hat{\varepsilon}_0 \tilde{\boldsymbol{E}}) = -\int k^2(\omega)\tilde{\boldsymbol{E}}_\omega \exp(-\mathrm{i}\omega t)\,\mathrm{d}\omega\ . \tag{A.11}$$

Expanding the function $k^2(\omega)$ into Taylor series around the carrier frequency $\omega_0$ and holding terms quadratic in frequency deviation $\omega - \omega_0$, we obtain

---

[1] We neglect the medium spatial dispersion.

$$\frac{1}{c^2} \frac{\partial^2}{\partial t^2} (\hat{\varepsilon}_0 \tilde{\boldsymbol{E}})$$

$$\approx -\mathrm{Re}\left[\left(k_0^2 \boldsymbol{E} + 2ik_0 \frac{1}{v_{\mathrm{gr}}} \frac{\delta \boldsymbol{E}}{\partial t} - D_2 \frac{\partial^2 \boldsymbol{E}}{\partial t^2}\right) \exp[i(k_0 z - \omega_0 t)]\right] . \quad \text{(A.12)}$$

*Group velocity* $v_{\mathrm{gr}}$ and *quadratic dispersion parameter* $D_2$ are introduced by the relations

$$\frac{1}{v_{\mathrm{gr}}} = \left[\frac{dk}{d\omega}\right]_{\omega=\omega_0} = \frac{1}{c} \frac{d}{d\omega}[\omega\sqrt{\varepsilon_0(\omega)}]_{\omega=\omega_0} , \quad \text{(A.13)}$$

$$D_2 = \frac{1}{2}\left(\frac{d^2 k^2}{d\omega^2}\right)_{\omega=\omega_0} = \frac{1}{2c^2} \frac{d^2}{d\omega^2}[\omega^2\varepsilon_0(\omega)]_{\omega=\omega_0} . \quad \text{(A.14)}$$

We transform the first term of (A.1) into the following form:

$$\nabla \times \nabla \tilde{\boldsymbol{E}} = \nabla(\nabla \cdot \tilde{\boldsymbol{E}}) - \Delta \tilde{\boldsymbol{E}} = \nabla(\nabla \cdot \tilde{\boldsymbol{E}}) - \Delta_\perp \tilde{\boldsymbol{E}} - \frac{\partial^2 \tilde{\boldsymbol{E}}}{\partial z^2} , \quad \text{(A.15)}$$

where

$$\Delta_\perp \tilde{\boldsymbol{E}} = \frac{\partial^2 \tilde{\boldsymbol{E}}}{\partial x^2} + \frac{\partial^2 \tilde{\boldsymbol{E}}}{\partial y^2} = \mathrm{Re}\left(\left(\frac{\partial^2 \boldsymbol{E}}{\partial x^2} + \frac{\partial^2 \boldsymbol{E}}{\partial y^2}\right) \exp[i(k_0 z - \omega_0 t)]\right) . \quad \text{(A.16)}$$

Under condition (A.8) one can neglect the term $\partial^2 \boldsymbol{E}/\partial z^2$:

$$\frac{\partial^2 \tilde{\boldsymbol{E}}}{\partial z^2} \approx \mathrm{Re}\left(\left(-k_0^2 \boldsymbol{E} + 2ik_0 \frac{\partial \boldsymbol{E}}{\partial z}\right) \exp[i(k_0 z - \omega_0 t)]\right) . \quad \text{(A.17)}$$

To evaluate the first term on the right-hand side of (A.15), we utilize (A.2). Then, under the condition of weak nonlinearity

$$|\delta\varepsilon| \ll \varepsilon_0(\omega_0) , \quad \text{(A.18)}$$

the following relation is true, if we neglect the operator character of $\delta\varepsilon$:

$$(\nabla \cdot \tilde{\boldsymbol{E}}) = -\frac{1}{\varepsilon_0(\omega_0)} \tilde{\boldsymbol{E}} \nabla \delta\varepsilon . \quad \text{(A.19)}$$

Therefore

$$\nabla\left(\nabla \cdot \{\boldsymbol{E} \exp[i(k_0 z - \omega_0 t)]\}\right)$$

$$\approx -\frac{ik_0}{\varepsilon_0(\omega_0)} \boldsymbol{E} \nabla(\delta\varepsilon) \exp[i(k_0 z - \omega_0 t)]\boldsymbol{e}_z , \quad \text{(A.20)}$$

where $\boldsymbol{e}_z$ is the unit vector along the $z$-axis. This term is small because the nonlinearity is weak and varies in space slowly. However, it would be more precise not to neglect it directly, but to put this vector projection to the transverse plane equal to zero. Introducing transverse projections of the electric field $\boldsymbol{E}_\perp = (E_x, E_y)$ and of nonlinear polarization $\delta\boldsymbol{P}_\perp$ and using (A.10)–(A.20), we obtain

$$
2ik_0 \left( \frac{\partial \boldsymbol{E}_\perp}{\partial z} + \frac{1}{v_{\mathrm{gr}}} \frac{\partial \boldsymbol{E}_\perp}{\partial t} \right) + \Delta_\perp \boldsymbol{E}_\perp - D_2 \frac{\partial^2 \boldsymbol{E}_\perp}{\partial t^2}
$$
$$
+ \frac{4\pi\omega_0}{c^2} \left( \omega_0 \delta\boldsymbol{P}_\perp + 2i \frac{\partial \delta\boldsymbol{P}_\perp}{\partial t} \right) = 0 . \tag{A.21}
$$

As for the longitudinal field component $E_z$, it can be expressed in $E_x$ and $E_y$, taking into account (A.19):

$$
E_z = \frac{i}{k_0} \left( \frac{\partial E_x}{\partial x} + \frac{\partial E_y}{\partial y} \right) . \tag{A.22}
$$

Hence it follows that this value is small:

$$
\left| \frac{E_z}{E} \right|^2 \propto \left( \frac{\lambda_0}{r_\parallel} \right)^2 \ll 1 . \tag{A.23}
$$

Paraxial equation (A.21) must be supplemented with a constitutive equation, which establishes a form of nonlinear polarization (or of the nonlinear part of dielectric permittivity $\delta\hat{\varepsilon}$) (see Appendix B). Then its solution expresses the field $\boldsymbol{E}_\perp(\boldsymbol{r}_n, z)$ in any section $z$ in terms of the field in some initial section, e.g., $z = 0$, $\boldsymbol{E}_\perp(\boldsymbol{r}_n, 0)$. It serves as the basic equation of paraxial optics and describes most of the known phenomena involved in nonlinear radiation propagation. Under certain conditions, in which that or other phenomenon is more pronounced, some of the terms in (A.21) can be neglected. It allows us to demonstrate clearly the meaning of different terms in the paraxial equation.

Neglecting diffraction $(\Delta_\perp \boldsymbol{E}_\perp = 0)$, dispersion $(D_2 = 0)$ and medium nonlinearity or inhomogeneity $(\delta\boldsymbol{P} = 0)$, the paraxial equation is transformed into a transport equation:

$$
\frac{\partial \boldsymbol{E}_\perp}{\partial z} + \frac{1}{v_{\mathrm{gr}}} \frac{\partial \boldsymbol{E}_\perp}{\partial t} = 0 , \tag{A.24}
$$

the solution of which is

$$
\boldsymbol{E}_\perp(z, t) = \boldsymbol{f} \left( t - \frac{z}{v_{\mathrm{gr}}} \right) . \tag{A.25}
$$

Here $\boldsymbol{f}$ is an arbitrary function; it is defined at specification of the field at $z = 0$: $\boldsymbol{f} = \boldsymbol{E}_\perp(z = 0, t)$. It corresponds to radiation pulse transport without any distortions with a constant group velocity $v_{\mathrm{gr}}$ determined by (A.13). Transverse coordinates enter into (A.25) only as parameters.

The equation of diffraction of monochromatic radiation $(\partial \boldsymbol{E}_\perp/\partial t = 0)$ in a linear homogeneous medium includes the term $\Delta_\perp \boldsymbol{E}_\perp$:

$$2ik_0 \frac{\partial \boldsymbol{E}_\perp}{\partial z} + \Delta_\perp \boldsymbol{E}_\perp = 0 .$$ (A.26)

If the transverse dimensions of a beam are of the order of $r_\perp$, (A.26) gives the characteristic length of diffraction:

$$l_{\mathrm{dfr}} = k_0 r_\perp^2 .$$ (A.27)

With a smaller path length ($z \ll l_{\mathrm{dfr}}$), diffraction effects can be neglected. More exactly, it is convenient to use in this near-field zone the formal solution of (A.26) and its expansion:

$$\boldsymbol{E}_\perp(\boldsymbol{r}_n, z) = \exp\left(\frac{iz}{2k_0} \Delta_\perp\right) \boldsymbol{E}_\perp(\boldsymbol{r}_n, 0)$$

$$= \sum_{j=0}^{\infty} \frac{1}{j!} \left(\frac{iz}{2k_0}\right)^j \Delta_\perp^j \boldsymbol{E}_\perp(\boldsymbol{r}_n, 0) .$$ (A.28)

The general solution of (A.26) is given in the integral form

$$\boldsymbol{E}_\perp(\boldsymbol{r}_n, z) = \left[\frac{k_0}{2\pi z} \exp\left(-i\frac{\pi}{2}\right)\right]^{n/2}$$

$$\times \int \boldsymbol{E}_\perp(\boldsymbol{r}_n', 0) \exp\left[\frac{ik_0}{2z}(\boldsymbol{r}_n - \boldsymbol{r}_n')^2\right] d\boldsymbol{r}_n' .$$ (A.29)

Here dimensionality $n = 1$ for slit beams (with essential field dependence on only one transverse coordinate), $n = 2$ for transversely two-dimensional field structures; the case $n = 3$ is also meaningful and corresponds to the taking into account of anomalous dispersion ($D_2 < 0$).

Note that the equation for pulsed radiation diffraction,

$$2ik_0 \left(\frac{\partial \boldsymbol{E}_\perp}{\partial z} + \frac{1}{v_{\mathrm{gr}}} \frac{\partial \boldsymbol{E}_\perp}{\partial t}\right) + \Delta_\perp \boldsymbol{E}_\perp = 0 ,$$ (A.30)

is also reduced to (A.26) after substitution of variables $t \to \tau = t - z/v_{\mathrm{gr}}$. Analogous to (A.26) is also the equation with the inclusion of the dispersion term:

$$2ik_0 \left(\frac{\partial \boldsymbol{E}_\perp}{\partial z} + \frac{1}{v_{\mathrm{gr}}} \frac{\partial \boldsymbol{E}_\perp}{\partial t}\right) - D_2 \frac{\partial^2 \boldsymbol{E}_\perp}{\partial t^2} = 0 .$$ (A.31)

The characteristic length of dispersion distortions of a pulse with front duration $\tau_{\mathrm{fr}}$ is

$$l_{\mathrm{disp}} = k_0 \tau_{\mathrm{fr}}^2 / |D_2| .$$ (A.32)

Let us proceed now to an analysis of the role of nonlinear terms in (A.21). Here we will give only some evaluations of the significance of the factors. For monochromatic radiation ($\partial \boldsymbol{E}_\perp/\partial t = 0$, $\partial \delta \boldsymbol{P}_\perp/\partial t = 0$), with diffraction neglected ($\Delta_\perp \boldsymbol{E} = 0$), (A.21) is reduced to the form

$$2ik_0\frac{\mathrm{d}\boldsymbol{E}_\perp}{\mathrm{d}z} + \left(\frac{\omega_0}{c}\right)^2 \delta\hat{\varepsilon}\boldsymbol{E}_\perp = 0\;. \tag{A.33}$$

It causes the effect of nonlinear rotation of the polarization ellipse, if the field polarization structure and the tensor character of the operator $\delta\hat{\varepsilon}$ are taken into account [220, 402]. With fixed (linear or circular) polarization of radiation, $\delta\hat{\varepsilon}$ can be considered as a scalar function of intensity $I = |E_\perp|^2$. When absorption or gain is present in the medium, this function is complex:

$$\delta\varepsilon(I) = \delta\varepsilon'(I) + \mathrm{i}\delta\varepsilon''(I)\;. \tag{A.34}$$

Separating real amplitude $A$ and phase $\Phi$ in the complex amplitude $E_\perp$,

$$E_\perp = A\exp(\mathrm{i}\Phi)\;, \tag{A.35}$$

and then passing to intensity $I = A^2$, we obtain from (A.33)

$$\frac{\mathrm{d}I}{\mathrm{d}z} = -\frac{k_0}{\varepsilon_0(\omega_0)}I\delta\varepsilon''(I)\;, \tag{A.36}$$

$$\frac{\mathrm{d}\Phi}{\mathrm{d}z} = \frac{k_0}{2\varepsilon_0(\omega_0)}\delta\varepsilon'(I)\;. \tag{A.37}$$

As is seen from (A.36-A.37), the imaginary part $\delta\varepsilon''$ is responsible for the intensity variation, and the real part $\delta\varepsilon'$ is responsible for the nonlinear phase change. At a given dependence $\delta\varepsilon(I)$ one finds the intensity longitudinal variation from (A.36), and then the phase change is determined by means of (A.37). For a transparent nonlinear medium $\delta\varepsilon'' = 0$ and

$$\boldsymbol{E}_\perp(x,y,z) = \boldsymbol{E}_\perp(x,y,0)\exp(\mathrm{i}B)\;. \tag{A.38}$$

Here the "break-up integral" is introduced:

$$B = \frac{k_0}{2\varepsilon_0(\omega_0)}\int_0^z \delta\varepsilon\,\mathrm{d}z \approx \frac{k_0\delta\varepsilon_0 z}{2\varepsilon_0(\omega_0)}\;. \tag{A.39}$$

Relations (A.38-A.39) correspond to the thin-layer approximation. The characteristic length of nonlinear distortions of the wave front $l_{\mathrm{nl}}$ for wide beams is determined by the condition $B \approx 1$, so

$$l_{\mathrm{nl}} \sim \lambda_0\varepsilon_0(\omega_0)/|\delta\varepsilon|\;. \tag{A.40}$$

Taking into account diffraction of monochromatic radiation and medium nonlinearity within the framework of the "nonlinear Schrödinger equation",

$$2ik_0\frac{\partial\boldsymbol{E}_\perp}{\partial z} + \Delta_\perp\boldsymbol{E}_\perp + \frac{\omega_0^2}{c^2}\delta\hat{\varepsilon}\boldsymbol{E}_\perp = 0\;, \tag{A.41}$$

leads to a series of nontrivial effects of the self-focusing type [220, 402, 206, 244]. A nonstationary equation

$$2ik_0 \left( \frac{\partial \boldsymbol{E}_\perp}{\partial z} + \frac{1}{v_{\mathrm{gr}}} \frac{\partial \boldsymbol{E}_\perp}{\partial t} \right) + \Delta_\perp \boldsymbol{E}_\perp + \frac{\omega_0^2}{c^2} \delta\hat{\varepsilon} \boldsymbol{E}_\perp = 0 \qquad (A.42)$$

is also reduced to this equation with the transform $t \to \tau = t - z/v_{\mathrm{gr}}$. Neglecting diffraction, but taking dispersion into account, we obtain a similar equation for the nonlinear propagation of pulses in a dispersive medium

$$2ik_0 \left( \frac{\partial \boldsymbol{E}_\perp}{\partial z} + \frac{1}{v_{\mathrm{gr}}} \frac{\partial \boldsymbol{E}_\perp}{\partial t} \right) - D_2 \frac{\partial^2 \boldsymbol{E}_\perp}{\partial t^2} + \frac{\omega_0^2}{c^2} \delta\hat{\varepsilon} \boldsymbol{E}_\perp = 0 \,. \qquad (A.43)$$

The last term in (A.21) is responsible for the existence of shock waves. If we hold it and neglect diffraction ($\Delta_\perp \boldsymbol{E}_\perp = 0$), we obtain, from (A.21), for a nonlinear medium without dispersion ($D_2 = 0$):

$$2ik_0 \left( \frac{\partial \boldsymbol{E}_\perp}{\partial z} + \frac{1}{v_{\mathrm{gr}}} \frac{\partial \boldsymbol{E}_\perp}{\partial t} \right) + 2i \frac{\omega_0}{c^2} \frac{\partial}{\partial t} (\delta\hat{\varepsilon} \boldsymbol{E}_\perp) + \frac{\omega_0^2}{c^2} \delta\hat{\varepsilon} \boldsymbol{E}_\perp = 0 \,. \qquad (A.44)$$

Now let us abstract ourselves from polarization effects (i.e., we assume linear or circular polarization of radiation) and assume that the medium nonlinearity response is fast, i.e., relaxation time $\tau_{\mathrm{rel}}$ is less than the duration of radiation pulse fronts $\tau_{\mathrm{fr}}$. Then we can consider $\delta\varepsilon$ as the given real function of intensity of type (A.34) with $\delta\varepsilon'' = 0$.

Again we separate the real amplitude $A$ and phase $\Phi$ in the complex amplitude $E_\perp$ [see (A.35)]. For the amplitude $A$ we obtain the transport equation, in which the transport velocity depends on intensity $u = u(I)$:

$$\frac{\partial I}{\partial z} + \frac{1}{u(I)} \frac{\partial I}{\partial t} = 0 \,, \qquad (A.45)$$

$$\frac{1}{u(I)} = \frac{1}{v_{\mathrm{gr}}} + \frac{1}{c\sqrt{\varepsilon_0}} \left( \delta\varepsilon(I) + 2I \frac{\mathrm{d}}{\mathrm{d}I} \delta\varepsilon \right) \,. \qquad (A.46)$$

The general solution of (A.45) can be written in the implicit form

$$I(z,t) = f \left( t - \frac{z}{u\big(I(z,t)\big)} \right) \,, \qquad (A.47)$$

where $f(t)$ is an arbitrary function; in our case it determines the input pulse shape [at $z = 0$ $f(t) = I(0,t)$]. As it follows from (A.47), parts of the laser pulse with different local intensity propagate with different velocities. Therefore, the intensity profile slope can increase. At some time moment corresponding to an infinite slope (in the framework of the assumed approximations), an *electromagnetic shock wave* arises. The length of the nonlinear path necessary for shock wave formation, $l_{\mathrm{sh}}$, can be estimated from the condition of equality of the nonlinear pulse front distortion and its width $\tau_{\mathrm{fr}}$, so

$$l_{\mathrm{sh}} \sim c\tau_{\mathrm{fr}} / |\delta\varepsilon| \,. \qquad (A.48)$$

We will consider the length of a nonlinear path to be sufficiently short, $z \ll l_{\mathrm{sh}}$, which permits us to neglect the really small term (under normal conditions) which contains $\partial(\delta\hat{\varepsilon}\boldsymbol{E}_\perp)/\partial t$ in paraxial equation (A.21).

The paraxial equation is also generalized for cases of media with smooth spatial variation of dielectric permittivity [367] and anisotropic media [379, 315]. Equation (A.21) corresponds to the propagation of a single laser beam close to a monochromatic plane wave (A.7). Nonlinear interaction of several beams is described by a set of coupled paraxial equations. A detailed analysis of the three-wave or three-frequency interactions of beams or pulses described by a set of coupled paraxial equations is given in [379], see also [315] .

**Nonparaxial Propagation.** The paraxial approximation is of limited usefulness being restricted to the case of sufficiently wide beams. But even weak nonparaxiality of wide beams is of basic importance in a number of degenerate situations of practical significance. Thus paraxial theory predicts the catastrophic collapse of beams with overcritical power in media with Kerr nonlinearity. However, close to the point of nonlinear focusing, the radiation beam width approaches the wavelength of light. Then the paraxial approximation is not valid anymore, and nonparaxial effects are of fundamental importance. Second, the standard paraxial approach does not take into account variations of the radiation polarization state, arising due to the effective inhomogeneity of the nonlinear medium. To find the complete polarization structure, it is also necessary to invoke more rigorous nonparaxial theory. Below we present the derivation of the corresponding propagation equation for the case of sufficiently wide beams and media with cubic nonlinearity (B.1).

The procedure is close to the derivation of the paraxial equation (A.21), but now we will take into account certain additional terms [347]. Starting with the same equations (A.1-A.2) and neglecting for simplicity frequency dispersion, we assume instead of (A.7)

$$\tilde{E} = \mathrm{Re}\{\boldsymbol{E}(\boldsymbol{r}_\perp, z)\exp[\mathrm{i}(\Gamma z - \omega_0 t)]\} \ . \tag{A.49}$$

The shift of the carrier wave number ($\Gamma$ instead of $k_0$) is convenient, e.g., for the description of steady-state spatial solitons with envelope $\boldsymbol{E} = \boldsymbol{E}_\mathrm{s}(\boldsymbol{r}_\perp)$, which depends on transverse coordinates $\boldsymbol{r}_\perp = x, y$ only. Now an exact consequence of (A.1) is

$$2\mathrm{i}\Gamma\frac{\partial\boldsymbol{E}_\perp}{\partial z} + \frac{\partial^2\boldsymbol{E}_\perp}{\partial z^2} + \Delta_\perp\boldsymbol{E}_\perp - (\Gamma^2 - k_0^2)\boldsymbol{E}_\perp + \frac{k_0^2}{\varepsilon_0}\boldsymbol{D}_\perp - \nabla(\nabla\cdot\boldsymbol{E}) = 0 \ . \tag{A.50}$$

This equation is not closed with respect to the transverse components of the envelope $\boldsymbol{E}_\perp = E_x, E_y$ because of the form of the last term on its left-hand side. However, in the approximation of weak nonparaxiality valid for sufficiently wide beams, a closed form of this equation can be derived, using an approximate expression for the field longitudinal component $\boldsymbol{E}_z$ via $\boldsymbol{E}_\perp$. To this end, let us introduce a small parameter of nonparaxiality,

$$\mu^2 = \frac{\Gamma^2 - k_0^2}{k_0^2} \approx 2\frac{\Gamma - k_0}{k_0} \ll 1 \,, \tag{A.51}$$

implying that for the field close to a steady-state soliton with a typical width $w$ we have

$$\mu^2 \sim (k_0 w)^{-2} \,, \quad \delta D \sim E(k_0 w)^{-2} \,. \tag{A.52}$$

Then, neglecting higher-order corrections with respect to $\mu$, we obtain the previous expression (A.22) for $\boldsymbol{E}_z$ and find a nonparaxial equation for $\boldsymbol{E}_\perp$:

$$2\mathrm{i}\Gamma\frac{\partial \boldsymbol{E}_\perp}{\partial z} + \Delta_\perp \boldsymbol{E}_\perp - (\Gamma^2 - k_0^2)\boldsymbol{E}_\perp$$
$$+ \frac{k_0^2}{\varepsilon_0}[\alpha(\boldsymbol{E}_\perp, \boldsymbol{E}_\perp^*)\boldsymbol{E}_\perp + \beta(\boldsymbol{E}_\perp, \boldsymbol{E}_\perp)\boldsymbol{E}_\perp^*] = \boldsymbol{Q}_\perp \,. \tag{A.53}$$

The right-hand side of (A.53) is a small nonparaxial correction term with $\boldsymbol{Q}_\perp = \boldsymbol{Q}_\mathrm{s} + \boldsymbol{Q}_z$ and

$$\boldsymbol{Q}_\mathrm{s} = -\frac{1}{\varepsilon_0}\left[\alpha|\nabla_\perp \cdot \boldsymbol{E}_\perp|^2\boldsymbol{E}_\perp - \beta(\nabla_\perp \cdot \boldsymbol{E}_\perp)^2\boldsymbol{E}_\perp^* + \nabla_\perp\left(\nabla_\perp \cdot (\alpha|\boldsymbol{E}_\perp|^2\boldsymbol{E}_\perp\right.\right.$$
$$\left.\left. + \beta\boldsymbol{E}_\perp^2\boldsymbol{E}_\perp^*)\right) - \alpha\nabla_\perp\left(|\boldsymbol{E}_\perp|^2(\nabla_\perp \cdot \boldsymbol{E}_\perp)\right) + \beta\nabla_\perp\left(\boldsymbol{E}_\perp^2(\nabla_\perp \cdot \boldsymbol{E}_\perp^*)\right)\right] \,,$$

$$\boldsymbol{Q}_z = -\frac{\partial^2 \boldsymbol{E}_\perp}{\partial z^2} = \frac{1}{k_0^2}\left(\Delta_\perp - (\Gamma^2 - k_0^2) + \frac{k_0^2}{\varepsilon_0}\alpha|\boldsymbol{E}_\perp|^2\right)$$
$$\times \left(\Delta_\perp \boldsymbol{E}_\perp - (\Gamma^2 - k_0^2)\boldsymbol{E}_\perp + \frac{k_0^2}{\varepsilon_0}[\alpha|\boldsymbol{E}_\perp|^2\boldsymbol{E}_\perp + \beta\boldsymbol{E}_\perp^2\boldsymbol{E}_\perp^*]\right)$$
$$+ \frac{\alpha}{4\varepsilon_0}[(\Delta_\perp \boldsymbol{E}_\perp, \boldsymbol{E}_\perp^*) - (\Delta_\perp \boldsymbol{E}_\perp^*, \boldsymbol{E}_\perp)]\boldsymbol{E}_\perp$$
$$+ \frac{\beta}{4\varepsilon_0}\left(2(\Delta_\perp \boldsymbol{E}_\perp, \boldsymbol{E}_\perp)\boldsymbol{E}_\perp^* - \boldsymbol{E}_\perp^2\Delta_\perp \boldsymbol{E}_\perp^* - (\Gamma^2 - k_0^2)\boldsymbol{E}_\perp^2\boldsymbol{E}_\perp^*\right.$$
$$\left. + \frac{k_0^2}{\varepsilon_0}(\alpha + 2\beta)|\boldsymbol{E}_\perp|^2\boldsymbol{E}_\perp^2\boldsymbol{E}_\perp^* - \frac{k_0^2\beta}{\varepsilon_0}\boldsymbol{E}_\perp^2\boldsymbol{E}_\perp^{*2}\boldsymbol{E}_\perp\right) \,. \tag{A.54}$$

In the lowest approximation $\boldsymbol{Q}_\perp = 0$. Then (A.53) takes the form of a paraxial equation. Its steady-state solutions have a continuous spectrum of the propagation constant $\Gamma$ connected with the beam width and are degenerate in power and polarization state. It is natural to take as the unperturbed solution the so-called Townes mode [64] with linear polarization ($E_y = 0$) and power equal to the critical power of self-focusing (independent of the propagation constant). In the next approximation there are corrections to the shape of this solution; the field polarization becomes elliptical and varies over the cross-section; and the radiation power increases with a decrease in the beam width. Because of the linear stability of this solution proved in [347], it represents a weakly nonparaxial spatial soliton. For a strongly nonlinear regime, numerical solution of the exact nonlinear Maxwell's equations shows the existence of extremely narrow (with a width less than the wavelength of light) spatial solitons – *optical needles* [359, 345]. They are stable and serve as the final stage of self-focusing of radiation beams with overcritical power.

# B. Constitutive Equations
# for Medium Nonlinear Polarization

The constitutive equations that set up the correspondence between the characteristics of light and medium polarization are prescribed phenomenologically or are determined from quantum consideration of some model of the medium. An explicit form of these equations is valid in a restricted range of parameters.

Generally speaking, the medium polarization can be determined not only by the electric field $\tilde{\boldsymbol{E}}$, but also by the magnetic field. However, the contribution of magnetic effects to the optical nonlinearity of medium is not essential in actual situations, when the speed of electrons is much less than the speed of light.

For weak fields and media with central symmetry, self-action is described by the lowest (cubic in $\tilde{\boldsymbol{E}}$) terms in the medium polarization or induction:

$$\delta\tilde{\boldsymbol{D}} = \alpha(\tilde{\boldsymbol{E}}, \tilde{\boldsymbol{E}}^*)\tilde{\boldsymbol{E}} + \beta(\tilde{\boldsymbol{E}}, \tilde{\boldsymbol{E}})\tilde{\boldsymbol{E}}^* . \tag{B.1}$$

In the case of fixed (linear or circular) polarization it corresponds to the simplest form of Kerr nonlinearity:

$$\delta\varepsilon = \varepsilon_2|\tilde{E}|^2 . \tag{B.2}$$

The nonlinearity coefficient $\varepsilon_2$ is positive for media with self-focusing and negative for media with self-defocusing. A generalization is the following dependence:

$$\delta\varepsilon = \delta\varepsilon_{\mathrm{nl}}(|\tilde{E}|^2) , \tag{B.3}$$

which is valid if the medium relaxation time $\tau_{\mathrm{rel}}$ is less than the characteristic time of variation of the field envelope.

Approximation of fast nonlinearity (B.3) follows at $\tau_{\mathrm{rel}} \to 0$ from a relaxation (Debye's) equation, which takes into account a transient process of polarization relaxation phenomenologically:

$$\tau_{\mathrm{rel}}\frac{\partial\delta\varepsilon}{\partial t} = -\delta\varepsilon + \delta\varepsilon_{\mathrm{nl}}(|\tilde{E}|^2) . \tag{B.4}$$

Nonlinearity relaxation plays an especially important role in the case of extremely short (femtosecond) laser pulses [9]. Diffusion in the medium is taken into account by introduction of the following additional term into (B.4):

$$\frac{\partial \delta\varepsilon}{\partial t} = -\frac{1}{\tau_{\mathrm{rel}}}\left(\delta\varepsilon + \delta\varepsilon_{\mathrm{nl}}(|\tilde{E}|^2)\right) + D\Delta\delta\varepsilon \,, \tag{B.5}$$

where $D$ is the diffusion coefficient and $\Delta$ is the Laplace operator.

A more detailed description of the nonlinear properties of the medium requires definition of its microstructure. The model of the medium consisting of two-level particles with states "a" and "b" with transition frequency $\omega_{\mathrm{ab}}$ close to the radiation frequency $\omega_0$ holds much significance [18]. In this case the medium response to the applied field has an oscillatory character. The contribution of two-level particles to the medium polarization $\delta P$ is expressed in terms of the nondiagonal elements of the density matrix $\varrho$:

$$\delta P = N\mu(\varrho_{\mathrm{ab}} + \varrho_{\mathrm{ba}}) \,, \quad \varrho_{\mathrm{ba}} = \varrho_{\mathrm{ab}}^* \,. \tag{B.6}$$

Here $N$ is the concentration of particles and $\mu$ is the matrix element of the electric dipole moment of transition between the two levels. In the case of inhomogeneous broadening, when there is a particle distribution in transition frequencies with a certain weight $g(\omega_{\mathrm{ab}})$, we have the following instead of (B.6):

$$\delta P = N\mu \int g(\omega)(\varrho_{\mathrm{ab}} + \varrho_{\mathrm{ba}}) \, \mathrm{d}\omega \,. \tag{B.7}$$

The diagonal elements of the density matrix $\varrho_{\mathrm{aa}}$ and $\varrho_{\mathrm{bb}}$ represent populations of the states "a" and "b" ($\varrho_{\mathrm{aa}} + \varrho_{\mathrm{bb}} = 1$). The *Bloch equations* for the elements of the density matrix have the form

$$\frac{\partial \varrho_{\mathrm{ab}}}{\partial t} = -\mathrm{i}\omega_{\mathrm{ab}}\varrho_{\mathrm{ab}} - \gamma_\perp \varrho_{\mathrm{ab}} + \mathrm{i}\tilde{V}(\varrho_{\mathrm{aa}} - \varrho_{\mathrm{bb}}) \,,$$

$$\frac{\partial \varrho_{\mathrm{aa}}}{\partial t} = -\gamma_\| \varrho_{\mathrm{aa}} + \mathrm{i}\tilde{V}(\varrho_{\mathrm{ab}} - \varrho_{\mathrm{ba}}) \,,$$

$$\frac{\partial \varrho_{\mathrm{bb}}}{\partial t} = \gamma_\| \varrho_{\mathrm{aa}} - \mathrm{i}\tilde{V}(\varrho_{\mathrm{ab}} - \varrho_{\mathrm{ba}}) \,, \tag{B.8}$$

where $\gamma_\perp$ is the rate of "phase" relaxation, $1/\gamma_\|$ is the lifetime of the upper (excited) level "a", and $\tilde{V} = -(\mu/\hbar)\tilde{E}$. These Bloch equations are naturally generalized to the case when the level "b" is also excited.

Transition to the slowly varying values $E$ and $\sigma$,

$$\tilde{E} = E\cos(\omega_0 t) \,, \quad \varrho_{\mathrm{ab}} = \sigma \exp(-\mathrm{i}\omega_0 t) \,, \tag{B.9}$$

corresponds to resonance approximation. In this case, after substitution of (B.9) into (B.8), the terms that include fast oscillating exponents (with frequencies multiple of $\omega_0$) should be neglected.

A more consistent approach involves consideration of the nature of relaxation caused by the interaction of medium particles with another sufficiently large system – a reservoir or a thermostat, in which the energy of the excited particles dissipates. It is essential that, besides the energy decrease, noise arises, caused by the return of part of the energy from the reservoir to particles [114]. The noise is determined both by temperature fluctuations depending on thermostat temperature $T$ and by purely quantum ones, which also remain at zero temperature $T = 0$.

The master equation for the density matrix $\varrho$ of the particle, which interacts with the reservoir (with a system of a large number of harmonic oscillators) has the form [114]

$$
\frac{\partial \varrho}{\partial t} = -\frac{i}{\hbar}[H_0, \varrho] + \frac{\gamma}{2}(1 + \bar{N})(2C\varrho C^\dagger - C^\dagger C\varrho - \varrho C^\dagger C)
$$
$$
+ \frac{\gamma}{2}\bar{N}(2C^\dagger \varrho C - CC^\dagger \varrho - \varrho CC^\dagger) . \tag{B.10}
$$

Here $H_0$ is the Hamiltonian of an isolated particle (in the absence of interaction with the reservoir); $C^\dagger$ and $C$ are the operators which coincide with operators of the creation and annihilation of states for popular models of a quantum particle – a two-level scheme and a harmonic oscillator; constant magnitude $\gamma$ determines the strength of the interaction of the particle with the reservoir and thus the rate of relaxation of excited states; and $\bar{N}$ is the average population of oscillators in the reservoir ($\bar{N} \to 0$ at $T \to 0$).

# C. Transverse Structures and Digital Optical Computing

**Introductory Notes.** In the main text of the book we described the specific features of optical patterns in different distributed nonlinear systems, which have sufficiently general character and therefore are inherent in systems of different physical nature. In this appendix we will discuss the possibilities for the application of transverse optical structures to the problem of optical computing, underlying new opportunities given by the transverse distributivity of the systems.

Depending on the form – continuous or discrete – of representation of the information processed, computers are separated into analogue and digital ones. Analogue computers are efficient in the solution of specialized problems. The more widely distributed now digital computers are much more universal and have considerably higher accuracy and reliability in computation. However, opportunities of rate increase for traditional (one-processor) computers are apparently almost exhausted at present.

If one does not take into account any principal advancements in electronics, for instance, on the basis of the wide application of high-temperature superconductivity, the progress here can be connected with multiprocessor complexes. However, organization of effective high-parallel computing in a computer is connected with certain principal difficulties, mainly when providing multiple interconnections of processors and elements of different levels by means of electronics [229].

The optical computers used now are analogous and specialized. The classical example of the efficiency of optics in parallel computing is the realization of a complex two-dimensional Fourier transform by a single lens during the time of the light trip from the entrance plane to the focal plane of the lens [127]. Realization of the same integral transform with the help of electronics is much more difficult.

An increase in the supercomputer performance by several orders of magnitude could cause revolutionary changes, e.g., in the problem of artificial intelligence, which is connected with attempts in the computer realization of such techniques for problem solution characteristic of intelligent human activity; various knowledge bases and expert systems allowing automatic solution of complex scientific and applied problems could become available.

Taking into account the history of computers, it is natural to suppose that universal optical computers will use a discrete representation of information. The usefulness of the application of separate optical elements inside a computer (e.g., of optical fibers for interconnections between computer blocks) does not cause any doubts. Currently, attempts are being made to design an electronic–optical computer, in which the demarcation line between electronics and optics would be chosen in a way that the advantages of both these approaches would be used in the most efficient way. As to the prospects of the construction of universal optical computers superior in their parameters to modern or designed electronic ones, they are controversial [135]. Among the nearest applications could be the use of driven nonlinear interferometers based on semiconductors [380] for the commutation of a large number of channels in telecommunications [377].

A potential advantage of optical computers is the availability of parallel information processing and interconnections inherent in optics; it is essential also that optical frequencies provide a considerably greater transmission band and a greater ultimate possible computation speed, as compared with the radio-frequency band. Transverse optical patterns – such as switching waves and localized structures – give us the opportunity to change from operations with separate information signals or units to operations with entire information arrays, i.e., to "spatial algebra" and "spatial logic". Below we will illustrate this idea with examples of a shift register and a full adder on the basis of a driven wide-aperture nonlinear interferometer.

**Analogous-Digital Information Processing.** Accuracy and reliability are inherent in digital methods of computation, but they give way to analogous methods in their potential and simplicity of fast processing of large informational arrays. This raises the question of the possibility for realizing a system of a new (hybrid) kind, which would provide a combination of advantages of digital and analogous methods, on the basis of wide-aperture nonlinear optical systems. This can be achieved by using transverse phenomena in driven nonlinear interferometers (Chap. 4), two-beam schemes of nonlinear reflection (Chap. 5), or bistable laser schemes (Chap. 6), taking into account the possibility of the schemes' effective reconstruction by means of radiation. In this case, the same device is maintained for a definite part of the duration of the computing clock in the discrete (digital) regime, the other part being in the continuous (analogous) regime [333, 311]. This potential is based on the phenomena of spatial hysteresis (see Chap. 2) and on the features of the switching waves and localized structures (Chaps. 2,4 and 6).

**Shift Register.** An important type of informational array transform is the shift operation. It is used, for example, in multiplication, image recognition, and in cellular automata. To organize an array of bistable cells we use the spatial modulation of the holding radiation intensity or phase across the driven nonlinear interferometer, and the array shift is achieved at oblique incidence of radiation (see Chap. 4). Spatial modulation of radiation intensity

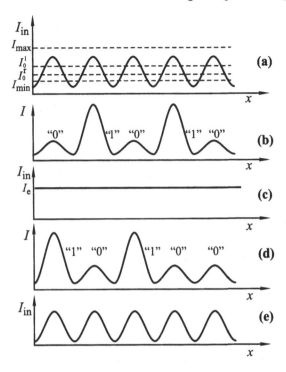

**Fig. C.1.** Field structures in a shift register scheme: coinciding transverse distributions of holding intensity at the beginning (**a**) and end (**e**) of computation cycle; the initial (**b**) and shifted (**d**) intensity distributions coding the information array and the transversely homogeneous intensity of holding radiation at the stage of information array shifting (**c**) [311]

$I_{in} = I_{in}(x)$ is presented in Fig. C.1a. Here $I_{min}$ and $I_{max}$ are the values of $I_{in}$ which are the boundaries of the bistability interval, $I_0^l$ and $I_0^r$ are the values of $I_{in}$ for which the velocities $v_l$ and $v_r$ of the "left" and "right" switching waves (see Sect. 4.4) are equal to zero, i.e., $v_l(I_0^l) = v_r(I_0^r) = 0$, and $x$ is the transverse coordinate with respect to the direction of radiation propagation.

As was indicated in Chap. 2, spatial modulation of the holding radiation allows us to organize the multichannel memory on the basis of an initially homogeneous bistable device due to the phenomenon of spatial hysteresis. In Fig. C.1b the transverse profile of intensity inside the interferometer is shown for the case of recording the informational array 01010 into the array of memory cells corresponding to the modulation with period $p$ presented in Fig. C.1a. The lower state of the bistable system ($I_1$) corresponds to the zero code, and the upper one ($I_2$) to the unity code.

To shift the information recorded in such a memory array by $n$ positions, one has to replace the coordinate dependence $I_{in}(x)$ depicted in Fig. C.1a by such a dependence with the period $p$ that $I_0^l < I_{in} < I_0^r$ during the time interval equal to

$$t_{sh} = n \int_0^p \frac{dx}{v_l(x)} = n \int_0^p \frac{dx}{v_r(x)} \, , \tag{C.1}$$

where $n$ is the value of the shift. A simple example of such dependence is the value $I_{in} = I_e$ independent of the transverse coordinate, the intensity $I_e$ being determined from the condition of equality of velocities of the "left" and "right" switching waves: $v_l(I_0) = v_r(I_0) = v_e(I_e)$. In this case $t_{sh} = np/v_e$.

During the time $t_{sh}$ the switched regions and the fronts of switching waves will shift in the $x$-direction by the distance $np$. As the initial (modulated) dependence $I_{in}(x)$ [see Fig. C.1a] confines the switched regions within the cells, the motion of switching waves will terminate when this dependence is restored, and the information is shifted by $n$ cells. The acceptable inaccuracy in the choice of time $t_{sh}$ depends on the system parameters and can reach half of the period $p$, so $\delta t \leq 0.5p/v_e$.

If the spatial modulation and the interval $t_{sh}$ are chosen so that

$$t_{sh} = (n+m) \int_0^p \frac{dx}{v_l(x)} = n \int_0^p \frac{dx}{v_r(x)} \, , \tag{C.2}$$

then the domain switched to the upper branch of the bistable curve will widen by $m$ cells during the interval $t_{sh}$. In this case the "right" and "left" switching waves could propagate in the same direction or in opposite directions. An operation like this can be applied, for instance, to separate contours in images.

Spatial discreteness and independence of parallel channels can be provided either by spatial modulation of the holding radiation intensity or phase or by the physical inhomogeneity of the system in the transverse direction; joint action of these two factors is possible as well. If spatial modulation of the holding radiation is two-dimensional, two-dimensional arrays of memory cells can be formed. The operation described above can be performed over digital information recorded in these cells in a similar way.

Now we estimate possible parameters of the shift register based on a wide-aperture Fabry–Perot interferometer homogeneous in the transverse direction and filled with a medium with fast nonlinearity. Propagation of the "left" and "right" switching waves in the same direction takes place usually at the angles of incidence at which the velocity of switching waves does not exceed $0.01c$ ($c$ is the speed of light). Using optical radiation, we can choose the dimensions for memory cells and the distances between them to be about $1\,\mu m$. If we assume the period of cell location to be equal to $3\,\mu m$, the shift by one position in such a scheme can be performed during a time of about $10^{-12}\,$s.

**Full Optical Adder.** One can also use localized structures (dissipative optical solitons, including moving asymmetric structures) within the framework of the analogue–digital approach stated above. Figure C.2 illustrates an example of a scheme for the summation of two numbers represented in binary

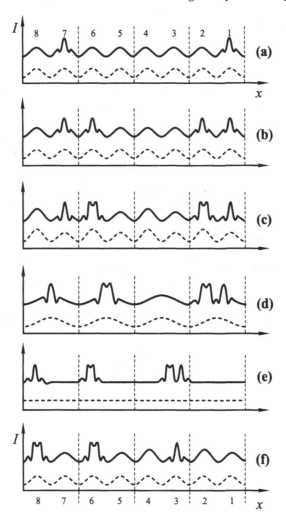

**Fig. C.2.** Field structures in a full adder scheme: $n = 4$, addition of two binary numbers 1001 (**a**) + 0101 (**b**, odd cells) = 1110 (**f**); *dashed lines* are intensity distributions for the holding radiation; *solid lines* are intensities of the output radiation [292]

code [292, 294]. A pair of cells of the adder – a wide-aperture nonlinear interferometer driven by external radiation – are allotted for one digit of the numbers.

The first stage is the same as in the scheme of the shift register. The cells are again formed using spatial hysteresis in conditions of spatial modulation of the external radiation (Fig. C.2a). First, an addend – the first number – is recorded into the odd cells. Then, during the second stage, the recorded array is shifted in the transverse dimension toward the left by one position, using, e.g., oblique incidence of radiation, and an augend – the second number – is recorded again into the odd cells (Fig. C.2b).

In the next stage, the profile of incident radiation intensity is changed, its maxima are made wider in the cells with even numbers. It allows us to form symmetric dissipative solitons, by transformation of field perturbations localized at the cells, after elimination of modulation. Solitons in cells with even numbers are wider than those in cells with odd numbers (Fig. C.2c). After that, the spatial modulation of the holding radiation, with the number of intensity maxima reduced by half, is switched on. As a result, either symmetric (upon addition of 1 and 0) or asymmetric (1 + 1) solitons are formed in the vicinity of these maxima (stage corresponding to Fig. C.2d).

Elimination of spatial modulation for certain time intervals causes the shift of asymmetric solitons by the dimension of the initial cells. At oblique incidence the entire picture is additionally shifted by a half of this dimension (Fig. C.2e). Then the spatial modulation of radiation with double period is used again. It provides now the transformation of soliton structures into narrow (in odd cells) and wide (in even ones) single solitons (Fig. C.2f). The latter stages – from Fig. C.2c to Fig. C.2f – should be repeated to organize the transfer in all digits. The main stages of the summation cycle are numerically simulated in [88]. The variant of hybrid architecture described above can be used not only in universal optical computers, but also in specialized systems of image processing (e.g., to separate image contours).

# D. Bistability of the Quantum Anharmonic Oscillator

**Introductory Notes.** A classical anharmonic oscillator is characterized by the nonlinearity of the quasi-elastic force arising when the oscillator deviates from the equilibrium position. An oscillator such as this serves as a standard example for the demonstration of bistability and hysteresis in a point (lumped) nonlinear system [185, 22]. A model of the classical anharmonic oscillator is the basis for the analysis of a great variety of nonlinear phenomena, including those in nonlinear optics [46] and in the problem of interaction of coherent radiation with molecules; estimations in [105] indicate the reality of the observation of hysteresis phenomena when we have laser excitation of molecular vibrations.

We naturally come to the problem of bistability in elementary (quantum) objects when determining the limits of the miniaturization of optical bistable elements. This problem is connected with the power and rate-limiting characteristics of these elements [117, 30]. It is interesting to note that quantum analysis of the nonlinear cavity, for example, in [76, 75], involves the Hamiltonian coinciding with the Hamiltonian of the quantum anharmonic oscillator. The problem of hysteresis in the quantum anharmonic oscillator is also of great importance from a purely theoretical standpoint. The argument in favour of the absence of bistability connected with linearity of the Schrödinger equation or the equation of motion for a density matrix, which describes quantum objects, is not convincing. Actually, as it has been discussed in Chap. 1, the meaningful requirement is nonlinearity with respect to a physical parameter – the amplitude of the external force here. Naturally, it takes place for quantum objects as well. Similarly, in connection with the stochastic behaviour of the classical anharmonic oscillator when the amplitude of external force exceeds some critical value [236, 145, 356], the model of the quantum anharmonic oscillator allows us to study the manifestation of stochasticity in the quantum region [356].

Let us recall that bistability in classical objects is possible only when we ignore fluctuations (see Chaps. 1 and 2). Even a weak noise results in metastability of the two stable (in the absence of noise) states of the classical oscillator [107, 78]. Fluctuations cause a finite probability of switchings between these two states, so that with the statistical description the unique distribution function forms with time, which does not depend on the initial conditions.

Bistability observation is possible only over periods shorter than the lifetime of the metastable state. In a quantum system, interaction with the reservoir causing fluctuations remains even at zero temperature $T \to 0$ (*quantum fluctuations*). Therefore only "bimetastability" and "dynamic hysteresis" would be possible for quantum objects under the condition of sufficiently large time of spontaneous switchings between the states.

Analytical solution of the problem of resonance excitation of the quantum anharmonic oscillator is extremely difficult and is achieved only at the cost of grave simplifications not always well founded. In connection with this, different opinions on the possibility of "quantum bistability" have been presented in the literature [76, 75, 353, 58, 326, 73, 79, 120, 56]. In the absence of fluctuations and anharmonicity, the quantum oscillator has a degenerate (equidistant) spectrum, and there is an exact solution of the Schrödinger nonstationary equation in the case of an arbitrary external force [37]. It is possible to neglect fluctuations for the time period limited, for example, by radiation lifetimes which are large enough for molecular vibrations (of order $10^{-8}$ s for the infra-red spectrum) [143]. In [326] the Schrödinger coherent states of the quantum harmonic oscillator [37] were used for more complete correspondence of the classical and quantum descriptions. Such states are described by the wave functions in the form of a wave packet whose centre of gravity moves along the classical trajectory of the oscillator. This analytical approach is in good agreement with numerical calculations [60] and leaves room for bistability.

Nevertheless, it is not easy to make final conclusions from a consideration of oscillator excitation neglecting relaxation – for the following reason: If an external force is absent, then an infinite number of states is possible in the quantum oscillator, which include both eigenstates with some definite energy and their linear superpositions. The influence of the external force adds "forced" oscillations to these "free" ones. However, under resonance conditions and at the infinite lifetime of the states, separation of free and forced oscillations is not unambiguous. Therefore, when the oscillator amplitude or frequency are scanned in the vicinity of the resonance, considerable distortions of the wave packet shape can arise, apart from its oscillations as a whole.

A more definite statement corresponds either to analysis of the kinetics of the isolated (with no relaxation) quantum oscillator, its initial state being given, and the consequent solution of the nonstationary problem (which is true for the time period less than the lifetime of the levels) or to consideration of relaxation (and fluctuations connected with it), which eliminates restrictions for the accessible time interval. In the second variant, the analytical approach is very complicated, and its conclusions are not unambiguous. At the same time, sufficiently convincing and complete results can be obtained by numerical solution of the problem. Therefore, we present below, following

[263], numerical results for the oscillator interacting with a reservoir, taking into account the scanning of frequency of the external force (radiation).

**Dynamic Hysteresis.** Below we differentiate the following: *free oscillator* – without relaxation and an external force; *isolated oscillator* – without relaxation, but with an external force; and *driven oscillator* – with relaxation and an external force. Anharmonicity is described by the Morse potential accepted in the theory of molecular vibrations [143, 186]. The Hamiltonian of the isolated oscillator has the form

$$H_0 = -\frac{1}{2}\frac{\partial^2}{\partial x^2} + D_a\{1 - \exp[-\alpha(x - x_0)]\}^2 - f(t)x \ . \tag{D.1}$$

In the assumed system of units, Planck's constant $\hbar = 1$ and the oscillator's mass $m = 1$. The constant of anharmonicity $\beta = -(1/3)(2D_a)^{-2/3}$; the dimensionless time used is that at which the oscillator frequency $\omega_0 = \sqrt{2D_a}\alpha = 1$. The discrete spectrum of a free oscillator is

$$\omega_n = (n + 1/2)\omega_0 + (3/2)\beta(n^2 + n + 1/2) \ , \quad n = 0, 1, \ldots, n_{max} \ . \tag{D.2}$$

Let the instant frequency of the external force be scanned with the constant rate $v_f$:

$$f(t) = A\cos\varphi(t) \ , \quad \dot\varphi = \omega_f(t) = \omega_{f0} + v_f t \ . \tag{D.3}$$

For scanning from low frequencies, $v_f > 0$, and from high frequencies, $v_f < 0$. Energy of oscillator excitation is determined by

$$E_{exc} = \sum_n n|a_n|^2 \ , \tag{D.4}$$

where $a_n$ are the amplitudes of the stationary states of the isolated anharmonic oscillator. The results of calculations of excitation energy $E_{exc}$ at frequency scanning with different rates $v_f$ are presented in Fig. D.1. Note that at $v_f < 0$ the excitation energy practically does not depend on the absolute value of scanning rate $|v_f|$. At large amplitudes of force $A > |\beta|$, the distribution of populations $|a_n|^2$ is close to Poisson's. At $v_f < 0$ the resonance response of the quantum anharmonic oscillator presented in Fig. D.1 approaches the classical resonance curve (dashed lines in Fig. D.1). At the same time estimates show that in the conditions of Fig. D.1 (comparatively small amplitude of force) only approximately half of the total energy of excitation is due to oscillations of the centre of mass of the wave packet, which is indicative of considerable distortions of the packet shape.

With scanning from low frequencies ($v_f > 0$), the excitation energy depends essentially on the scanning rate. Slow scanning leads to consecutive quasi-two-level multiphoton transitions, so that excitation is reduced practically completely to variation of the shape of the wave packet. The oscillator state at some instant force frequency $\omega_f(t)$ in the resonance region is

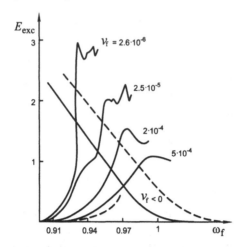

**Fig. D.1.** Energy of excitation of isolated quantum (*solid lines*) and classical (*dashed lines*) anharmonic oscillators as dependent on the instantaneous frequency of the external force: $A = 0.03$, $\beta = -0.01$

determined essentially by its previous history, so that there is a strongly pronounced dynamic hysteresis.

Now we include a consideration of relaxation in our description. As explained above, it means the simultaneous presence both of damping and fluctuations in the oscillator. Following [114], we simulate relaxation by interaction of the quantum anharmonic oscillator with the reservoir of harmonic oscillators. As we are mainly interested in quantum fluctuations, we assume that temperature $T = 0$. Therefore, the equation for the density matrix of the quantum anharmonic oscillator $\varrho$ takes the form

$$\partial\varrho/\partial t = -i[H_0, \varrho] + (\gamma/2)(2C\varrho C^+ - C^+C\varrho - \varrho C^+C) . \qquad (D.5)$$

Here $[H_0, \varrho] = H_0\varrho - \varrho H_0$ is the commutator, $C^+$ and $C^-$ are the operators of creation and annihilation of oscillator states, and a constant $\gamma$ determines the intensity of oscillator interaction with the reservoir. Master equation (D.5) is obtained from the more general equation (B.10) (Appendix B) with zero temperature of the reservoir $T \to 0$.

Let the frequency of the external force be constant ($\omega_f = $ const., $v_f = 0$, scanning is absent). Then in the resonance approximation (D.5) is reduced to the set of linear differential equations for amplitudes $a_n$ with constant coefficients. Its general solution is the superposition of the different eigensolutions with exponential (decaying) dependence on time, damping factors (exponentials) $\alpha_n$ being eigenvalues. There is a zero eigenvalue $\alpha_0 = 0$ corresponding to the only stationary solution of (D.5).

It can be shown that at $\gamma \to 0$ all eigenvalues approach zero: $\alpha_n \to 0$. In this way we come to the case of an isolated (no reservoir) oscillator. Thus, the

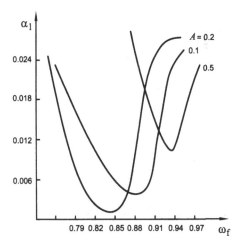

**Fig. D.2.** Dependence of the minimum eigenvalue $\alpha_1$ on frequency $\omega_f$: $\gamma = 0.03$, $\beta = -0.01$

only source of instability of the properly chosen states of quantum anharmonic oscillator excited in resonance is its interaction with the reservoir.

Calculation of $\alpha_n$ shows that, when approaching resonance, one of the eigenvalues ($\alpha_1$) decreases essentially in value, whereas the rest of them remain of the order of $\gamma$. Figure D.2 shows the dependence of the minimum (nonzero) eigenvalue $\alpha_1$ on the excitation frequency $\omega_f$ at different force amplitudes $A$. With an increase in $A$ the depth of the gap in this dependence increases. At $A = 0.2$, $\beta = -0.01$ and $\gamma < 0.03$ the eigenvalue $\alpha_1 = 0.03\gamma$.

The region of existence of a small eigenvalue characterized by the frequency range of external force (of exciting radiation),

$$\omega_{min} < \omega_f < \omega_{max} , \tag{D.6}$$

is remarkable in relation to one more thing. The analysis of the stationary distribution of the driven oscillator populations in the stationary state ($\alpha_0 = 0$) over the levels of free oscillator $\varrho_{nn}$ for different force frequencies $\omega_f$ shows that sharp changes in the character of this dependence occur in the vicinity of the edges of the interval (D.6) (Fig. D.3). For frequencies $\omega_f < \omega_{min}$, the distributions $\varrho_{nn}(\omega_f)$ have maxima at $n = 0$. For $\omega_f > \omega_{max}$ these distributions are close to Poisson's, with the only maximum at $n > 1$. However, in the region (D.6) the distribution of the populations has two maxima (at $n = 0$ and $n > 1$). In the conditions of Fig. D.3 $\omega_{min} \approx 0.85$ and $\omega_{max} \approx 0.87$.

The presence of an anomalously small eigenvalue $\alpha_1 \ll \gamma$ in the resonance region allows us to speak about the existence of metastable states with the lifetime $\tau_1 = \alpha_1^{-1} \gg \gamma^{-1}$. Note that the nonstationary eigensolutions of (D.5) with the fixed eigenvalue $\alpha_n \neq 0$ have no physical sense by themselves.

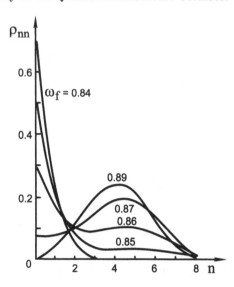

**Fig. D.3.** Stationary distributions of the population over levels of free oscillator $\varrho_{nn}$ for different frequencies $\omega_f$; for convenience, points in discrete distributions are connected by solid lines

This can be seen from the fact that for them $\mathrm{Sp}\,\varrho = \sum_n \varrho_{nn} = 0$. Therefore some of the diagonal elements of the density matrix $\varrho_{nn}$, which should correspond to population of levels, are negative. However, linear combinations of the eigensolutions do have physical sense. Due to a substantial difference in the eigenvalues, we can speak about two metastable states which are the linear combinations of the stationary ($\alpha_0 = 0$) and slowly decaying ($\alpha_1 \ll \gamma$) eigensolutions.

It is these two states that determine the dynamics of excitation of the driven oscillator if the scanning time $\tau_{sc}$ is in the intensity range $\gamma^{-1} < \tau_{sc} < \tau_1$. In this case frequency values $\omega_f = \omega_{min}$ and $\omega_{max}$ serve as natural boundaries for the bistability zone. Actually, at moderate-rate scanning of the frequency from outside range (D.6) to inside it, it is at these frequency values that we can expect the appearance (along with the stationary eigenstate) of considerable additions of the slowly decaying state. Calculations show that for scanning of the frequency in range (D.6) taking into account of only two proper driven oscillator states (with the eigenvalues $\alpha_0 = 0$ and $\alpha_1 \ll \gamma$) provides high accuracy.

Figure D.4 shows instant distributions of the populations (over the levels of the free oscillator) of the quantum anharmonic oscillator excited by external radiation and interacting with the reservoir with radiation frequency scanning at the rate $v_f = \pm 10^{-4}$. In the case of scanning from the low-frequency region (Fig. D.4a), population distributions have only one maximum in the normal state up to the frequency $\omega'_{max} = 0.87$. Upon scanning from

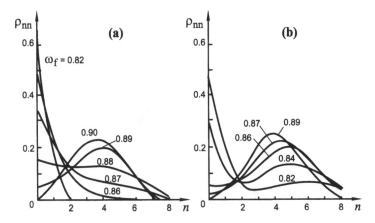

**Fig. D.4.** Instantaneous distributions of populations at frequency scanning at the rates $v_f = 10^{-4}$ (**a**) and $-10^{-4}$ (**b**)

the high-frequency region down up to $\omega'_{min} = 0.86$ (Fig. D.4b), excited states are maximally populated. Thus, regions of instant frequencies at which two different types of population distribution form overlap. These distributions correspond to two metastable states with the lifetime $\tau_1 \propto \alpha_1^{-1}$.

Figure D.5 presents the energy of a quantum oscillator in dependence on the frequency of external force $\omega_f$ for scanning rates $v_f = \pm 10^{-4}$ and $\pm 10^{-5}$. In the second case the scanning time in the bistability zone is close to $\tau_1$; therefore the resonance curve is close to the stationary limit. With sufficiently fast scanning, dynamic hysteresis takes place, but it is noticeably weakened as compared with classical hysteresis (in the absence of noise), if the amplitude of the exciting force is small ($A < 1$). The availability of metastable states whose lifetime is considerably larger than the relaxation time is a necessary condition for hysteresis.

Comparing the energy of excitation of the isolated quantum anharmonic oscillator (Fig. D.1) and that of the oscillator connected with the reservoir (Fig. D.5), one can conclude that interaction with the reservoir, which is the source of relaxation and quantum fluctuations, considerably suppresses multiphoton transitions (frequency scanning also acts in the same direction). This brings the dynamics of quantum and classical oscillators closer together.

The results presented allow us to trace the transition from the quantum oscillator to the classical one. With growth of force amplitude $A$, the lifetime of metastable states increases (therefore one can decrease correspondingly the scanning velocity of the exciting radiation frequency while retaining hysteresis), the bistability zone widens, and the frequency dependence of the energy of excitation of the quantum anharmonic oscillator approaches the resonance curve of the classical anharmonic oscillator.

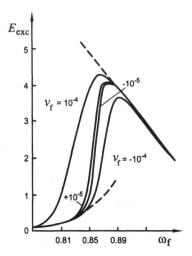

**Fig. D.5.** Excitation energy and dynamic hysteresis in quantum (*solid lines*) and classical (*dashed lines*) anharmonic oscillators

# E. Transverse Effects
# for Squeezed States of Light

In all of the above we considered radiation as classical, obeying Maxwell's equations (see Appendix A). In the quantum description, field amplitudes become operators with definite commutation relations [41, 72]. Neglecting their noncommutativity and, accordingly, quantum effects, is justified for large numbers of photons, i.e., for sufficiently strong fields. In the opposite case of weak fields no nonlinear effects necessary for bistability would take place.

At the same time, the quantum nature of light is important in optical bistability for the analysis of quantum noise absent in the classical treatment. Quantum noise of radiation and photodetectors serve as a factor limiting the opportunities of optical information processing and optical computing. It is essential that photodetectors respond to the photon number (radiation intensity). Therefore fluctuations of photocurrent in them can be suppressed (without violation of quantum-mechanic uncertainty relations) for radiation in *squeezed states*, in which intensity fluctuations are decreased at the cost of an increase in phase fluctuations. One can become acquainted with modern achievements in quantum optics in [363, 381, 168, 413, 217]. Note that such states can also be formed by means of a driven nonlinear interferometer [270], which plays the role of a distinctive "filter" of the initial radiation from photonic noise in a definite spectral range.

For the problem of information parallel processing it is very important to depress not only temporal, but also spatial (transverse) photonic fluctuations over the entire aperture of sufficiently wide light beams. Therefore here we will consider the generation of such light states at the exit of a wide-aperture nonlinear interferometer excited by external radiation. The basic opportunity for photonic fluctuation suppression in optical images was theoretically predicted and studied in detail in the works by Kolobov and Sokolov [177, 176, 175]. When such images are recorded by a two-dimensional array of photodetectors, one can reduce the quantum threshold of the maximum information density determined by the level of detector shot noise and considered earlier insuperable. To generate radiation beams in squeezed states, Kolobov and Sokolov [177, 176] considered the use of a nonlinear optical parametric travelling-wave oscillator. Easier for realization, we think, is the driven wide-aperture nonlinear interferometer proposed as a source of multimode light in a squeezed state in [39], which we follow here. An additional factor greatly increasing the squeezing efficiency is the opportunity to obtain differential gain (i.e., steep slope of the transfer function) in nonlinear interferometers (and other nonlinear feedback systems).

It is more convenient in analysis to consider a ring interferometer (see Fig. 4.1). We assume that its input mirror has a small transmission coefficient, and the rest of the mirrors reflect ideally. The interferometer is characterized by unlimited (wide) aperture and sufficiently small longitudinal dimension; so the model of purely transverse distributivity is valid (see Sect. 4.1). The interferometer is filled with a Kerr medium (fast nonlinearity, absorption absent). The radiation incident on the interferometer is a sum of a coherent regular component – a classical plane wave – and a small fluctuating component.

It is remarkable that in the statement of such a problem (at weak fluctuations), its solution can be split into two stages. First one needs to determine the response of a nonlinear interferometer to a small perturbation of the incident radiation within the classical consideration. Then one uses known general relations, which describe the characteristics of quantum fluctuations in terms of the response functions obtained at the first stage.

Here we will describe in more detail the first stage, which is close to the analysis of transverse instability (see Sect. 4.3). The results of Chap. 4, including the conditions for low-threshold bistability (4.1.45-4.1.46), are true for the classical field. The field amplitude $E$ is described by (4.1.23), an equation of the transversely distributed model (although the same results can be obtained without field averaging in the longitudinal direction, as was done in Sect. 4.3). For normal incidence of external radiation and a nonlinear medium without dispersion ($\theta = 0$, $D_2 = 0$, $\partial \langle \delta P \rangle / \partial t = 0$), this equation acquires the form

$$\frac{1}{v_{\mathrm{gr}}} \frac{\partial E}{\partial t} - \frac{i}{2k_0} \Delta_\perp E - i k_0 \frac{\delta \varepsilon}{\varepsilon_0} E$$
$$+ \frac{1}{d} \{ [1 - R \exp(i \Delta_{\mathrm{ph}})] E - \tau E_{\mathrm{in}} \} = 0 \, . \qquad (\mathrm{E.1})$$

Here we explicitly introduced the amplitude transmission coefficient $\tau$ of the entrance mirror. We will consider that the unperturbed field state inside the interferometer corresponds to a transversely homogeneous regime. We denote the amplitudes of the corresponding plane waves at the entrance, inside, and at the exit of the interferometer as $E_{\mathrm{in}}^{(0)}$, $E^{(0)}$, and $E_{\mathrm{out}}^{(0)}$, the relative perturbations as $\delta E_{\mathrm{in}}$ and $\delta E$, and the absolute ones as $\Delta E_{\mathrm{in}} = E_{\mathrm{in}}^{(0)} \delta E_{\mathrm{in}}$, $\Delta E = E^{(0)} \delta E$ and $\Delta E_{\mathrm{out}} = E_{\mathrm{out}}^{(0)} \delta E_{\mathrm{out}}$.

We present the field amplitudes in the form

$$E_{\mathrm{in}}(\boldsymbol{r}_\perp, t) = E_{\mathrm{in}}^{(0)} [1 + \delta E_{\mathrm{in}}(\boldsymbol{r}_\perp, t)] \, , \quad |\delta E_{\mathrm{in}}|^2 \ll 1 \, ,$$
$$E(\boldsymbol{r}_\perp, t) = E_{\mathrm{out}}^{(0)} [1 + \delta E(\boldsymbol{r}_\perp, t)] \, , \qquad |\delta E|^2 \ll 1 \, , \qquad (\mathrm{E.2})$$

where $r_\perp$ is the two-dimensional vector of transverse coordinates $(x, y)$. Paraxial equation (E.1) linearized with respect to $\delta E$ is

$$\frac{1}{v_{gr}}\frac{\partial \delta E}{\partial t} - \frac{i}{2k_0}\Delta_\perp \delta E - ik_0 I_0 \frac{\varepsilon_2}{2\varepsilon_0}(2\delta E + \delta E^*)$$

$$+ \frac{1}{d}\left([1 - R\exp(i\Delta_{ph})]\delta E - \tau \frac{E_{in}^{(0)}}{E^{(0)}}\delta E_{in}\right) = 0 , \tag{E.3}$$

where $I^{(0)} = |E^{(0)}|^2$. The solution of the linear inhomogeneous differential equation (E.3) is combined from a general solution of the homogeneous equation ("free oscillations," $\delta E_{in} = 0$) and a partial solution of the inhomogeneous equation ("forced oscillation"). Consider first the homogeneous equation.

As (E.3) is linear and homogeneous in the transverse direction, it is convenient to expand the initial (at $t = 0$) perturbation into a spectrum in plane waves, or in spatial frequencies $q_\perp = (q_x, q_y)$:

$$\delta E(r_\perp, 0) = \int \delta E(q_\perp)\exp(iq_\perp r_\perp)\,dq_\perp . \tag{E.4}$$

Equation (E.3) relates in pairs the perturbation components with opposite frequencies $q_\perp$ and $-q_\perp$. Therefore we seek a solution of (E.3) for the perturbation components with frequencies $\pm q_\perp$ in the form

$$\delta E = \alpha \exp[i(q_\perp r_\perp + \Omega t)] + \beta^* \exp[-i(q_\perp r_\perp + \Omega^* t)] . \tag{E.5}$$

Substitution of (E.5) into (E.3) leads to a set of two linear equations with respect to $\alpha$ and $\beta$:

$$d_1\alpha - i\varphi_{nl}\beta = 0 , \quad i\varphi_{nl}\alpha + d_2\beta = 0 , \tag{E.6}$$

where

$$\varphi_{nl} = k_0 d\frac{\varepsilon_2 I_0}{2\varepsilon_0} ,$$

$$d_{1,2}(q_\perp, \Omega) = i\left[\frac{\Omega d}{v_{gr}} \pm \left(\frac{q_\perp^2 d}{2k_0} - 2\varphi_{nl}\right)\right] + [1 - R\exp(i\Delta_{ph})] . \tag{E.7}$$

The condition of existence for a nontrivial solution of (E.6) gives a dispersion equation for the determination of the complex frequency $\Omega = \Omega(q_\perp)$ in the form

$$Z(q_\perp, \Omega) = 0 ,$$
$$Z(q_\perp, \Omega) = d_1(q_\perp, \Omega)d_2(q_\perp, \Omega) - \varphi_{nl}^2 . \tag{E.8}$$

Since dispersion equation (E.8) is quadratic with respect to $\Omega$, we find two values of $\Omega$ at fixed $q_\perp^2$. We can then determine coefficients $\alpha$ and $\beta$ and perturbation $\delta E(r_\perp, t)$, using initial conditions. At $\text{Im}\,\Omega < 0$ perturbations $\delta E$

will grow with time; so an initial transversely homogeneous field distribution will be unstable (radiation filamentation, if the corresponding value $q_\perp \neq 0$). At the stability boundary, $\mathrm{Im}\,\Omega = 0$, i.e., $\Omega$ is real. There, by separating the real and imaginary parts of the dispersion equation, we obtain $\Omega = 0$ and

$$\left(\frac{q_\perp^2 d}{2k_0} - 2\varphi_{\mathrm{nl}} - R\sin\Delta_{\mathrm{ph}}\right)^2 + (1 - R\cos\Delta_{\mathrm{ph}})^2 = \varphi_{\mathrm{nl}}^2 . \qquad (E.9)$$

At $\Delta_{\mathrm{ph}}^2 \ll 1 - R \ll 1$, which we assume below, (E.9) is transformed into an equation for the stability boundary (4.3.30). This stresses once more the validity of employing the transverse distributivity model and (E.1). Then, in accordance with the results in Sect. 4.3, we choose conditions for which the transversely homogeneous state is stable with respect to small transverse perturbations. Then "free oscillations" will be damped with time, and "forced oscillations" will form. To determine them we use again a spectral expansion with real $q_\perp$ and $\Omega$,

$$\delta E_{\mathrm{in}}(r_\perp, t) = \int \delta E_{\mathrm{in}}(q_\perp, \Omega)\exp(iq_\perp r_\perp + i\Omega t)\,dq_\perp\,d\Omega , \qquad (E.10)$$

and similar expansions for perturbations $\delta E$, $\Delta E$ and $\Delta E_{\mathrm{out}}$. For a "monochromatic" perturbation at the entrance

$$\delta E_{\mathrm{in}} = \gamma\exp(iq_\perp r_\perp + i\Omega t) , \qquad (E.11)$$

the solution of (E.3) has the form

$$\delta E = \tau\gamma\alpha\exp(iq_\perp r_\perp + i\Omega t) + (\tau\gamma\beta)^*\exp(-iq_\perp r_\perp - i\Omega t) . \qquad (E.12)$$

For coefficients $\alpha$ and $\beta$, we obtain the inhomogeneous set

$$d_1\alpha - i\varphi_{\mathrm{nl}}\beta = E_{\mathrm{in}}^{(0)}/E^{(0)} , \quad i\varphi_{\mathrm{nl}}\alpha + d_2\beta = 0 . \qquad (E.13)$$

Coefficients of this set are determined by (E.7). The determinant of (E.13) equals $Z$ [see (E.8)] and differs from zero under the stability conditions assumed. Solution of (E.13) has the form

$$\alpha(q_\perp, \Omega) = \frac{d_2(q_\perp, \Omega)}{Z(q_\perp, \Omega)}\frac{E_{\mathrm{in}}^{(0)}}{E^{(0)}} ,$$

$$\beta(q_\perp, \Omega) = -i\frac{\varphi_{\mathrm{nl}}}{Z(q_\perp, \Omega)}\frac{E_{\mathrm{in}}^{(0)}}{E^{(0)}} . \qquad (E.14)$$

Hence absolute perturbations are

$$\Delta E(q_\perp, \Omega) = \tau\frac{d_2(q_\perp, \Omega)}{Z(q_\perp, \Omega)}\Delta E_{\mathrm{in}}(q_\perp, \Omega)$$

$$+ i\tau^*\frac{\varphi_{\mathrm{nl}}}{Z(-q_\perp, -\Omega)}\Delta E_{\mathrm{in}}^*(-q_\perp, -\Omega) . \qquad (E.15)$$

The field at the exit is the sum of radiation reflected by the entrance mirror and the radiation going out from the interferometer through the same mirror:

$$E_{\text{out}} = RE_{\text{in}} + \tau E \,, \tag{E.16}$$

A similar relation is true for Fourier spectra of absolute perturbations. Therefore, taking into account (E.15), we obtain

$$\Delta E_{\text{out}}(\boldsymbol{q}_\perp, \Omega) = \mu(\boldsymbol{q}_\perp, \Omega)\Delta E_{\text{in}}(\boldsymbol{q}_\perp, \Omega) + \nu(\boldsymbol{q}_\perp, \Omega)\Delta E_{\text{in}}^*(-\boldsymbol{q}_\perp, -\Omega) \,, \tag{E.17}$$

· where

$$\mu(\boldsymbol{q}_\perp, \Omega) = 1 - (1 - R) + \tau^2 \frac{d_2(\boldsymbol{q}_\perp, \Omega)}{Z(\boldsymbol{q}_\perp, \Omega)} \,,$$

$$\nu(\boldsymbol{q}_\perp, \Omega) = \mathrm{i}|\tau|^2 \frac{\varphi_{\text{nl}} \exp(2\mathrm{i}\Theta)}{Z^*(-\boldsymbol{q}_\perp, -\Omega)} \,,$$

$$E^{(0)} = |E^{(0)}|^2 \exp(\mathrm{i}\Theta) \,. \tag{E.18}$$

Generality unrestricted, one may consider that the phase of the unperturbed state $\Theta = 0$, and the transmission coefficient of the exit mirror is purely imaginary (absorption is absent):

$$\tau = \mathrm{i}|\tau| = \mathrm{i}\sqrt{1 - R^2} \approx \mathrm{i}\sqrt{2(1 - R)} \,. \tag{E.19}$$

Relations (E.17) and (E.18) solve the problem of the calculation of the nonlinear interferometer's response to weak perturbations in the classical statement. Now we proceed to the quantum description of the field. The operator for the positive-frequency part of the electric field in the Heisenberg representation $E^{(+)}$ is related to the slowly varying amplitude of the classical electric field $E$. The Kerr nonlinearity of the medium is described by the product of operators:

$$\delta\varepsilon = \varepsilon_2 E^{(-)} E^{(+)} \,, \tag{E.20}$$

where $E^{(-)}$ is the operator of the negative-frequency part of the field. The fluctuational field component (additional to the coherent one with amplitude $E_{\text{in}}^{(0)}$) corresponds to a multimode vacuum.

As in the preceding classical analysis, we consider small deviations of the field inside the interferometer and of the field at its exit from the corresponding regular (classical) component. In this formulation, there is complete correspondence between the classical and quantum descriptions [270, 169]. Namely, photonic annihilation and creation operators $a$ and $a^\dagger$ in the spectral representation at the entrance and exit of the interferometer are connected by a relation similar to (E.17)

$$a_{\text{out}}(\boldsymbol{q}_\perp, \Omega) = \mu(\boldsymbol{q}_\perp, \Omega)a_{\text{in}}(\boldsymbol{q}_\perp, \Omega) + \nu(\boldsymbol{q}_\perp, \Omega)a_{\text{in}}^*(-\boldsymbol{q}_\perp, -\Omega) \,, \tag{E.21}$$

with coefficients $\mu(\boldsymbol{q}_\perp, \Omega)$ and $\nu(\boldsymbol{q}_\perp, \Omega)$ determined by (E.18). Field transform (E.21) enters the class of Bogolyubov's transforms [72]. It converts the vacuum field state into a squeezed state, with suppressed fluctuations in one of the quadrature field components, noise suppression taking place only in a definite interval of spatial and temporal frequencies. A decrease in the level of shot noise in radiation detection is reached by its heterodyning. In the regime of direct light detection, it is realized by mixing the incident radiation directly reflected by the entrance mirror with that exiting the interferometer [see (E.16)]. According to [363], the spectral density of the photocurrent noise characterizing the squeezing efficiency has the following form upon registration of the exit radiation:

$$\langle \delta i^2(q_\perp, \Omega) \rangle = \bar{i}\{1 - \eta + \eta[(|\mu(q_\perp, \Omega)| + |\nu(q_\perp, \Omega)|)^2 \cos^2 \psi$$
$$+ (|\mu(q_\perp, \Omega)| - |\nu(q_\perp, \Omega)|)^2 \sin^2 \psi]\} . \qquad (E.22)$$

Here we introduced the phase parameter

$$\psi(q_\perp, \Omega) = \frac{1}{2} \arg[\mu(q_\perp, \Omega)\nu(-q_\perp, -\Omega)] - \arg \bar{a}_{\text{out}} , \qquad (E.23)$$

where $\bar{i}$ is the average photocurrent and $\eta$ is the quantum efficiency of the photodetector. The first two summands in (E.22) determine the shot noise, i.e., the standard limit of noise with a classical treatment of light, and the last term (with phase parameter $\psi$) reflects the opportunities for its suppression.

An example of the noise spectrum is presented in Fig. E.1. For its calculation we chose a working point at the lower branch of the hysteresis curve. This regime is stable with respect to small transverse perturbations, while the states at the upper branch are unstable. Quantum efficiency of the photodetector is $\eta = 1$. The level of unity corresponds to shot noise upon registration of the coherent radiation. In the particular case of $q_\perp = 0$ the calculations coincide with those presented in [270] for a purely temporal problem. According to Fig. E.1, shot noise is not suppressed at quite low frequencies $\Omega$. However, this is not so important practically, as main technical fluctuations are concentrated in the low-frequency range, and depression of quantum noise gives no advantage because of their background. Another picture is observed at higher frequencies: the photonic statistics, especially at small $q_\perp$, becomes deeply subpoissonic. The efficiency of noise suppression can be greater than one order of magnitude. And it is not the limiting value. The point is that radiation exiting from an interferometer can be subjected to additional heterodyning. The ultimate degree of squeezing will be reached at the optimal phase modulation of the local oscillator [177, 368]. One can widen the frequency interval in which squeezing is effective by means of additional optical elements, for instance, a lens [368].

The presence of bistability is not necessary for squeezing, as a high value of differential gain is sufficient to increase the efficiency of noise suppression. In the bistable regime, differential gain increases when the working point is close to the edge of a branch of the hysteresis curve. However, a regime such as this is unreliable, as the corresponding states are metastable. Also, the

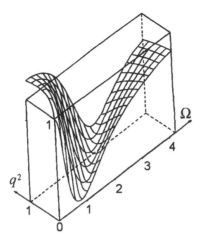

**Fig. E.1.** Spectral density of photocurrent noise in the mode of direct detection: $\Delta_{\mathrm{ph}} = -5(1 - R)$, $\mathrm{d}I/\mathrm{d}I_{\mathrm{in}} = 3.3/(1 - R)$, $\eta = 1$ [39]

probability of spontaneous switching into the state corresponding to the other branch, which has lower differential gain and can even appear unstable with respect to small transverse perturbations, increases under these conditions (see Chap. 2). Instead, a monostable regime close to the critical conditions of bistability onset ("nascent bistability") may be more convenient [in this case, at $\varepsilon_2 < 0$ and $\Delta_{\mathrm{ph}}^2 \approx 3(1 - R)^2$]. Indeed, in this case one can combine stability and the absence of spontaneous switching of the single stationary regime with a high value of differential gain in a comparatively wide range of perturbations. However, the requirements of optical homogeneity of the wide-aperture interferometer scheme and of the wide external radiation beam remain here.

Considerable attention has recently been focussed also on other manifestations of the quantum nature of the electromagnetic field in macroscopic optical structures, including attempts to unify nonclassical effects with classical transverse nonlinear phenomena [203, 116]. This, however, is the subject of another book.

# References

1. E. Abraham: Opt. Lett. **11**, 689 (1986)
2. E. Abraham, C. Rae: J. Opt. Soc. Am. B **4**, 490 (1987)
3. E. Abraham, S.D. Smith: Rep. Prog. Phys. **45**, 815 (1982)
4. A.B. Aceves, P. Varatharajah, A.C. Newell, E.M. Wright, G.I. Stegeman, D.R. Heatley, J.V. Moloney, H. Adachihara: J. Opt. Soc. Am. B **7**, 963 (1990)
5. H.A. Adachihara, D.W. McLaughin, J.V. Moloney, A.C. Newell: J. Math. Phys. **29**, 63 (1988)
6. V.M. Agranovich, D.L. Mills (Eds): *Surface Polaritons: Electromagnetic Waves at Surfaces and Interfaces*, Modern Problems in Condensed Matter Sciences 1 (North-Holland, Amsterdam 1982)
7. S.A. Akhmanov, M.A. Vorontsov, V.Yu. Ivanov: JETP Lett. **47**, 707 (1988)
8. S.A. Akhmanov, M.A. Vorontsov, V.Yu. Ivanov, A.V. Larichev, N.I. Zheleznykh: J. Opt. Soc. Am. B **9**, 78 (1992)
9. S.A. Akhmanov, V.A. Vyslouh, S.A. Chirkin: *Optics of Femtosecond Laser Pulses* (Am. Inst. Phys., New York 1992)
10. N.N. Akhmediev, A. Ankiewicz: *Solitons* (Chapman and Hall, London 1997)
11. N. Akhmediev, A. Ankiewicz: "Solitons of the Complex Ginzburg–Landau Equation", In: *Spatial Solitons*, ed. by S. Trillo, W.E. Torruellas (Springer, Berlin 2001) p. 311–341
12. N.N. Akhmediev, A. Ankiewicz, J.M. Soto-Crespo: Phys. Rev. Lett. **79**, 4047 (1997); J. Opt. Soc. Am. B **15**, 515 (1998)
13. N.N. Akhmediev, V.M. Eleonskiĭ, N.E. Kulagin: Sov. Phys. JETP **62**, 894 (1985)
14. N.N. Akhmediev, V.I. Kornev, Yu.V. Kuz'menko: Sov. Phys. JETP **61**, 62 (1985)
15. Yu.R. Alanakyan: Sov. Phys. Tech. Phys. **12**, 587 (1967)
16. E.B. Alexandrov, A.A. Ansel'm, A.N. Moskalev: Sov. Phys. JETP **62**, 680 (1985)
17. Yu.M. Aliev, A.D. Boardman, A.I. Smirnov, K. Xie, A.A. Zharov: Phys. Rev. E **53**, 5409 (1996)
18. L. Allen, J.H. Eberli: *Optical Resonance and Two-Level Atoms* (Wiley, New York 1975)
19. G.B. Al'tshuler, V.S. Ermolaev: Sov. Phys. Dokl. **28**, 146 (1983)
20. G.B. Al'tshuler, M.V. Inochkin, A.A. Manenkov: Sov. J. Quantum Electron. **17**, 362 (1987)
21. Yu.A. Anan'ev: *Laser Resonators and the Beam Divergence Problem* (Adam Hilger, Bristol 1992)
22. A.A. Andronov, A.A. Vitt, S.E. Khaiken: *Theory of Oscillators* (Pergamon, Oxford 1966)
23. S.P. Apanasevich, F.V. Karpushko, G.V. Sinitsyn: Sov. J. Quantum Electron. **15**, 251 (1985)

296    References

24. I.S. Aranson, K.A. Gorshkov, A.S. Lomov, M.I. Rabinovich: Physica D **43**, 435 (1990)
25. F.T. Arecchi, S. Bocaletti, P. Ramazza: Phys. Rep. **318**, 1 (1999)
26. I.P. Areshev, N.N. Rosanov, V.K. Subashiev et al.: Semiconductors **22**, 674 (1988)
27. V.I. Arnold: *Geometrical Methods in the Theory of Ordinary Differential Equations* (Springer, Heidelberg 1983)
28. M.L. Asquini, L.A. Lugiato, H.J. Carmichael, L.M. Narducci: Phys. Rev. A **33**, 360 (1986)
29. G.J. Assanto: J. Mod. Opt. **5**, 855 (1990)
30. D.H. Auston, T.K. Gustafson, A.E. Kaplan et al.: Appl. Opt. **26**, 231 (1987)
31. Yu.I. Balkarei, A.S. Kogan: JETP Lett. **57**, 286 (1993)
32. Yu.I. Balkarei, M.G. Evtikhov, M.I. Elinson, A.S. Kogan, V.S. Posvyanskii: Quantum Electron. **25**, 641 (1995)
33. Yu.I. Balkarei, M.G. Evtikhov, A.S. Kogan: Quantum Electron. **29**, 347 (1999)
34. Yu.I. Balkarey, M.G. Evtikhov, J.V. Moloney, Yu.A. Rzhanov: J. Opt. Soc. Am. B **7**, 1298 (1990)
35. Yu.I. Balkarey, A.V. Grigor'yants, Yu.A. Rzhanov: Sov. J. Quantum Electron. **17**, 72 (1987)
36. Yu.I. Balkarey, A.V. Grigor'yants, Yu.A. Rzhanov, M.I. Elinson: Opt. Comm. **66**, 161 (1988)
37. A.I. Baz', Ya.B. Zel'dovich, A.M. Perelomov: *Scattering, Reactions and Decay in Nonrelativistic Quantum Mechanics* (Israel Program for Scientific Transl., Jerusalem 1969)
38. V.Yu. Bazhenov, M.S. Soskin, V.B. Taranenko: Izv. Akad. Nauk Beloruss. SSR No. 1, 67 (1989)
39. A.V. Belinskii, N.N. Rosanov: Opt. Spectrosc. **70**, 89 (1992)
40. M.G. Benedict, A.M. Ermolaev, V.A. Malyshev, I.V. Sokolov, E.D. Trifonov: *Super-radiance: Multiatomic Coherent Emission* (Institute of Physics, Bristol 1996)
41. V.B. Berestetskii, E.M. Lifshitz, L.P. Pitaevkii: "Quantum Electrodynamics", In: *Course in Theoretical Physics, Vol. 4*, ed. by L.D. Landau and E.M. Lifshitz (Butterworth–Heinemann, Oxford 1990)
42. V.I. Bespalov, V.I. Talanov: Sov. Phys. JETP Lett. **3**, 307 (1966)
43. A.Y. Bigot, A. Daunois, P. Mandel: Phys. Lett. A **123**, 123 (1987)
44. J.E. Bjorkholm, P.W. Smith, W.J. Tomlinson et al.: IEEE J. Quantum Electron. **17**, 118 (1981)
45. J.E. Bjorkholm, P.W. Smith, W.J. Tomlinson, A.E. Kaplan: Opt. Lett. **6**, 345 (1981)
46. N. Bloembergen: *Nonlinear Optics* (Benjamin, New York 1965)
47. A.D. Boardman, S.A. Nikitov, Q. Wang: IEEE Trans. Magnetics **30**, 1 (1994)
48. A.D. Boardman, K. Xie: J. Opt. Soc. Am. B **14**, 3102 (1997)
49. A.D. Boardman, M. Xie: "Nonlinear Magnetooptic Solitons", In: *Spatial Solitons*, ed. by S. Trillo, W.E. Torruellas (Springer, Berlin 2001) p. 415–430
50. Ya.L. Bogomolov, A.V. Kochetov, A.G. Litvak, V.A. Mironov: Sov. Phys. JETP **71**, 667 (1990)
51. B.B. Boiko, N.S. Petrov: *Light Reflection from Amplifying and Nonlinear Media* (Nauka i Tekhnika, Minsk 1988) (in Russian)
52. B.B. Boiko, I.Z. Dzhilavdari, N.S. Petrov: J. Appl. Spectrosc. **23**, 1511 (1975)
53. L.G. Bol'shinskiĭ, A.I. Lomtev: Sov. Phys. Tech. Phys. **31**, 499 (1986)
54. R. Bonifacio, L.A. Lugiato: Lett. Nuovo Cim. **21**, 510 (1978)
55. R. Bonifacio, L.A. Lugiato, J.D. Farina, L.M. Narducci: IEEE J. Quantum Electron. **17** 357 (1981)

56. D. Bortman, A. Ron: Phys. Rev. A **52**, 3316 (1995)
57. C.M. Bowden, C.C. Sung, J.W. Haus, J.M. Cook: J. Opt. Soc. Am. B **5**, 11 (1988)
58. P.A. Braun: Theor. Math. Phys. **37**, 1070 (1978)
59. L.M. Brechovskich: *Waves in Layered Media* (Academic, New York 1960)
60. J. Brickmann, P.J. Russegger: Chem. Phys. **75**, 5744 (1981)
61. F.V. Bunkin, N.A. Kirichenko, B.S. Lukjyanchuk: Sov. Phys. Uspekhi **25**, 662 (1982)
62. V.S. Butylkin, A.E. Kaplan, Yu.G. Khronopulo, E.I. Yakubovich: *Resonant Nonlinear Interaction of Light with Matter* (Springer, Berlin 1989)
63. H.J. Carmichael: Opt. Comm. **53**, 122 (1985)
64. R.Y. Chiao, E. Garmire, C.H. Townes: Phys. Rev. Lett. **13**, 479 (1964)
65. J.D. Cole: *Perturbation Methods in Applied Mathematics* (Blaisdell, Toronto 1968)
66. P. Collet, J.P. Eckmann: *Iterated Maps of the Intervals as Dynamical Systems* (Birkhäuser, Boston 1980)
67. L.-C. Crasovan, B.A. Malomed, D. Mihalache: Phys. Rev. E **63**, 016605 (2000)
68. M.C. Cross, P.C. Hohenberg: Rev. Mod. Phys. **65**, 851 (1993)
69. M. Dagenais, W.F. Sharfin: Appl. Phys. Lett. **45**, 210 (1984)
70. R. Daisy, B. Fisher: J. Opt. Soc. Am. B **11**, 1059 (1994)
71. G. D'Alessandro, W.J. Firth: Phys. Rev. A **46**, 537 (1992)
72. A.S. Davydov: *Quantum Mechanics* (Pergamon Press, Oxford 1966)
73. A.P. Dmitriev, M.I. D'yakonov: Sov. Phys. JETP **63**, 838 (1986)
74. T.S. Dneprovskaya: Phys. Status Solidi B **52**, 39 (1972)
75. P.D. Drummond, D.F. Walls: J. Phys. A **13**, 725 (1980)
76. P.D. Drummond, C.W. Gardiner, D.F. Walls: Phys. Rev. A **24**, 914 (1981)
77. M.I. Dykman, M.A. Krivoglas: Physica A **104**, 480 (1980)
78. M.I. Dykman, M.A. Krivoglaz: Sov. Phys. JETP **50**, 30 (1979)
79. M.I. Dykman, D.N. Smelyanskiĭ: Sov. Phys. JETP **67**, 1769 (1988)
80. M.I. Dykman, F.L. Velikovich, G.P. Golubev et al.: Sov. Phys. JETP Lett. **53**, 193 (1991)
81. V.F. Elesin, Yu.V. Kopaev: Sov. Phys. JETP **36**, 767 (1973)
82. L.E. El'sgol'c, S.B. Norkin: *Introduction to Theory and Applications of Differential Equations with Deviating Arguments* (Academic, New York 1973)
83. R.H. Enns: Phys. Rev. A **36**, 5441 (1987)
84. E.M. Epshtein: Sov. Phys. Tech. Phys. **23**, 983 (1978)
85. E. Etrich, U. Peschel, F. Lederer: Phys. Rev. Lett. **79**, 2454 (1997)
86. I.L. Fabelinskij: *Molecular Scattering of Light* (Plenum, New York 1968)
87. A.V. Fedorov: Opt. Spectrosc. **77**, 396 (1994)
88. A.V. Fedorov, S.V. Fedorov, G.V. Khodova, N.N. Rosanov, V.A. Smirnov, N.V. Vyssotina: in *Optical Computing*, ed. B.S. Wherrett (Institute of Physics, Bristol, 1995) p. 653
89. S.V. Fedorov, G.V. Khodova, N.N. Rosanov: Proc. SPIE **1840**, 208 (1992)
90. S. Fedorov, D. Michaelis, U. Peschel, C. Etrich, N. Rosanov, F. Lederer: Phys. Rev. E **64**, 036610 (2001)
91. S.V. Fedorov, A.G. Vladimirov, G.V. Khodova, N.N. Rosanov: Phys. Rev. E **61**, 5814 (2000)
92. S.V. Fedorov, N.N. Rosanov, A.N. Shatsev, N.A. Veretenov, A.G. Vladimirov: "Oscillating and rotating states for laser solitons", Proc. SPIE (in press)
93. W.J. Firth: J. Phys. **49**, Coll. C2, Suppl. No 6, C2-451 (1988)
94. W.J. Firth: J. Mod. Opt. **37**, 151 (1990)
95. W.J. Firth: "Theory of Cavity Solitons", In: Soliton-driven Photonics, ed. by A.D. Boardman, A.P. Suchorukov (Kluwer, Dordrecht 2001) p. 459-485

96. W.J. Firth, E. Abraham, E.M. Wright: Philos. Trans. R. Soc. Lond. A **313**, 299 (1984)
97. W.J. Firth, I. Galbraith: IEEE J. Quantum Electron. **21**, 1399 (1985)
98. W.J. Firth, G.K. Harkness: Asian J. Phys. **7**, 665 (1998)
99. W.J. Firth, G.K. Harkness: "Existence, Stability and Properties of Cavity Solitons", In: *Spatial Solitons*, ed. by S. Trillo, W.E. Torruellas (Springer, Berlin 2001) p. 343–358
100. W.J. Firth, A.J. Scroggie: Europhys. Lett. **26**, 521 (1994)
101. W.J. Firth, S.W. Siclair: J. Mod. Opt. **35**, 431 (1990)
102. W.J. Firth, E.M. Wright: Opt. Comm. **40**, 233 (1982)
103. W.J. Firth, E.M. Wright: Phys. Lett. A **92**, 211 (1982)
104. W.J. Firth, I. Galbraith, E.M. Wright: J. Opt. Soc. Am. B **2**, 1005 (1985)
105. C. Flytzanis, C.L. Tang: Phys. Rev. Lett. **45**, 441 (1980)
106. W. Forysiak, R.G. Flesch, J.V. Moloney, E.M. Wright: Phys. Rev. Lett. **76**, 3695 (1996)
107. M.I. Frejdlin, A.D. Wentzell: *Random Perturbations of Dynamic Systems* (Springer, Berlin 1984)
108. A.V. Gainer, G.I. Surdutovich: Phys. Status Solidi B **150**, 539 (1988)
109. A.V. Gaĭner, G.I. Surdutovich: Sov. J. Quantum Electron. **18**, 629 (1988)
110. A.V. Gaĭner, G.I. Surdutovich: Autometry No. 3, 3 (1989)
111. I. Galbraith, H. Haug: J. Opt. Soc. Am. B **4**, 1116 (1987)
112. I. Ganne, G. Slekys, I. Sagnes, R. Kuszelewicz: Phys. Rev. B **63**, 075318 (2001)
113. A.V. Gaponov-Grekhov, M.I. Rabinovich, I.M. Starobinets: JETP Lett. **39**, 688 (1984)
114. C.W. Gardiner: *Handbook of Stochastic Methods for Physics, Chemistry and the Natural Sciences*, 2nd edn. (Springer, Berlin 1990)
115. E. Garmire, J.H. Marburger, S.D. Allen: Appl. Phys. Lett. **32**, 320 (1978)
116. A. Gatti, L.A. Lugiato, L. Spinelli, G. Tissoni, M. Brambilla, P. Di Trapani, F. Prati, G.L. Oppo, A. Berzanskis: Chaos Solitons Fractals **10**, 875 (1999)
117. H.M. Gibbs: *Optical Bistability: Controlling Light with Light* (Academic, Orlando 1985)
118. H.M. Gibbs, G.R. Albright, N. Peyghambarian H.E. Schmidt, S.W. Koch, H. Haug: Phys. Rev. A **32**, 692 (1985)
119. H.M. Gibbs, F.A. Hopf, D.L. Kaplan, R.L. Shoemaker: Phys. Rev. Lett. **46**, 474 (1981)
120. M.E. Goggin, P.W. Milloni: Phys. Rev. A **37**, 796 (1988)
121. Ch. Golaire, P. Mandel, P. Laws: Opt. Comm. **57**, 297 (1986)
122. L.L. Golik, A.V. Grigor'yants, M.I. Elinson: Tech. Phys. Lett. **7**, 51 (1981)
123. L.L. Golik, A.V. Grigor'yants, M.I. Elinson, Yu.I. Balkarei: Opt. Comm. **46**, 51 (1983)
124. K.A. Gorshkov, L.A. Ostrovsky: Physica D **3**, 428 (1981)
125. A.V. Grigor'yants, L.L. Golik, Yu.A. Rzhanov, M.I. Elinson, Yu.I. Balkarei: Sov. J. Quantum Electron. **14**, 714 (1984)
126. A.V. Grigor'yants, L.L. Golik, Yu.A. Rzhanov et al.: Bull. Russ. Acad. Sci., Ser. Phys. **48**, 133 (1984)
127. J. Goodman: *Introduction to Fourier Optics* (McGraw–Hill, San Francisco 1968)
128. V.S. Grigor'yan, A.I. Maimistov, Yu.M. Sklyarov: Sov. Phys. JETP **67**, 530 (1988)
129. A.V. Grigor'yants, I.N. Dyuzhikov: J. Opt. Soc. Am. B **7**, 1303 (1990)
130. A.V. Grigor'yants, I.N. Dyuzhikov: JETP **74**, 784 (1992); Quantum Electron. **24**, 469 (1994)

131. A.V. Grigor'yants, L.L. Golik, M.I. Elinson, Yu.I. Balkareĭ: Sov. J. Quantum Electron. **13**, 1135 (1983)
132. A.V. Grigor'yants, L.L. Golik, Yu.A. Rzhanov, Yu.I. Balkarei, M.I. Elinson: Sov. J. Quantum Electron. **17**, 793 (1987)
133. J. Grohs: *Der Einfluss von Rauschen und Rueckkopplung auf die Dynamik eines optisch nichtlinearen Halbleiterelements*, Ph.D. Thesis, University of Kaiserslautern (1993)
134. J. Guckenheimer, P. Holmes: *Nonlinear Oscillations, Dynamical Systems and Bifurcations of Vector Fields* (Springer, Heidelberg 1983)
135. P.S. Guilfoyle, D.S. McCallum: Opt. Eng. **35**, A3 (1996)
136. Yu.V. Gulyaev, A.V. Grigor'yants, Yu.A. Rzhanov, Yu.I. Balkarei, L.L. Golik, M.I. Elinson: Sov. Phys. Dolk. **28**, 740 (1983)
137. J. Hajto, I. Jánossy: Phyl. Mag. B **47**, 347 (1983)
138. H. Haken: *Synergetics. An Introduction* (Springer, Berlin 1978)
139. H. Haken (Ed.): *Evolution of Order and Chaos* (Springer, Berlin 1982)
140. H. Haken: *Advanced Synergetics. Instability Hierarchies of Self-Organizing Systems and Devices* (Springer, Berlin 1983)
141. J.W. Haus, C.C. Sung, C.M. Bowden, J.M. Cook: J. Opt. Soc. Am. B **2**, 1920 (1985)
142. F. Henneberger, H. Rossmann: Phys. Status Solidi B **121**, 685 (1984)
143. G. Herzberg, K.P. Huber: *Molecular Spectra and Molecular Structure* (Van Nostrand Reinhold, New York), *Vol. 1: Spectra of diatomic molecules, 2nd edn.* (1950), *Vol. 4: Constants of diatomic molecules* (1979)
144. M. Hoyuelos, P. Colet, M. San Miguel, D. Walgraef: Phys. Rev. E **58**, 2992 (1998)
145. B.A. Huberman, J.P. Crutchfield: Phys. Rev. Lett. **43**, 1743 (1979)
146. K. Ikeda: Opt. Comm. **30**, 257 (1979)
147. K. Ikeda, H. Daido, O. Akimoto: Phys. Rev. Lett. **45**, 709 (1980)
148. H. Issler: *Raeumliche und zeitliche Strukturbildung in der nichtlinearen Optik am Beispiel induziert absorptiver und rueckgekoppelter Systeme*, Ph.D. Thesis, University of Kaiserslautern (1994)
149. I. Jannosy, M.R. Taghizaden, G.H. Mathew, S.D. Smith: IEEE J. Quantum Electron. **21**, 1447 (1985)
150. E.C. Jarque, V.A. Malyshev: Opt. Comm. **142**, 66 (1997)
151. J.L. Jewell, Y.H. Lee, J.F. Duffy et al.: Appl. Phys. Lett. **48**, 1342 (1986)
152. R. Jin, D. Richardson, S.W. Koch, H.M. Gibbs: Opt. Eng. **28**, 344 (1989)
153. N.A. Kaliteevskii, N.N. Rosanov: Opt. Spectrosc. **89**, 569 (2000)
154. N.A. Kaliteevskii, N.N. Rosanov, S.V. Fedorov: Opt. Spectrosc. **85**, 485 (1998)
155. E. Kamke: *Differential Gleichungen. Lösungsmethoden und Lösungen* (Butterworth–Heinemann, Oxford 1985)
156. N.G. van Kampen: *Stochastic Processes in Physics and Chemistry* (North-Holland, Amsterdam 1984)
157. A.E. Kaplan: Sov. Phys. JETP Lett. **24**, 114 (1976)
158. A.E. Kaplan: Radiophys. Quantum Electron. **22**, 229 (1979)
159. A.E. Kaplan: Opt. Lett. **6**, 360 (1981)
160. A.E. Kaplan: IEEE J. Quantum Electron. **17**, 118 (1981)
161. A.E. Kaplan: Phys. Rev. Lett. **55**, 1291 (1985)
162. N.V. Karlov, N.A. Kirichenko, B.S. Luk'yanchuk: *Laser Thermochemistry* (Nauka, Moscow 1992) (in Russian)
163. P.I. Khadzhi, K.D. Lyakhomskaya: Quantum Electron. **10**, 881 (1999)
164. P.I. Khadzhi, G.D. Shibarshina: Semiconductors **21**, 1089 (1987)

165. P.I. Khadzhi, G.D. Shibarshina, A.H. Rotaru: *Optical Bistability in a System of Coherent Excitons and Biexcitons in Semiconductors* (Shtiintsa, Kishinev 1988) (in Russian)
166. B.S. Kerner, V.V. Osipov: *Autosolitons: A New Approach to Problems of Self-Organization and Turbulence* (Kluwer, Dordrecht 1994)
167. V.G. Kisilev: Phys. Status Solidi B **152**, 667 (1989)
168. D.N. Klyshko: *Photons and Nonlinear Optics* (Gordon and Breach, New York 1988)
169. D.N. Klyshko: Phys. Lett. A **137**, 334 (1989)
170. S.W. Koch, E.M. Wright: Phys. Rev. A **35**, 2542 (1987)
171. V.A. Kochelap, A.V. Kuznetsov: Phys. Rev. B **42**, 7497 (1990)
172. V.A. Kochelap, L.Yu. Melnikov, V.N. Sokolov: Semiconductors **16**, 746 (1982)
173. A.V. Kochetov, A.G. Litvak, V.A. Mironov, E.M. Sher: Physica D **87**, 342 (1995); Radiophys. Quantum Electron. **38**, 160 (1995)
174. A.N. Kolmogorov, I.G. Petrovskii, N.S. Piskunov: Bull. Moscow Univ., Sect. A **1**, No. 6, 1 (1937)
175. M.I. Kolobov: Rev. Mod. Phys. **71**, 1539 (1999)
176. M.I. Kolobov, I.V. Sokolov: Phys. Lett. A **140**, 101 (1989)
177. M.I. Kolobov, I.V. Sokolov: Sov. Phys. JETP **69**, 1097 (1989)
178. A.A. Kolokolov, A.I. Sukov: Radiophys. Quantum Electron. **21**, 909 (1978)
179. A.A. Kolokolov, A.I. Sukov: Radiophys. Quantum Electron. **21**, 1013 (1978)
180. M. Krezer, W. Balzer, T. Tschudi: Appl. Opt. **29**, 579 (1990)
181. R. Kuszelewicz, I. Ganne, I. Sagnes, G. Slekys: Phys. Rev. Lett. **84**, 6006 (2000)
182. J.N. Kutz, T. Erneux, S. Trillo, M. Haelterman: J. Opt. Soc. Am. B **16**, 1936 (1999)
183. A.V. Kuznetsov: Semiconductors **22**, 1143 (1988)
184. A.V. Kuznetsov: Phys. Status Solidi B **159**, 223 (1990)
185. L.D. Landau, E.M. Lifshitz: *Course in Theoretical Physics. Vol. 1. Mechanics* (Butterworth–Heinemann, Oxford 1996)
186. L.D. Landau, E.M. Lifshitz: *Course in Theoretical Physics. Vol. 3. Quantum Mechanics* (Butterworth–Heinemann, Oxford 2000)
187. L.L. Landau, E.M. Lifshitz: *Course in Theoretical Physics. Vol. 5. Statistical Physics* (Butterworth–Heinemann, Oxford 1999)
188. L.D. Landau, E.M. Lifshitz: *Course in Theoretical Physics. Vol. 8. Electrodynamics of Continuous Media* (Butterworth–Heinemann, Oxford 1995)
189. L. Larger, V.S. Udaltsov, J.P. Goedgebuer, W.T. Rhodes: Electron. Lett. **36**, 199 (2000)
190. N.M. Lawandy, W.S. Rabinovich: IEEE J. Quantum Electron. **20**, 458 (1984)
191. M. Le Berre, E. Ressayre, A. Tallet, H.M. Gibbs: Phys. Rev. Lett. **56**, 274 (1986)
192. M. Le Berre, D. Leduc, E. Ressaire, A. Tallet: J. Opt. B **1**, 153 (1999)
193. B.D. Levitan, A.V. Subashiev: Sov. Phys. Tech. Phys. **34**, 1194 (1989)
194. E.M. Lifshitz, L.P. Pitaevskii: Physical Kinetics, in *Course in Theoretical Physics, Vol. 10*, ed. by L.D. Landau, E.M. Lifshitz (Pergamon Press, Oxford 1981)
195. M. Lindberg, S.W. Koch, H. Haug: Phys. Rev. A **33**, 407 (1986)
196. V.N. Lisitsyn, V.P. Chebotaev: Sov. Phys. JETP Lett. **7**, 1 (1978)
197. A.G. Litvak, G.M. Fraiman: Radiophys. Quantum Electron. **15**, 1341 (1972)
198. A.G. Litvak, V.A. Mironov: Radiophys. Quantum Electron. **11**, 1096 (1968)
199. P. Lodahl, M. Saffman: Phys. Rev. A **60**, 3251 (1999)
200. S. Longhi: Phys. Scripta **56**, 611 (1997)
201. S. Longhi: Opt. Lett. **23**, 346 (1998)

202. L.A. Lugiato: Prog. Opt. **21**, 69 (1984)
203. L.A. Lugiato, M. Brambilla, A. Gatti: "Optical Pattern Formation", In *Advances in Atomic, Molecular, and Optical Physics*, Vol. 40, ed. by B. Bederson, H. Walther (Academic, Boston 1999) p. 229
204. L.A. Lugiato, R. Lefever: Phys. Rev. Lett. **58**, 2209 (1987)
205. V.N. Lugovoĭ: Sov. J. Quantum Electron. **9**, 1207 (1979)
206. V.N. Lugovoĭ, A.M. Prokhorov: Sov. Phys. Uspekhi **16**, 658 (1974)
207. K.D. Lyakhomskaya, L.Yu. Nad'kin, P.I. Khadzhi: Quantum Electron. **31**, 67 (2001)
208. A.V. Lykov: *Heat and Mass Transfer* (Mir, Moscow 1980)
209. T. Maggipinto, M. Brambilla, G.K. Harkness, W.J. Firth: Phys. Rev. E **62**, 8726 (2000)
210. A.I. Maimistov, A.M. Basharov: *Nonlinear Optical Waves* (Kluwer, Dordrecht 1999)
211. A.A. Mak, L.N. Soms, V.A. Fromsel, V.E. Yashin: *Neodim-Glass Lasers* (Nauka, Moscow 1990) (in Russian)
212. B.A. Malomed: Physica D **29**, 155 (1987)
213. B.A. Malomed: Phys. Rev. E **58**, 7928 (1998)
214. B.A. Malomed, A.G. Vladimirov, G.V. Khodova, N.N. Rosanov: Phys. Lett. A **274**, 111 (2000)
215. V. Malyshev, E.C. Jarque: J. Opt. Soc. Am. B **12**, 1868 (1995)
216. V.A. Malyshev, E.C. Jarque: Opt. Spectrosc. **82**, 630 (1997)
217. L. Mandel, E. Wolf: *Optical Coherence and Quantum Optics* (Cambridge University Press, New York 1995)
218. P. Mandel: *Theoretical Problems in Cavity Nonlinear Optics* (Cambridge University Press, Cambridge, England, 1997)
219. P. Manneville, Y. Pomeau: Phys. Lett. A **75**, 1 (1979)
220. J.H. Marburger: Prog. Quantum Electron. **4**, 35 (1975)
221. D. Marcuse: Appl. Opt. **19**, 3130 (1980)
222. F. Marquis, P. Dobiasch, P. Meystre, E.M. Wright: J. Opt. Soc. Am. B **3**, 50 (1986)
223. S.L. McCall: Appl. Phys. Lett. **32**, 284 (1978)
224. S.L. McCall, E.L. Hahn: Phys. Rev. Lett. **18**, 908 (1967)
225. S.L. McCall, H.M. Gibbs, G.G. Churchill, T.N.C. Venkatesan: Bull. Am. Phys. Soc. **20**, 636 (1975)
226. S.L. McCall, H.M. Gibbs, T.N.C. Venkatesan: J. Opt. Soc. Am. **65**, 1184 (1975)
227. B.P. McGinnis, E.M. Wright, S.W. Koch, N. Peyghambarian: Opt. Lett. **15**, 258 (1990)
228. D.W. McLaughlin, J.V. Moloney, A.C. Newell: Phys. Rev. Lett. **51**, 75 (1983)
229. J.D. Meindl: Proc. IEEE **83**, 619 (1995)
230. D. Michaelis, U. Peschel, F. Lederer: Phys. Rev. A **56**, R3366 (1997)
231. D. Michaelis, U. Peschel, F. Lederer: Opt. Lett. **23**, 337 (1998)
232. F. Mitschke, C. Boden, W. Lange, P. Mandel: Opt. Comm. **71**, 385 (1989)
233. J.V. Moloney: Opt. Acta **29**, 1503 (1982)
234. J.V. Moloney: IEEE J. Quantum Electron. **21**, 1393 (1985)
235. J.V. Moloney, F.A. Hopf, H.M. Gibbs: Phys. Rev. A **25**, 3442 (1982)
236. A.D. Morozov: Differential Equations **12**, 164 (1976)
237. T.A. Murina, N.N. Rosanov: Sov. J. Quantum Electron. **11**, 711 (1981)
238. T.A. Murina, N.N. Rosanov: Sov. Phys. Tech. Phys. **29**, 100 (1984)
239. M. Nakazawa, K. Suzuki, H.A. Haus: IEEE J. Quantum Electron. **25**, 2036 (1989)

240. M. Nakazawa, K. Suzuki, H. Kubota, H.A. Haus: IEEE J. Quantum Electron. **25**, 2045 (1989)
241. P. Nardone, P. Mandel, R. Kapral: Phys. Rev. A **33**, 2465 (1986)
242. L.A. Nesterov: Opt. Spectrosc. **64**, 694 (1988)
243. A.C. Newell: *Solitons in Mathematics and Physics* (SIAM, Philadelphia 1987)
244. A.C. Newell, J.V. Moloney: *Nonlinear Optics* (Addison–Wesley, Redwood City 1992)
245. H.X. Nguyen, V.D. Egorov: Phys. Status Solidi B **150**, 519 (1988)
246. H.X. Nguyen, V.D. Egorov, A. Harendt, E.V. Nazvanova: Phys. Status Solidi B **148**, 407 (1988)
247. G. Nikolis, I. Progogine: *Self-Organization in Nonequilibrium Systems* (Wiley-Interscience, New York 1977)
248. J.F. Nye, M.V. Berry: Proc. R. Soc. A **336**, 165 (1974)
249. A.Yu. Okulov, A.N. Oraevskii: Trudy FIAN USSR **187**, 202 (1988)
250. U. Olin: J. Opt. Soc. Am. B **7**, 35 (1990)
251. U. Olin, O. Sahlen: Opt. Lett. **14**, 566 (1989)
252. G.L. Oppo, A. Scroggie, W.J. Firth: J. Opt. B **1**, 133 (1999)
253. K. Otsuka, K. Ikeda: Phys. Rev. Lett. **59**, 194 (1987)
254. E. Ott, C. Grebogi, J.A. Yorke: Phys. Rev. Lett. **64**, 1196 (1990)
255. A. Ouarzeddini, H. Adachihara, J.V. Moloney et al.: J. Phys. **49**, Coll. C2, Suppl. No. 6, C2-455 (1988)
256. F. Papoff, G. D'Alessandro, W.J. Firth: Phys. Rev. A: **48**, 634 (1993)
257. U. Parlitz, L.O. Chua, L. Kocarev, K.S. Halle, A. Shang: Int. J. Bifurcation and Chaos **2**, 973 (1992)
258. L.M. Pecora, T.L. Carroll: Phys. Rev. Lett. **64**, 821 (1990)
259. U. Peschel, D. Michaelis, C. Etrich, F. Lederer: Phys. Rev. E **58**, 2735 (1998)
260. N.S. Petrov, V.A. Shakin: Sov. Phys. Tech. Phys. **30**, 443 (1985)
261. F. Pi, C. Schmidt, G. Orriols: Appl. Phys. B **42**, 85 (1987)
262. D. Pieroux, S.V. Fedorov, N.N. Rosanov, P. Mandel: Europhys. Lett. **49**, 322 (2000)
263. P.N. Pigurnov, N.N. Rosanov, V.A. Smirnov: Opt. Spectrosc. **68**, 119 (1990)
264. Yu.P. Raiser: Sov. Phys. Uspekhi **23**, 789 (1980)
265. A.N. Rakhmanov: Opt. Spectrosc. **74**, 701 (1993)
266. A.N. Rakhmanov, V.I. Shmalgausen: Sov. J. Quantum Electron. **22**, 1020 (1992)
267. A.N. Rakhmanov, V.I. Shmalhausen: Proc. SPIE **2108**, 428 (1993)
268. A.N. Rakhmanov, M.I. Vorontsov, A.P. Popova, V.I. Shmalgausen: Sov. J. Quantum Electron. **22**, 593 (1992)
269. V.P. Reshetin: Sov. J. Quantum Electron. **15**, 180 (1985)
270. S. Reynaud, E. Giacobino: J. Phys. **49**, No. 6, C2-477 (1988)
271. H. Richardson, E. Abraham, W.J. Firth: Opt. Comm. **63**, 199 (1987)
272. N.N. Rosanov: Tech. Phys. Lett. **3**, 239 (1977)
273. N.N. Rosanov: Tech. Phys. Lett. **4**, 30 (1978)
274. N.N. Rosanov: Opt. Spectrosc. **47**, 335 (1979)
275. N.N. Rosanov: Tech. Phys. Lett. **6**, 77 (1980)
276. N.N. Rosanov: Tech. Phys. Lett. **6**, 335 (1980)
277. N.N. Rosanov: Sov. Phys. JETP **53**, 47 (1981)
278. N.N. Rosanov: Sov. Phys. Tech. Phys. **26**, 1249 (1981)
279. N.N. Rosanov: Bull. Russ. Acad. Sci., Ser. Phys. **46**, 1886 (1982)
280. N.N. Rosanov: Opt. Spectrosc. **52**, 326 (1982)
281. N.N. Rosanov: Opt. Spectrosc. **55**, 125 (1983)
282. N.N. Rosanov: Sov. Phys. Tech. Phys. **29**, 957 (1984)
283. N.N. Rosanov: J. Phys., **49**, Coll. C2, C2-429 (1988)

284. N.N. Rosanov: Sov. Phys. Dokl. **33**, 924 (1988)
285. N.N. Rosanov: Opt. Spectrosc. **66**, 526 (1989)
286. N.N. Rosanov: Opt. Spectrosc. **67**, 787 (1989)
287. N.N. Rosanov: Opt. Spectrosc. **68**, 250 (1990)
288. N.N. Rosanov: Opt. Spectrosc. **68**, 558 (1990)
289. N.N. Rosanov: Sov. J. Quantum Electon. **20**, 1250 (1990)
290. N.N. Rosanov: Opt. Spectrosc. **70**, 784 (1991); **71**, 475 (1991)
291. N.N. Rosanov: Proc. SPIE **1840**, 130 (1992)
292. N.N. Rosanov: Opt. Spectrosc. **72**, 243 (1992)
293. N.N. Rosanov: Opt. Spectrosc. **73**, 243 (1992)
294. N.N. Rosanov: Proc. SPIE **1840**, 130 (1992)
295. N.N. Rosanov: JETP **76**, 991 (1993); JETP **86**, 284 (1998)
296. N.N. Rosanov: Opt. Spectrosc. **76**, 555 (1994); **77**, 555 (1994)
297. N.N. Rosanov: Opt. Spectrosc. **78**, 78 (1995)
298. N.N. Rosanov: Opt. Spectrosc. **78**, 914 (1995)
299. N.N. Rosanov: Opt. Spectrosc. **80**, 772 (1996)
300. N.N. Rosanov: Prog. Opt. **35**, 1 (1996)
301. N.N. Rosanov: Opt. Spectrosc. **81**, 248 (1996); **81**, 856 (1996); **82**, 151 (1997)
302. N.N. Rosanov: Opt. Spectrosc. **82**, 396 (1997)
303. N.N. Rosanov: Opt. Spectrosc. **86**, 559 (1999)
304. N.N. Rosanov: Opt. Spectrosc. **88**, 238 (2000)
305. N.N. Rosanov: Opt. Spectrosc. **88**, 721 (2000)
306. N.N. Rosanov: Opt. Spectrosc. **89**, 897 (2000)
307. N.N. Rosanov: Physics – Uspekhi **43**, 421 (2000)
308. N.N. Rosanov: Quantum Electron. **30**, 1005 (2000)
309. N.N. Rosanov, A.V. Fedorov: Opt. Spectrosc. **64**, 817 (1988)
310. N.N. Rosanov, A.V. Fedorov: Bull. Acad. Sci. USSR, Ser. Phys. **52**, No. 3, 101 (1988)
311. N.N. Rosanov, A.V. Fedorov: Opt. Spectrosc. **68**, 565 (1990)
312. N.N. Rosanov, S.V. Fedorov: Opt. Spectrosc. **72**, 782 (1992)
313. N.N. Rosanov, S.V. Fedorov: Opt. Spectrosc. **80**, 478 (1996)
314. N.N. Rosanov, S.V. Fedorov: Opt. Spectrosc. **84**, 767 (1998)
315. N.N. Rosanov, S.V. Fedorov: Phys. Pev. E **63**, 066601 (2001)
316. N.N. Rosanov, G.V. Khodova: Sov. J. Quantum Electron. **16**, 241 (1986)
317. N.N. Rosanov, G.V. Khodova: Opt. Spectrosc. **61**, 128 (1986)
318. N.N. Rosanov, G.V. Khodova: Opt. Spectrosc. **65**, 449 (1988)
319. N.N. Rosanov, G.V. Khodova: Opt. Spectrosc. **65**, 828 (1988)
320. N.N. Rosanov, G.V. Khodova: Sov. J. Quantum Electron. **19**, 512 (1989)
321. N.N. Rosanov, G.V. Khodova: J. Opt. Soc. Am. B **7**, 1057 (1990)
322. N.N. Rosanov, G.V. Khodova: Opt. Spectrosc. **72**, 1394 (1992)
323. N.N. Rosanov, V.E. Semenov: Opt. Spectrosc. **48**, 59 (1980)
324. N.N. Rosanov, V.E. Semenov: Opt. Comm. **38**, 435 (1981)
325. N.N. Rosanov, V.A. Smirnov: Sov. Phys. JETP **43**, 1075 (1976); Sov. J. Quantum Electron. **10**, 232 (1980)
326. N.N. Rosanov, V.A. Smirnov: JETP Lett. **33**, 488 (1981); Sov. Phys. JETP **59**, 689 (1984)
327. N.N. Rosanov, V.A. Smirnov: Sov. J. Quantum Electron. **10**, 232 (1980)
328. N.N. Rosanov, A.V. Fedorov, S.V. Fedorov: Opt. Spectrosc. **82**, 151 (1997)
329. N.N. Rosanov, A.V. Fedorov, S.V. Fedorov, G.V. Khodova: JETP **80**, 199 (1995)
330. N.N. Rosanov, A.V. Fedorov, S.V. Fedorov, G.V. Khodova: Opt. Spectrosc. **79**, 795 (1995)

331. N.N. Rosanov, A.V. Fedorov, S.V. Fedorov, G.V. Khodova: Physica D **96**, 272 (1996)
332. N.N. Rosanov, A.V. Fedorov, G.V. Khodova: Phys. Status Solidi B **150**, 499 (1988)
333. N.N. Rosanov, A.V. Fedorov, G.V. Khodova: Bull. Acad. Sci. USSR Phys. **53**, No. 6, 57 (1989)
334. N.N. Rosanov, A.V. Fedorov, G.V. Khodova: "Spatiotemporal Structures in Bistable Optical Schemes". In: *OSA Proceedings on Nonlinear Dynamics in Optical Systems*, Vol. 7, Opt. Soc. Am., Washington, USA. Ed. by N.B. Abraham, E.M. Garmire, P. Mandel (1990) p. 196
335. N.N. Rosanov, A.V. Fedorov, V.V. Shashkin: J. Opt. Soc. Am. B. **8**, 1471 (1991)
336. N.N. Rosanov, S.V. Fedorov, N.A. Kaliteevskii, D.A. Kirsanov, P.I. Krepostnov, V.O. Popov: Nonlinear Opt. **23**, 221 (2000)
337. N.N. Rosanov, S.V. Fedorov, G.V. Khodova: Opt. Spectrosc. **81**, 896 (1996)
338. N.N. Rosanov, S.V. Fedorov, G.V. Khodova: Opt. Spectrosc. **88** 790 (2000)
339. N.N. Rosanov, S.V. Fedorov, G.V. Khodova, A.A. Zinchik: Opt. Spectrosc. **83**, 370 (1997)
340. N.N. Rosanov, S.V. Fedorov, A.N. Shatsev: Opt. Spectrosc. **90**, 261 (2001)
341. N.N. Rosanov, S.V. Fedorov, A.N. Shatsev: Opt. Spectrosc. **91**, 232 (2001)
342. N.N. Rosanov, D.V. Liseev, A.D. Boardman, K. Xie: "Magneto-optic Cavity Multistability and Polarized Dissipative Solitons: Dissitons", In: *Nonlinear Guided Waves and Their Applications*, Dijon, France, OSA Technical Digest Series (, DC: Optical Society of America, Washington, DC 1999) p. 139-141
343. N.N. Rosanov, V.E. Semenov, G.V. Khodova: Sov. J. Quantum Electron. **12**, 193, 198 (1982)
344. N.N. Rosanov, V.E. Semenov, G.V. Khodova: Sov. J. Quantum Electron. **13**, 1534 (1983)
345. N.N. Rosanov, V.E. Semenov, N.V. Vyssotina: J. Opt. B: Quantum Semiclass. Opt. **3**, S96 (2001)
346. N.N. Rosanov, A.N. Sutyagin, G.V. Khodova: Bull. Russ. Acad. Sci. Phys. **48**, 188 (1984)
347. N.N. Rosanov, N.V. Vyssotina, A.G. Vladimirov: JETP **91**, 1130 (2000)
348. L. Roso-Franco: Phys. Rev. Lett. **55**, 2149 (1985); J. Opt. Soc. Am. B **4**, 1878 (1987)
349. D. Ruelle, F. Takens: Comm. Math. Phys. **20**, 167 (1971)
350. S.M. Rytov, Yu.A. Kravtsov, V.I. Tatarskii: *Principles of Statistical Radiophysics. Random Processes. Vol. 1. Elements of Random Process Theory* (Springer, Berlin 1987)
351. Yu.A. Rzhanov, A.V. Grigor'yants, Yu.I. Balkarei, M.I. Elinson: Sov. J. Quantum Electron. **20**, 419 (1990)
352. O. Sahlen: Opt. Comm. **59**, 238 (1986)
353. V.N. Sazonov: Theor. Math. Phys. **31**, 349 (1977); **35**, 514 (1978)
354. B. Schäpers, M. Feldman, T. Ackemann, W. Lange: Phys. Rev. Lett. **85**, 748 (2000)
355. A. Schülzgen, N. Peyghambarian, S. Hughes: Phys. Status Solidi B **206**, 125 (1995)
356. H.G. Schuster: *Deterministic Chaos: An Introduction* (Physik., Weinheim 1984)
357. B. Segar, B. Macke: Phys. Rev. Lett. **60**, 412 (1988)
358. B. Segard, J. Zemmouri, B. Macke: Opt. Comm. **63**, 339 (1987)
359. V.E. Semenov, N.N. Rosanov, N.V. Vyssotina: JETP **89**, 243 (1999)
360. V.P. Silin: Sov. Phys. JETP **26**, 955 (1968)

361. D.V. Skryabin, A.R. Champneys, W.J. Firth: Phys. Rev. Lett. **84**, 463 (2000)
362. V.I. Smirnov: *A Course of Higher Mathematics, Vol. IV* (Pergamon, Oxford 1964)
363. D.F. Smirnov, A.S. Troshin: Sov. Phys. Uspekhi **30**, 851 (1987)
364. P.W. Smith, W.J. Tomlinson: IEEE J. Quantum Electron. **20**, 30 (1984)
365. P.W. Smith, J.-P. Hermann, W.J. Tomlinson, P.J. Maloney: Appl. Phys. Lett. **35**, 846 (1979)
366. P.W. Smith, W.J. Tomlinson, P.J. Maloney, J.-P. Hermann: IEEE J. Quantum Electron. **17**, 340 (1981)
367. M.S. Sodha, L.A. Patel, S.L. Kaushik: Plasma Physics **21**, 1 (1979)
368. I.V. Sokolov: Opt. Spectrosc. **73**, 689 (1992)
369. L. Spinelli, G. Tissoni, M. Brambilla, F. Prati, L.A. Lugiato: Phys. Rev. A **58**, 2542 (1998)
370. V.A. Stadnik: Sov. Phys. JETP Lett. **45**, 175 (1987)
371. A.A. Stadnik: Sov. Phys. JETP Lett. **49**, 731 (1989)
372. V.A. Stadnik: Phys. Solid State **29**, 2059 (1987)
373. V.A. Stadnik: Opt. Comm. **68**, 445 (1988)
374. V.A. Stadnik: Phys. Status Solidi B **159**, 241 (1990)
375. K. Staliunas: Phys. Rev. A **48**, 1573 (1993)
376. K. Staliunas, V.J. Sánchez-Morcillo: Phys. Rev. A **57**, 1454 (1998)
377. N. Streible, K.-H. Brenner, A. Huang, et al.: Proc. IEEE **77**, 1954 (1989)
378. A.F. Suchkov: Sov. Phys. JETP **22**, 1026 (1966)
379. A.P. Sukhorukov: *Nonlinear Wave Interaction in Optics and Radiophysics* (Nauka, Moscow 1988) (in Russian)
380. V.B. Taranenko, I. Ganne, R.J. Kuszelewicz, C.O. Weiss: Phys. Rev. A **61**, 063818 (2000)
381. M.C. Teich, B.E.A. Saleh: Quantum Opt. **1**, 153 (1989)
382. E.C. Titchmarsh: *Eigenfunction Expansions Associated with Second-Order Differential Equations* (Clarendon Press, Oxford 1969 – Part 1 and 1970 – Part 2)
383. M. Tlidi, M. Le Berre, E. Ressayre, A. Tallet, L. Di Menza: Phys. Rev. A **61**, 043806 (2000)
384. M. Tlidi, P. Mandel, R. Lefever: Phys. Rev. Lett. **73**, 640 (1994)
385. M. Tlidi, P. Mandel, R. Lefever: Phys. Rev. Lett. **81**, 979 (1998)
386. M. Tlidi, P. Mandel, M. Le Berre, E. Ressayre, A. Talet, L. Di Menza: Opt. Lett. **25**, 487 (2000)
387. S. Trillo, W.E. Torruellas (Eds.): *Spatial Solitons* (Springer, Berlin 2001)
388. W.J. Tomlinson, J.P. Gordon, P.W. Smith, A.E. Kaplan: Appl. Opt. **21**, 2041 (1982)
389. S. Trillo, M. Haelterman, A. Sheppard: Opt. Lett. **22**, 970 (1997)
390. V.A. Trofimov: Proc. SPIE **3733**, 281 (1999)
391. U. Trutschel, F. Lederer, M. Golz: IEEE J. Quantum Electron. **25**, 194 (1989)
392. N. Tsukada, T. Nakayama: Phys. Rev. A **25**, 964 (1982)
393. E.A. Vanin, A.I. Korytin, A.M. Sergeev et al.: Phys. Rev. **49**, 2806 (1994)
394. G.D. VanWiggeren, R. Roy: Science **279**, 1198 (1998)
395. V.A. Vasil'ev, Yu.M. Romanovskii, V.G. Yakhno: *Autowave Processes in Kinetic Systems: Spatial and Temporal Self-Organization in Physics, Chemistry, Biology, and Medicine* (Reidel, Dordrecht 1987)
396. N.A. Veretenov, A.G. Vladimirov, N.A. Kaliteevskii, N.N. Rosanov, S.V. Fedorov, A.N. Shatsev: Opt. Spectrosc. **89**, 380 (2000)
397. A.G. Vladimirov, S.V. Fedorov, N.A. Kaliteevskii, G.V. Khodova, N.N. Rosanov: J. Opt. B: Quantum Semiclass. Opt. **1**, 101 (1999)

306    References

398. A.G. Vladimirov, G.V. Khodova, N.N. Rosanov: Phys. Rev. E **63**, 056607 (2001)
399. A.G. Vladimirov, N.N. Rosanov, S.V. Fedorov, G.V. Khodova: Quantum Electron. **27**, 949 (1997); **28**, 55 (1998)
400. S.N. Vlasov: Sov. J. Quantum Electron. **6**, 245 (1976)
401. S.N. Vlasov: Sov. J. Quantum Electron. **14**, 1233 (1984)
402. S.N. Vlasov, V.I. Talanov: *Wave Self-Focusing* (Inst. Appl. Phys., Nizhni Novgorod 1997) (in Russian)
403. M. Volmer: *Kinetik der Phasenbildung* (Steinkopff, Dresden 1939)
404. M.A. Vorontsov: J. Opt. B: Quantum Semiclass. Opt. **1**, R1 (1999)
405. M.A. Vorontsov, W.J. Firth: Phys. Rev. A **49**, 2891 (1994)
406. M.A. Vorontsov, W.B. Miller (Eds.): *Self-Organization in Optical Systems and Applications to Information Technology* (Springer, Berlin 1995)
407. M.A. Vorontsov, G.W. Carhart, R. Dou: J. Opt. Soc. Am. B **17**, 266 (2000)
408. M.A. Vorontsov, A.N. Rakhmanov, V.I. Shmalgausen: Sov. J. Quantum Electron. **22**, 56 (1992)
409. N.V. Vyssotina, L.A. Nesterov, N.N. Rosanov, V.A. Smirnov: Opt. Spectrosc. **85**, 218 (1998)
410. N.V. Vyssotina, N.N. Rosanov, V.A. Smirnov: Tech. Phys. Lett. **10**, 510 (1984)
411. N.V. Vyssotina, N.N. Rosanov, V.A. Smirnov: Sov. Phys. Tech. Phys. **32**, 104 (1987)
412. N.V. Vyssotina, N.N. Rosanov, V.A. Smirnov: "Formation of a High-Frequency Pulse Sequence in a Cavity with a Passive Nonlinear Medium with Anomalous Dispersion ", In: *Proc. VI All-Russian Conf. "Laser Optics"*, (S.I.Vavilov State Optical Institute, Leningrad 1990) p. 259 (in Russian)
413. D.F. Walls, G.J. Milburn: *Quantum Optics* (Springer, Berlin 1994)
414. C.O. Weiss, G. Slekys, V.B. Taranenko, K. Staliunas, R. Kuszelewicz: "Spatial Solitons in Resonators", In: *Spatial Solitons*, ed. by S. Trillo, W.E. Torruellas (Springer, Berlin 2001) p. 393–414
415. C.O. Weiss, M. Vaupel, K. Staliunas, G. Slekys, V.B. Taranenko: Appl. Phys. B: Laser Opt. **68**, 151 (1999)
416. S.G. Wenden, H.A. Adachihara, J.V. Moloney: "Encoded and Spontaneous Dark and Bright Solitary Wave Patterns in a Nonlinear Optical Feedback System". In: *OSA Proceedings on Nonlinear Dynamics in Optical Systems*, Vol. 7, Opt. Soc. Am., Washington, USA, ed. by N.B. Abraham, E.M. Garmire, P. Mandel, (1990) p. 188
417. E. Yanke, F. Emde, F. Lösch: *Tafelen höhenerer Funktionen* (Teubner, Stuttgart 1960)
418. J. Young, H. Richardson, H.A. MacKenzie, E. Abraham: J. Opt. Soc. Am. B **5**, 3 (1988)
419. V.E. Zakharov, S.V. Manakov, S.P. Novikov, L.P. Pitaevskii: *Theory of Solitons: The Inverse Scattering Method* (Plenum, New York 1984)
420. J.B. Zel'dovič, G.N. Barenblatt, V.B. Librovich, G.M. Makhviladze: *The Mathematical Theory of Combustion and Explosions* (Consultants Bureau, New York 1985)
421. B.Ya. Zel'dovich, N.F. Pilipetsky, V.V. Shkunov: *Principles of Phase Conjugation* (Springer, Berlin 1985)

# Index

Adjoint operator 120
Autocorrelation function 79

Bifurcation 72
– Andronov–Hopf 96, 160, 218, 220
– period doubling 72, 73
– saddle-node 160, 218
Bistability 1
– absolute 3, 164
– all-optical 6
– convective 3, 164, 230
– hybrid 6, 7, 69
– increasing absorption 6, 17
– laser scheme 8, 207
– spatial 17, 202
Bloch equations 270
Break-up integral 103, 264

Componency 156
Constitutive equation 269
Critical nuclei 35, 37

Debye's equation 269
Density of probability 27
Dissipative optical soliton 114, 276
– polarized 162
Drag coefficient 113
Dynamic chaos 75, 78

Feigenbaum's relation 74
Fluctuation 3, 27, 271, 280
Fokker–Planck equation 28

Galilean invariance 214
Galilean transform 113, 164
Galilean transformation 234

Heat conductivity equation 19
– reduced 21
Hysteresis
– angle 2
– dynamic 282

– frequency 2
– intensity 2
– polarization 2
– spatial 18

Inhomogeneity 118, 135, 242
– large-scale 118
– small-scale 118
Instability
– Bonifacio–Lugiato 169
– Ikeda 96, 99, 148, 169, 171
– McCall 148
– modulation 98, 161, 167, 213, 225
– transverse 101, 155
Interferometer
– Fabry–Perot 7, 172, 276
– magneto-optic 157
– ring 84, 157, 288
– two-beam 163
– two-frequency 164

Kerr nonlinearity 269, 288

Lamerey's diagram 70, 71
Laser bullet 246
Laser soliton 214
Local perturbation 17, 126
Lyapunov exponent 75, 80, 81, 94

Maxwell's rule 54
Maxwell's value 17, 34, 39, 51, 129, 136, 214, 252
Mechanical analogy 18, 31, 39, 56, 179, 191
Mechanics
– Aristotelian 143
– Newtonian 143
Metastability 285
Multistability 1, 22, 42, 124, 160, 195

Neutral mode 37, 112, 120, 226, 233
Noise 3

Noise suppression   292
Nonlinear reflection   7, 177
- beam filamentation   200
- beam lateral shift   200
- continuum problem   187, 191, 192
- hybrid regime   186, 192, 193
- oscillation regime   194
- radiation beam   196
- total internal   181, 184, 188, 192
- transmission regime   182, 184, 192, 193
- two-beam scheme   205
Nonlinear Schrödinger equation   264
Nonlinear surface wave   180, 189
Nonlinearity
- Kerr   92, 100, 157, 167, 179, 200, 291
- threshold   108
- two-level scheme   148
Nonparaxial propagation   266

Optical adder   276

Paraxial approximation   259
Paraxial equation   85, 199, 262
Period doubling   171, 222
Point mapping   70
Principle of limiting absorption   177, 194

Quantum anharmonic oscillator   279

Saddle   35
Shift register   274
Shock wave   265
Slowing down
- critical   25
- uncritical   25
Spatial frequency   102, 107, 161, 213
Spatial hysteresis   137, 140, 147
Squeezed state   287
Switching wave   17, 33, 37, 38, 109, 136, 148, 172, 204
- diffractive   124
- stable   17, 42, 47
- unstable   42, 46

Topological index   145
Trajectory
- heteroclinic   35, 110, 122, 215
- homoclinic   35, 122, 127, 174
Transfer function   1, 108
- isolated loop   66

Vortex   145